Protected Cultivation of Horticultural Crops

THE AUTHORS

Dr. Karuna Shanker is working as an Assistant Professor-cum-Junior Scientist (Horticulture) Tilka Manjhi Agriculture College, Punsiya, Godda, (Birsa Agricultural University, Kanke, Ranchi) Jharkhand India. He obtained his B.Sc. Ag from Udai Pratap Autonomous College Varanasi (UP), He did his M.Sc. (Ag) Horticulture from Acharya Narendra Deva University of Agriculture and Technology, Kumarganj, Ayodhya (UP) and Ph.D. in Horticulture from Birsa Agricultural University, Kanke, Ranchi, Jharkhand, India. He has qualified ICAR-NET in Fruit Science & Floriculture and Landscaping conducted by ASRB, New Delhi. He has Published 18 research paper in national and international journals of repute, 2 book and 9 book chapter, several popular /semi-Scientific articles and organized/ Participated various Training programmes Seminar, symposia and workshops.

Dr. Awadhesh Kumar did his graduation, post-graduation and doctorate from Acharya Narendra Deva University of Agriculture and Technology, Kumarganj, Ayodhya Uttar Pradesh India. He has qualified National Eligibility Test in Fruit Science conducted by Agricultural Scientists Recruitment Board, New Delhi. He was awarded Rajiv Gandhi National Fellowship from University Grant Commission for his doctorate degree. Dr. Kumar started his carrier as Technical Assistant, Department of Agriculture, Government of Uttar Pradesh. At present he is working as Subject Matter Specialist in Horticulture, Krishi Vigyan Kendra Ballia, Acharya Narendra Deva University of Agriculture and Technology, Kumarganj, Ayodhya. Where he is looking protected cultivation of vegetable crops in poly house and green house. The author has published several Hindi and English articles, research paper in various magazine, national and international journals.

Protected Cultivation of Horticultural Crops

(As Per ICAR VIth Deans Committee Syllabus)

– Authors –

Dr. Karuna Shanker

Dr. Awadhesh Kumar

2026

Daya Publishing House®
A Division of
Astral International Pvt. Ltd.
New Delhi – 110 002

ISBN 978-93-5919-354-0

Publisher's Note:

Every possible effort has been made to ensure that the information contained in this book is accurate at the time of going to press, and the publisher and author cannot accept responsibility for any errors or omissions, however caused. No responsibility for loss or damage occasioned to any person acting, or refraining from action, as a result of the material in this publication can be accepted by the editor, the publisher or the author. The Publisher is not associated with any product or vendor mentioned in the book. The contents of this work are intended to further general scientific research, understanding and discussion only. Readers should consult with a specialist where appropriate.

Every effort has been made to trace the owners of copyright material used in this book, if any. The author and the publisher will be grateful for any omission brought to their notice for acknowledgement in the future editions of the book.

Published by : **Daya Publishing House®**
A Division of
Astral International Pvt. Ltd.
– ISO 9001: 2015 Certified Company –
4736/23, Ground Floor, Ansari Road,
Darya Ganj, New Delhi-110 002
Ph. 011-43549197, 8130496929
e-mail: info@astralint.com
Website: www.astralint.com

Printed at : **Replika Press Pvt Ltd**

Preface

Protected cultivation of horticultural crops has become an essential practice in modern agriculture due to increasing demand for high-quality fruits, vegetables, flowers, and medicinal plants throughout the year. It enables the cultivation of crops under controlled environmental conditions, allowing farmers to overcome the limitations of climate, seasonality, and adverse weather, such as frost, heavy rain, hailstorms, and extreme temperatures. By providing an optimal microclimate, protected cultivation enhances crop growth, development, and yield, ensuring consistent and superior quality produce. This method also helps in extending the growing season or producing crops off-season, which can fetch higher market prices and improve farmers' income. Protected cultivation structures, such as greenhouses, polyhouses, and shade net houses, provide physical barriers that protect crops from pests, diseases, and weeds, reducing dependency on chemical pesticides and promoting safer, more sustainable production systems. Efficient use of water through controlled irrigation and fertigation systems in protected cultivation also minimizes water wastage, ensuring water-use efficiency and resource conservation. Furthermore, the ability to precisely regulate light, temperature, humidity, and CO_2 concentration improves photosynthesis, nutrient uptake, and overall plant health, leading to better growth rates, higher yields, and improved quality of fruits, vegetables, flowers, and ornamental plants. Protected cultivation also plays a vital role in research, education, and training, enabling scientists and students to study plant responses to controlled environments, test new cultivars, and develop innovative crop management techniques. The method supports year-round seedling production, nursery management, and propagation of disease-free planting material, which are critical for maintaining crop health and productivity. Moreover, it contributes to food and nutritional security by enabling the production of nutrient-rich vegetables and fruits in all seasons.

With increasing climate variability, urbanization, and shrinking cultivable land, protected cultivation has emerged as a strategic solution to meet the growing demand for horticultural produce. It ensures the production of quality crops with high aesthetic, nutritional, and medicinal value while conserving natural resources. Overall, protected cultivation of horticultural crops is of immense importance as it promotes sustainable, high-yielding, and profitable agriculture, ensures environmental protection, enhances resource efficiency, and provides an opportunity for farmers, entrepreneurs, and researchers to adopt modern, technology-driven practices in horticulture. Protected cultivation has emerged as a key technology in modern horticulture, enabling the production of high-quality crops throughout the year while mitigating the risks posed by adverse climatic conditions. This book, "Protected Cultivation of Horticultural Crops", has been prepared to provide a comprehensive and systematic understanding of this important field, in line with the ICAR VI Deans Committee syllabus. It addresses all facets of protected cultivation, from the basic principles, types of structures, and cladding materials, to greenhouse design, environmental control, soil and substrate management, irrigation and fertigation, propagation, and crop-specific cultivation techniques. Special emphasis is also given to off-season production, pest and disease management, and the cultivation of economically important medicinal and aromatic plants. This book is designed to serve both as an academic resource and a practical guide. It is especially useful for undergraduate and postgraduate students of horticulture, researchers, and faculty members in agricultural universities. At the same time, it caters to the needs of farmers, greenhouse entrepreneurs, and extension personnel, providing practical insights into the design, management, and optimization of protected cultivation systems. It is our hope that this book will not only enhance knowledge but also support sustainable and high-value horticultural production under protected conditions.

We would like to express our sincere gratitude to all the individuals, as well as the various books, manuals, periodicals, newsletters, magazines, journals, and other resources, that contributed to the preparation of this book. Despite our best efforts, we acknowledge that some errors may have occurred during the compilation and editing process. We welcome any further queries, constructive suggestions, and feedback aimed at improving the book, and we will be deeply appreciative of such contributions.

Dr. Karuna Shanker

Dr. Awadhesh Kumar

Contents

Preface v

1. **Introduction to Protected Cultivation** 1

1.0 Protected Cultivation: An Introduction – 1.1 Importance of Protected Cultivation – 1.2 Scope of Protected Cultivation – 1.3 Status of Protected Cultivation in India – 1.4 Status of Protected Cultivation in the World.

2. **Types of Protected Structures** 23

2.0 Introduction – 2.1 Classification Based on Site – 2.2 Classification Based on Climate – 2.3 Types of Structures: Greenhouse, Polyhouse, Shade Net House.

3. **Cladding Materials for Protected Structures** 43

3.0 Introduction: Cladding Materials for Protected Structures – 3.1 Types of Cladding Materials – 3.2 Properties and Selection Criteria.

4. **Greenhouse Design and Environment Control** 65

4.0 Introduction – 4.1 Principles of Greenhouse Design – 4.2 Structural Components – 4.3 Environmental Factors Affecting Growth – 4.4 Methods of Environmental Control – 4.5. Use of Artificial Lighting – 4.6 Automation in Greenhouses.

5. Soil and Substrate Management 115

5.0 Introduction: An Overview – 5.1 Soil Preparation Techniques – 5.2 Soil Management Practices – 5.3 Types of Substrates – 5.4 Substrate Preparation and Management.

6. Benches and Containers in Protected Cultivation 159

6.0 Introduction: An Overview – 6.1 Protected Structures – 6.2. Benches – 6.3 Containers and their Uses.

7. Irrigation and Fertigation Management 189

7.0 Introduction – 7.1 Irrigation Systems in Protected Cultivation – 7.2 Fertigation: Concepts and Practices – 7.3 Scheduling and Management.

8. Propagation and Production of Quality Planting Material 223

8.0 Introduction – 8.1 Methods of Propagation – 8.2 Production of Quality Planting Material – 8.3 Nursery Management under Protected Conditions.

9. Greenhouse Cultivation of Important Horticultural Crops 249

9.0 Major Flower Crops and their Varieties Cultivated under Greenhouses – 9.1 Flowers – 9.2 Vegetables – 9.3 Cultivation of Pot Plants.

10. Cultivation of Economically Important Medicinal and Aromatic Plants 315

10.1. Introduction – 10.2. Importance of Protected Cultivation of MAPs – 10.3. Suitable Structures for MAP Cultivation – 10.4. Propagation Techniques under Protected Conditions – 10.5. Growing Media and Soil Management – 10.6. Nutrient and Water Management – 10.7. Crop Management Practices – 10.8. Pest and Disease Management – 10.9. Harvesting and Post-Harvest Handling – 10.10. Economically Important Medicinal Plants in Protected Cultivation – 10.11. Economically Important Aromatic Plants in Protected Cultivation – 10.12. Challenges in Protected MAP Cultivation.

11. Off-Season Production of Flowers and Vegetables 331

11.1 Off-season Production of Vegetables – 11.2 Categories of Protected Cultivation Structures in Vegetables – 11.3 Off-season Production of Flowers – 11.4 Common Practices in Protected Off-Season Flowers and Vegetable Production – 11.5. Advantages of Protected Off-Season Production Flowers and Vegetables – 11.6. Challenges in Off-Season Flowers and Vegetables Production under Protected Cultivation.

12. Insect Pest and Disease Management in Protected Cultivation **343**

12.1 Introduction: An Overview – 12.2 Preventive Measures in Pest and Disease Management – 12.3 Monitoring and Early Detection – 12.4 Control Methods.

Glossary **353**

References **373**

Index **381**

Introduction to Protected Cultivation

1.0 Protected Cultivation: An Introduction

Protected cultivation refers to the practice of growing crops in a controlled environment that safeguards plants from adverse climatic conditions and optimizes growth factors. Structures such as greenhouses, polyhouses, shade nets, and plastic tunnels create a microclimate that enables crops to grow throughout the year, irrespective of external weather. By regulating temperature, humidity, light intensity, and ventilation, protected cultivation ensures better productivity, quality, and uniformity of produce. This method has become increasingly important in regions prone to unpredictable rainfall, temperature extremes, or strong winds. The concept has gained momentum globally due to the rising demand

The latest technique for cultivating horticultural crops in regulated environments is termed protected culture. The demand for greenhouses has increased markedly in recent decades to enhance the qualitative and quantitative production of vegetable crops. Various challenges are currently being faced, including increasing urbanization, population expansion, land ownership limitations, and diminishing agricultural productivity. Advanced technology is employed in protected cultivation, or controlled environment agriculture (CEA), where plants are cultivated in soilless systems or on a limited area of land. Utilizing a novel technology known as hydroponics, plants can flourish without soil and yield greater output.

for high-value horticultural crops and the need to maximize production per unit area. In India, protected cultivation has become a key strategy for producing vegetables, flowers, fruits, and nursery plants with improved quality and extended shelf life. It also allows off-season production, which fetches higher market prices for farmers. The technology is particularly beneficial in peri-urban areas where space is limited, but the demand for fresh produce is high. Protected cultivation is not limited to temperature control; it integrates modern techniques like drip irrigation, fertigation, automated climate control, and artificial lighting. These measures ensure precise input application, minimize wastage, and enhance resource-use efficiency. The method is also suitable for organic farming, as it reduces the risk of pest and disease infestation, thereby minimizing chemical pesticide use. With proper management, farmers can achieve higher yields and consistent quality, making the investment economically viable in the long run (Figure 1.1).

Figure 1.1: Protected Cultivation.
(*Source*: https://media.architecturaldigest.com)

In the face of climate change and growing food security concerns, protected cultivation offers a sustainable pathway to agricultural development. Government policies, subsidies, and training programs are further encouraging farmers to adopt this technology. Research institutions are also working to develop cost-effective designs suitable for diverse agro-climatic zones in India. As consumer awareness of food quality and safety increases, protected cultivation is set to play a pivotal role in modern agriculture by ensuring year-round supply, higher profitability, and reduced environmental impact.

1.1 Importance of Protected Cultivation

Protected cultivation of horticultural crops offers numerous benefits, including higher productivity by enabling year-round crop production regardless of external weather conditions. It ensures better quality produce through controlled temperature, humidity, and light management, reducing blemishes and improving uniformity. The system minimizes pest and disease incidence, lowering the need for chemical pesticides and promoting safer food. Water and nutrient use efficiency is greatly enhanced through drip irrigation and fertigation. It allows the cultivation of high-value and exotic crops that may not thrive in open-field conditions. Farmers can fetch higher prices due to off-season production and improved quality standards. Protected structures shield crops from adverse climatic events like heavy rain, frost, hail, and strong winds. It supports export-oriented production by meeting stringent international quality norms. The technology promotes efficient land use, making small areas highly productive. Overall, it enhances profitability, sustainability, and resilience in horticulture.

1.1.1 Ensures Year-Round Production

Protected cultivation systems create a microclimate that remains favourable for plant growth regardless of external weather conditions. By using structures like greenhouses, shade nets, or polyhouses, parameters such as temperature, humidity, light intensity, and air circulation can be precisely regulated. This allows farmers to grow crops continuously, eliminating the dependency on specific sowing and harvesting seasons. For example, tomatoes, cucumbers, and leafy greens can be harvested multiple times a year, ensuring constant supply in local and distant markets. Year-round production also means that farmers can target periods of high demand, such as festive seasons, to fetch better prices. In areas with cold winters or scorching summers where open-field farming is difficult, protected cultivation provides a stable platform for uninterrupted crop cycles. This approach not only secures the farmer's income throughout the year but also benefits consumers by making fresh produce available consistently. It reduces the risk of market fluctuations caused by seasonal shortages and helps maintain price stability. Additionally, continuous production supports agro-industries such as food processing units by ensuring a steady flow of raw materials.

1.1.2 Enhances Crop Yield and Quality

Protected cultivation significantly boosts both the quantity and quality of agricultural produce by minimizing environmental stress on plants. With carefully controlled irrigation, nutrient application, and climate parameters, crops can achieve optimal growth and higher productivity compared to open-field systems. This controlled approach reduces losses from pests, diseases, and weather fluctuations, resulting in greater efficiency per unit area. The produce often has superior physical attributes such as uniform size, attractive colour, smooth texture, and improved flavour, making it highly appealing to consumers. For example, greenhouse-grown capsicum and roses are known to have more uniform shape and colour compared to open-field varieties. The absence of damage from environmental factors also means that the crops have better shelf life, which is essential for transportation over long distances and for export markets. Higher quality produce commands better prices in both domestic and international markets, providing farmers with increased profit margins. Moreover, uniformity in quality helps build brand reputation and market reliability. Protected cultivation also allows farmers to meet stringent quality standards set by retail chains and export agencies, enhancing market access.

1.1.3 Promotes Resource-Use Efficiency

Protected cultivation promotes optimal use of critical agricultural resources such as water, nutrients, and energy. Through systems like drip irrigation and fertigation, water and fertilizers are delivered directly to the root zone, ensuring that plants receive the exact amount they need without wastage. This targeted delivery reduces nutrient leaching into the soil and groundwater, minimizing environmental pollution. Controlled environmental systems also reduce the overuse of pesticides and fungicides, cutting costs while promoting eco-friendly farming practices. Energy-efficient greenhouse designs, combined with solar-powered irrigation or heating systems, further reduce operational expenses. By reducing dependency on large volumes of inputs, protected cultivation ensures sustainability in resource-scarce regions. The efficient use of resources also enables farmers to cultivate crops in non-traditional areas, such as arid zones, saline lands, or urban rooftops, where conventional farming would be unsustainable. Additionally, this efficiency helps in lowering the carbon footprint of agricultural activities. Farmers benefit not only from reduced input costs but also from increased yields per unit of water and fertilizer applied, improving overall farm profitability.

1.1.4 Protects Crops from Adverse Weather

One of the primary advantages of protected cultivation is its ability to safeguard crops from unpredictable and extreme weather conditions. Polyhouses, shade nets, and other protective structures act as physical barriers against heavy rains, hailstorms, strong winds, frost, and extreme heat. This protection reduces crop loss risk, especially for high-value crops such as strawberries, orchids, or exotic vegetables, where even minor damage can lead to significant financial loss. In areas affected by frequent climatic uncertainties due to climate change, such protection ensures the stability of agricultural operations. The controlled conditions inside protected structures also prevent excessive waterlogging or drought stress, both of which can severely impact yields. Farmers are thus able to plan their production cycles with greater confidence, knowing that external weather events will have minimal impact. For regions experiencing early or late frost, protected cultivation provides a thermal buffer that prevents plant injury. This stability in production not only safeguards farmer livelihoods but also contributes to regional food security. The ability to withstand extreme weather is becoming increasingly crucial in the era of global climate variability.

1.1.5 Reduces Pest and Disease Incidence

Protected cultivation minimizes pest and disease problems by creating a controlled environment that is less conducive to their spread. The use of insect-proof nets, physical barriers, and carefully regulated humidity and temperature levels reduces the likelihood of infestations. In addition, the closed nature of the structures prevents the entry of common pests such as aphids, thrips, and whiteflies. Disease spread is also curtailed by limiting excess moisture on leaves and ensuring proper ventilation. This leads to a significant reduction in pesticide usage, lowering both production costs and chemical residues on crops. Consequently, the produce is safer for consumption and meets food safety standards more easily. Farmers benefit from reduced crop loss and better market acceptance. Additionally, early detection and localized treatment of any outbreaks are easier under protected cultivation conditions. This method aligns with Integrated Pest Management (IPM) strategies, promoting environmentally responsible agriculture. The lower dependency on chemical inputs also helps preserve beneficial insects and pollinators, contributing to a balanced ecosystem.

1.1.6 Facilitates Off-Season and Exotic Crop Production

Protected cultivation enables farmers to grow crops outside their normal growing seasons, allowing them to capture premium market prices during scarcity. By controlling environmental parameters, farmers can simulate the ideal growing conditions for crops that would otherwise be impossible to cultivate in a given region or season. For instance, bell peppers, gerberas, orchids, and strawberries can be grown year-round, including in off-seasons, making them highly profitable. Moreover, exotic crops that require specific temperature and humidity levels–such as broccoli, lettuce, and cherry tomatoes–can be successfully cultivated in regions where open-field conditions are unsuitable. This market advantage is especially valuable in urban and peri-urban areas where consumer demand for exotic vegetables and flowers is growing rapidly. Off-season and exotic crop production also allows farmers to diversify their product range, reducing dependency on a single crop and mitigating market risks. By targeting high-value niche markets, farmers can enhance income stability while catering to changing consumer preferences.

1.1.7 Supports Export-Oriented Agriculture

Global markets demand consistent quality, uniform size, and chemical residue-free produce, all of which can be achieved through protected cultivation. By controlling the environment and limiting pesticide use, farmers can meet the strict sanitary and phytosanitary (SPS) requirements of international buyers. Protected cultivation also helps extend shelf life, as crops grown in controlled environments are generally healthier and less prone to post-harvest spoilage. This is especially important for perishable items like cut flowers, fresh vegetables, and soft fruits intended for export. Moreover, the ability to produce year-round ensures continuous supply to overseas markets, strengthening trade relationships. Many export-focused agricultural clusters in India and other countries use protected cultivation as a standard practice. By meeting international quality benchmarks, farmers can command premium prices and establish a strong reputation in global markets. This integration into export supply chains contributes to foreign exchange earnings and boosts the agricultural economy.

1.1.8 Encourages Diversification and Innovation

Protected cultivation offers a platform for introducing new crops, farming techniques, and value-added products, thereby encouraging diversification. Farmers can experiment with hydroponics, aeroponics, vertical farming, and specialized seedling nurseries within these structures. The controlled conditions also support high-value ornamental plant production, herbal crops, and medicinal plants, opening new avenues for farm income. Diversification not only reduces risks associated with market fluctuations but also makes agriculture more resilient and adaptable. Additionally, innovation in protected cultivation often leads to technology adoption, such as automated climate control systems, renewable energy integration, and sensor-based irrigation. These advancements improve efficiency, reduce labour requirements, and enhance productivity. As markets evolve and consumer preferences shift, farmers engaged in protected cultivation can quickly adapt to new trends. This dynamism makes agriculture more sustainable in the long term and positions farmers as competitive players in both local and global markets (Table 1.1).

Table 1.1: Importance of Protected Cultivation at a Glance

Sl.No.	*Aspect*	*Detailed Explanation*
1.	Ensures Year-Round Production	Creates a controlled microclimate that enables crop production in all seasons, regardless of external weather. Regulates temperature, humidity, and light for uninterrupted supply, reducing market shortages and price fluctuations.
2.	Enhances Crop Yield and Quality	Minimizes environmental stress, leading to higher productivity. Produces uniform, visually appealing, and nutritionally rich crops, fetching better market and export prices, especially for high-value crops.
3.	Promotes Resource-Use Efficiency	Uses drip irrigation and fertigation to deliver water and nutrients directly to roots, reducing wastage. Controlled systems minimize fertilizer and pesticide use, cutting costs and environmental pollution.
4.	Protects Crops from Adverse Weather	Shields plants from heavy rain, wind, hail, frost, and extreme heat. Structures like polyhouses and shade nets prevent physical damage and reduce the risk of crop failure.
5.	Reduces Pest and Disease Incidence	Enclosed systems with insect-proof nets and environmental control reduce pest and disease attacks. Less chemical pesticide use improves food safety and lowers costs.
6.	Facilitates Off-Season and Exotic Crop Production	Allows cultivation of crops outside their natural season and exotic varieties like bell peppers, cherry tomatoes, and orchids, which command premium prices.

Sl.No.	*Aspect*	*Detailed Explanation*
7.	Supports Export-Oriented Agriculture	Meets international quality standards with uniform produce and minimal pesticide residues. Longer shelf life reduces post-harvest losses during transport.
8.	Encourages Diversification and Innovation	Enables farmers to explore hydroponics, vertical farming, nurseries, and ornamental plants. Supports experimentation with new crops and market-driven production.

1.2 Scope of Protected Cultivation

1.2.1. Year-Round Production

Protected cultivation creates a controlled microclimate that allows crops to be grown throughout the year, independent of seasonal variations. Structures such as greenhouses, polyhouses, and shade nets maintain ideal temperature, humidity, and light levels, ensuring uninterrupted plant growth. Farmers can plan multiple cropping cycles annually, increasing land use efficiency and income. Crops like tomatoes, cucumbers, capsicum, and strawberries can be produced even during extreme summers, cold winters, or heavy monsoons. This ensures a constant supply to markets, reducing the risk of seasonal shortages and price spikes. By scheduling harvests to match peak demand periods, farmers can secure better prices. Continuous production also supports supply commitments with supermarkets, processing units, and exporters. Year-round availability enhances consumer satisfaction and strengthens market linkages. Moreover, it helps in stabilizing rural employment, as labour is required in all seasons. In peri-urban areas, this system ensures a steady flow of fresh produce to cities, reducing dependence on imports during off-seasons. Overall, year-round production through protected cultivation is a key driver of agricultural sustainability and profitability.

1.2.2. Off-Season and High-Value Crop Production

Protected cultivation enables the successful growth of crops outside their natural growing seasons, giving farmers access to premium markets. By regulating temperature, light, humidity, and photoperiod, it is possible to produce exotic vegetables such as lettuce, broccoli, cherry tomato, and seedless cucumber during months when open-field cultivation is not feasible. Similarly, high-value cut flowers like gerbera, roses, orchids, and carnations can be grown to meet demand during festivals, weddings, and export seasons. Off-season production commands higher market prices due to limited supply and high consumer demand. This advantage allows farmers to target niche markets and secure better profit margins. By using staggered planting schedules, produce can be supplied over extended periods without market saturation. Protected cultivation also supports the growing of medicinal and aromatic plants, which have year-round industrial demand. Such diversification helps reduce dependence on traditional field crops and enhances farm income stability. This system is particularly profitable for peri-urban farming, where high-value produce can quickly reach urban consumers. Exporters also benefit as they can meet strict quality and delivery schedules for international buyers.

1.2.3. Climate Control for Sensitive Crops

Protected cultivation systems are designed to offer precise control over environmental factors such as temperature, humidity, ventilation, and light intensity, which are critical for the growth of climate-sensitive crops. This allows successful cultivation of crops like collared capsicum, seedless cucumber, strawberry, carnation, and orchids in regions where open-field conditions would be unsuitable. Automated systems such as foggers, exhaust fans, shade nets, and thermal screens help maintain optimal growing conditions throughout the crop cycle. By preventing exposure to extreme heat, frost, heavy rain, or strong winds, protected structures reduce plant stress and ensure uniform growth. This technology also makes it possible to introduce exotic and high-value varieties from other climatic zones, expanding the diversity of crops grown locally. In addition, regulated environments extend the growing season of certain crops, allowing production even in periods of climatic adversity. Climate control enhances both yield and quality, improving colour, size, and shelf life of produce. The ability to fine-tune conditions also facilitates research trials, seed production, and specialty crop cultivation. Such adaptability is particularly valuable in areas facing climate change and unpredictable weather patterns.

1.2.4. Increased Productivity

Protected cultivation significantly boosts crop productivity by providing an environment where plants can perform to their full genetic potential. The use of controlled temperature, humidity, light, and irrigation ensures that plants face minimal stress, leading to better growth, flowering, and fruit set. Yields under protected conditions are often two to five times higher than those in open-field farming. This system also allows multiple cropping cycles in a year, enabling more output from the same piece of land. High-density planting techniques can be applied without compromising crop health, further enhancing land use efficiency. By minimizing losses from pests, diseases, and adverse weather, overall production becomes more reliable. Improved root development in controlled substrates ensures efficient nutrient uptake and stronger plants. The uniformity of produce also increases its market value, as size, shape, and quality meet buyer specifications. Higher productivity reduces the cost per unit of produce, making the system more profitable. Moreover, consistent yields support long-term supply agreements with retailers and exporters.

1.2.5 Water and Nutrient Efficiency

Protected cultivation incorporates advanced irrigation methods like drip systems and fertigation units, which deliver water and nutrients directly to the plant root zone with precision. This targeted approach minimizes evaporation losses and ensures that every drop of water is effectively utilized. Fertilizers are applied in dissolved form through irrigation water, allowing accurate nutrient delivery based on crop stage and requirements. Such precision reduces nutrient leaching into the soil, preventing wastage and environmental pollution. Studies have shown that water savings of 40–60 per cent are achievable compared to traditional flood irrigation methods. By maintaining optimal soil moisture levels, the risk of waterlogging and root diseases is minimized. Nutrient use efficiency is greatly improved, resulting in healthier plants and higher yields. Real-time monitoring systems can adjust irrigation and nutrient schedules automatically, further enhancing resource use. Lower water and fertilizer inputs reduce production costs and make farming more sustainable. This is particularly beneficial in water-scarce regions, where efficient water management is essential for agricultural viability.

1.2.6. Pest and Disease Management

Protected cultivation offers a natural defence against pests and diseases by using physical barriers such as insect-proof nets, double-door entry systems, and controlled ventilation. These measures limit the entry of harmful insects, thereby reducing pest pressure on crops. The enclosed environment also allows better control of temperature and humidity, which can be adjusted to make conditions unfavourable for the development of certain pathogens. As a result, the need for chemical pesticides is greatly reduced, leading to safer, residue-free produce. This is especially important for crops intended for export, where strict residue limits apply. Integrated Pest Management (IPM) practices, such as the use of biological control agents, sticky traps, and pheromone traps, are more effective under controlled conditions. Regular monitoring and sanitation help in early detection and suppression of outbreaks before they become severe. By reducing pesticide use, production costs decrease, beneficial insect populations are preserved, and environmental pollution is minimized. Overall, pest and disease management in protected cultivation ensures healthier crops, higher yields, and improved market acceptability.

1.2.7. Quality Improvement

Protected cultivation provides a stable and optimized growing environment that enhances the physical and nutritional quality of produce. Uniform temperature, humidity, and light conditions promote consistent fruit size, shape, and colour across the entire crop. This uniformity increases market value and meets the grading requirements of both domestic and export buyers. Controlled irrigation and nutrient supply improve texture and firmness, which are crucial for consumer appeal and transportability. By reducing mechanical damage from wind, rain, or hail, the system ensures blemish-free produce. Protected structures also help in maintaining optimal ripening conditions, resulting in better flavour and aroma. The reduced incidence of pests and diseases further enhances quality by minimizing blemishes and pesticide residues. Extended shelf life is achieved because produce harvested from a healthy, stress-free environment stays fresher for longer. Such quality attributes allow farmers to access premium markets, negotiate better prices, and establish long-term relationships with buyers. For export crops, these improvements ensure compliance with stringent international standards.

1.2.8. Seedling and Nursery Production

Protected cultivation offers an ideal environment for producing healthy, vigorous, and disease-free seedlings of vegetables, flowers, and fruit crops. Structures such as polyhouses, net houses, and shade houses allow precise control over temperature, humidity, and light, creating optimal germination and growth conditions. By preventing exposure to pests, heavy rains, frost, and strong winds, seedlings develop with minimal stress and stronger root systems. The use of sterilized growing media and trays reduces the risk of soil-borne diseases and pest infestations. This ensures uniform growth and higher survival rates when seedlings are transplanted to the field. Controlled environments also allow year-round nursery production, enabling farmers to supply planting material whenever needed. High-quality seedlings mature faster, leading to earlier harvests and increased productivity. Specialized nursery facilities can also produce grafted seedlings for disease resistance or enhanced vigour. Such nurseries support commercial farming, contract production, and large-scale plantation projects. By ensuring a steady supply of healthy planting material, protected seedling production plays a crucial role in achieving high yields and maintaining crop quality.

1.2.9 Adaptation to Harsh Climates

Protected cultivation enables successful crop production in regions that face extreme climatic conditions such as scorching summers, freezing winters, high rainfall, or strong winds. Structures like polyhouses, net houses, and greenhouses act as protective barriers, shielding plants from adverse weather while maintaining optimal growing conditions inside. In arid and semi-arid regions, they reduce water loss and heat stress, allowing crops to flourish despite limited rainfall. In high-altitude or frost-prone areas, heating systems and thermal screens protect plants from low temperatures. In coastal or cyclone-prone zones, robust greenhouse designs and windbreaks minimize storm damage. This adaptability makes it possible to grow high-value and climate-sensitive crops in areas where open-field cultivation would be risky or impossible. Farmers can thus utilize otherwise marginal lands for profitable agriculture. Adaptation to harsh climates also helps in stabilizing local food supplies, reducing dependency on imports during extreme seasons. It further supports resilience against climate change by offering a secure production environment.

1.2.10 Export-Oriented Production

Protected cultivation creates produce that meets the strict quality, appearance, and safety standards required for international markets. Controlled environmental conditions ensure uniform size, colour, shape, and texture, which are essential for export crops like roses, gerbera, chrysanthemums, capsicum, cherry tomato, and strawberries. Reduced pesticide uses and adherence to Good Agricultural Practices (GAP) help meet maximum residue limit (MRL) requirements of importing countries. High-quality packaging and extended shelf life achieved under protected cultivation improve transportability over long distances. Year-round production capability allows exporters to fulfil supply contracts consistently, enhancing reliability and buyer trust. Farmers can also plan harvests to align with high-price export windows. Specialized varieties bred for export can be grown successfully, even if their native climate differs from the production site. Export-oriented production through protected cultivation brings foreign exchange earnings, strengthens the horticulture export sector, and opens global market opportunities for Indian farmers.

1.2.11 Employment Generation

Protected cultivation is labour-intensive during its setup, operation, and maintenance, thereby generating employment opportunities, especially in rural areas. Skilled labour is needed for greenhouse construction, installation of irrigation and climate-control systems, and regular operation. Crop management tasks such as pruning, trellising, training, harvesting, grading, and packaging require continuous workforce involvement. The sector also creates opportunities for semi-skilled and unskilled workers in activities like transplanting, weeding, and post-harvest handling. Women workers often play a significant role in greenhouse operations, particularly in tasks demanding precision and care. Employment generation is not limited to farming; it extends to nursery production, equipment manufacturing, marketing, and distribution. Seasonal and contractual work under protected cultivation helps reduce rural unemployment and underemployment. Additionally, training programs for youth in protected farming techniques enhance skill development and open entrepreneurial opportunities. By creating sustainable jobs, this sector contributes to rural economic stability and reduces migration to urban areas.

1.2.12 Research and Development Opportunities

Protected cultivation offers a controlled environment ideal for conducting research on new crop varieties, production techniques, and agricultural technologies. Researchers can manipulate environmental parameters like temperature, humidity, and light to study their effects on plant growth, yield, and quality. It is also valuable for breeding programs aimed at developing varieties with improved traits such as disease resistance, higher productivity, and better adaptability. The system enables testing of new irrigation methods, nutrient formulations, and integrated pest management strategies with minimal external interference. Universities, research institutes, and private companies use protected structures for seed production and quality assurance. The predictable conditions also support experimentation with exotic crops and specialty produce for niche markets. Insights gained from such research can be transferred to farmers, helping them adopt innovative and efficient practices. By fostering continuous innovation, protected cultivation plays a vital role in advancing horticultural science and boosting overall agricultural productivity.

1.2.13 Integration with Modern Technologies

Protected cultivation can be seamlessly integrated with advanced technologies like hydroponics, aeroponics, vertical farming, and automation to maximize productivity and resource efficiency. In hydroponics and aeroponics, plants are grown without soil, receiving nutrients directly through water or mist, which complements the controlled environment of greenhouses. Automation systems can regulate irrigation, fertigation, temperature, and humidity based on sensor data, reducing labour costs and human error. Vertical farming within protected structures optimizes space usage, enabling high-density production in limited areas. LED lighting systems can supplement natural light, extend the photoperiod and enhance photosynthesis. Smart monitoring tools, including IoT devices and mobile apps, allow real-time decision-making. Renewable energy sources like solar panels can be integrated to reduce operational costs and environmental impact. Such technology integration transforms protected cultivation into a high-tech, sustainable farming solution suitable for urban, peri-urban, and rural applications.

1.2.14 Contribution to Food and Nutritional Security

Protected cultivation plays a crucial role in ensuring consistent availability of fresh, high-quality, and nutrient-rich produce throughout the year. By enabling year-round production of vegetables, fruits, and herbs, it helps stabilize food supply and reduce seasonal gaps in markets. Consumers gain access to diverse and fresh foods, improving dietary variety and nutrition. In urban areas, protected farming supports local supply chains, reducing transportation time and preserving nutrient content. It also encourages the production of biofortified and specialty crops with enhanced health benefits. By minimizing crop losses from pests, diseases, and extreme weather, protected cultivation safeguards yield and ensures reliable supply. In regions facing food scarcity, this system can provide a steady source of essential vitamins and minerals. For government nutrition programs, it can serve as a dependable source of quality produce. Overall, it strengthens both household-level and national food security while supporting sustainable agricultural practices.

1.3 Status of Protected Cultivation in India

Protected cultivation in India has expanded steadily over the past two decades, supported by government schemes and technological advancements. The total area under protected cultivation is estimated to be between 40,000 and 110,000 hectares, depending on the source and the definition of structures included. States such as Maharashtra, Andhra Pradesh, Karnataka, Chhattisgarh, and Gujarat are leaders in adoption. For instance, under the Mission for Integrated Development of Horticulture (MIDH) during 2020–21, Andhra Pradesh reported 5,142 hectares, Karnataka 4,152 hectares, Chhattisgarh 3,666 hectares, and Gujarat 3,075 hectares. Maharashtra alone has recorded around 4,478 hectares, with a majority of the area under shade nets and naturally ventilated polyhouses. Between 2005–06 and 2017–18, government initiatives like the National Horticulture Mission (NHM) helped bring over 215,000 hectares under improved horticultural practices, a significant share of which was protected cultivation. Despite this progress, the national penetration remains low–around 0.2 per cent of the total cultivated area–indicating substantial untapped potential. The expansion is driven by high-value crops such as capsicum, cucumber, tomato, strawberry, and cut flowers like roses and gerbera, which benefit from better yields and quality under controlled environments. Government subsidies, typically covering up to 50 per cent of infrastructure and planting material costs, have been crucial in encouraging adoption, especially for small and medium farmers. Institutions like ICAR's Precision Farming Development Centres have played a vital role in promoting polyhouses, net houses, and climate control technologies. Farmers adopting these systems report earnings of ₹2–5 lakh per hectare per season, along with the creation of employment for 4–5 rural labourers per structure. However, challenges such as high initial investment, maintenance costs, and the need for skilled manpower remain key constraints. Even so, protected cultivation is increasingly recognized as a strategic approach to increase productivity, improve quality, and ensure year-round supply, making it an important component of India's future horticultural growth.

Here's a detailed overview of the status of protected cultivation in India:

1.3.1. Growth and Adoption

- ✰ ***Increasing Area:*** The area under protected cultivation in India is experiencing continuous growth, driven by various factors like increasing population and demand for off-season and high-value produce.
- ✰ ***Current estimates:*** Current estimates indicate that approximately 30,000 hectares are currently under protected cultivation in India.
- ✰ ***Key States:*** Major states leading the adoption include Maharashtra, Karnataka, Himachal Pradesh, and the North Eastern states.
- ✰ ***Crop Focus:*** Protected cultivation in India primarily focuses on high-value crops like flowers (roses, gerbera, carnation, chrysanthemum, lilium, orchids) and vegetables (tomato, capsicum, melons).

1.3.2. Benefits of Protected Cultivation

- ✰ ***Enhanced Yield and Quality:*** Protected cultivation enables precise regulation of vital growth factors such as temperature, humidity, and light intensity, ensuring optimal crop development. These controlled conditions lead to uniform growth, better colour, size, and taste of produce. Nutrient management becomes more efficient, resulting in superior nutritional value and market appeal. Yield levels can increase drastically, often reaching 10–12 times that of open-field farming. This makes protected cultivation an attractive option for commercial horticulture and export-quality production.

- ***Year-round Production and Diversification:*** By creating a controlled microclimate, protected structures make it possible to cultivate crops in all seasons, irrespective of external weather conditions. Farmers can plan multiple crop cycles in a year, maximizing land productivity. This flexibility allows cultivation of exotic or off-season vegetables, fruits, and flowers that otherwise fail in local climates. It opens up new opportunities in high-value niche markets such as hotels, supermarkets, and export chains. Ultimately, this reduces dependence on seasonal markets and stabilizes income throughout the year.
- ***Resource Efficiency:*** Advanced irrigation techniques like drip and micro-sprinklers, combined with hydroponics or soilless media, ensure targeted delivery of water and nutrients to plant roots. This drastically cuts water wastage, often saving 40–70 per cent compared to open-field methods. Fertilizers are applied precisely through fertigation, reducing overuse and environmental contamination. Protected cultivation also reduces weed growth, thereby minimizing herbicide application. The result is a more sustainable farming approach that conserves scarce natural resources while lowering input costs.
- ***Pest and Disease Control:*** The use of insect-proof nets, UV-stabilized films, and other barriers in greenhouses greatly limits pest invasion. This physical exclusion reduces dependency on harmful chemical pesticides, leading to safer and residue-free produce. A stable microclimate also reduces plant stress, lowering susceptibility to certain diseases. Integrated pest management techniques can be implemented more effectively in such environments. Overall, this ensures healthier crops, improved food safety, and higher consumer confidence in the produce.
- ***Climate Resilience:*** Protected structures act as a buffer against extreme and unpredictable weather events, which are becoming more frequent due to climate change. They safeguard crops from heavy rains, hailstorms, frost, and excessive heat, which could otherwise wipe out entire fields. By maintaining consistent growing conditions, they prevent sudden yield losses and protect farmers' investments. This resilience allows farmers to commit to contracts and supply chains with greater confidence. In the long run, it contributes to food security by stabilizing production despite climatic uncertainties. Benefits of protected cultivation in agriculture is summarised in Table 1.2.

1.3.3. Challenges and Obstacles

- ***High Initial Investment:*** The establishment of greenhouses, polyhouses, or other protected cultivation structures requires a considerable financial outlay. This includes costs for construction materials, climate control systems, irrigation setups, and sometimes automation technologies. For small and marginal farmers, arranging such capital without adequate subsidies or credit support is difficult. Even when loans are available, repayment risks can discourage adoption. High investment often leads to hesitation, delaying the adoption of these technologies. Without affordable financing models, many farmers remain excluded from the benefits of protected cultivation.

Table 1.2: Benefits of Protected Cultivation in Agriculture

Benefit	*Key Points*	*Outcome*
Enhanced Yield and Quality	Controlled temperature, humidity, light, and nutrients; uniform growth; better colour, size, taste; efficient nutrient use.	Yield 10–12 times higher; superior quality; high market value; export potential.
Year-round Production and Diversification	Controlled microclimate; multiple crop cycles; cultivation of exotic/off-season crops.	Continuous supply; stable income; entry into niche markets (hotels, supermarkets, export).
Resource Efficiency	Drip/micro irrigation, hydroponics; precise fertigation; reduced weed growth.	40–70 per cent water saving; lower fertilizer use; reduced herbicides; sustainable farming.
Pest and Disease Control	Insect-proof nets, UV films, physical barriers; reduced plant stress; effective IPM.	Less pesticide use; residue-free produce; healthier crops; food safety ensured.
Climate Resilience	Protection from rains, frost, hail, heat; stable microclimate.	Reduced crop loss; secured investment; reliable supply chains; improved food security.

- ***Lack of Awareness and Technical Know-How:*** Despite the clear advantages, many farmers are unaware of the full potential of protected cultivation. Even those who are aware often lack the technical knowledge required for proper design, construction, and maintenance of structures. Advanced practices like fertigation scheduling, microclimate control, and pest management under protected conditions require specialized skills. Inadequate extension services and limited availability of practical training programs worsen the problem. Without continuous technical support, farmers may experience crop failures or reduced productivity. This knowledge gap is a major bottleneck in scaling the practice.
- ***Market Challenges:*** Producing high-quality crops under protected cultivation is only half the battle–selling them profitably is the other half. Farmers often struggle with market access, facing middlemen who exploit information gaps to pay lower prices. Price volatility, especially for perishable horticultural produce, can result in financial losses even after achieving high yields. Lack of organized supply chains and cold storage further compounds the issue. Without reliable market linkages, farmers

cannot fully capitalize on their investment in protected cultivation. This uncertainty discourages long-term commitment to the practice.

- ***Quality of Materials:*** The success of protected cultivation depends heavily on the durability and performance of construction materials like cladding films, nets, and structural frames. Unfortunately, in many regions, farmers end up using substandard or counterfeit materials due to cost-cutting or lack of awareness. Such materials degrade faster under UV exposure or weather stress, leading to frequent repairs and replacements. This not only increases maintenance costs but also disrupts crop cycles. The absence of strict quality standards and certification mechanisms adds to the problem. Ultimately, poor-quality inputs undermine the sustainability of the system.
- ***Availability of Specific Varieties:*** Protected cultivation often requires crop varieties that are specifically bred for controlled environments, such as those with resistance to humidity-related diseases or the ability to thrive in low-light conditions. However, access to such varieties remains limited in many regions. Farmers may have to rely on imported seeds or planting material, which can be costly and subject to supply delays. Lack of localized breeding programs further limits options. As a result, many farmers end up using open-field varieties that perform poorly under protected conditions. This reduces the potential benefits and profitability of the system.

1.3.4. Government Support and Initiatives

- ***Subsidies and Financial Assistance:*** The Government of India, through programs like the National Horticulture Mission (NHM) and the National Horticulture Board (NHB), offers significant subsidies to reduce the financial burden of adopting protected cultivation. These subsidies often cover 40–60 per cent of the cost of polyhouses, shade net houses, plastic mulching, and related infrastructure. Such support makes the technology more affordable for small and marginal farmers who might otherwise be unable to invest. In addition, some state governments run their own schemes, offering top-up subsidies. This financial push has been instrumental in accelerating the adoption of protected cultivation across different regions.
- ***Training and Skill Development:*** Recognizing that technology alone is not enough, the government and private sector have jointly initiated training programs for farmers. These include hands-on workshops, exposure visits, and demonstration plots to teach practical skills in greenhouse management, fertigation, pest control, and climate regulation. Krishi Vigyan Kendras (KVKs) and agricultural universities also play a key role in providing localized training. Such programs help bridge the knowledge gap and ensure that farmers can operate and maintain

protected cultivation systems efficiently. As a result, farmers are better equipped to achieve higher yields and maintain consistent quality.

- ✰ ***Focus on Research and Development*** The government, in collaboration with research institutions, is actively working on developing region-specific protected cultivation solutions. This includes designing structures that can withstand local weather extremes and optimizing microclimate control for different crop varieties. Efforts are also directed toward breeding crop varieties suited for greenhouse conditions and resistant to prevalent pests and diseases. By aligning research with farmers' needs, these initiatives make protected cultivation more practical and profitable. Such advancements ensure that the technology remains sustainable and adaptable to India's diverse agro-climatic zones.

Protected farming represents a promising avenue for the future of Indian agriculture, with substantial benefits in production, quality, resource efficiency, and climate resilience. Despite existing hurdles, governmental assistance, technical progress, and heightened awareness among farmers are facilitating broader adoption and fostering a more sustainable and profitable future for Indian farmers and the agricultural sector overall.

1.4 Status of Protected Cultivation in the World

Protected cultivation has emerged as a transformative technology in global agriculture, enabling year-round production of high-quality crops irrespective of seasonal or climatic constraints. This method involves using structures like greenhouses, shade nets, and tunnels to create controlled environments for plants. In recent decades, rapid urbanization, population growth, and the demand for safe, pesticide-free produce have accelerated the adoption of these systems. It is especially vital for regions facing climate change impacts such as droughts, heat waves, and erratic rainfall.

1.4.1. Area Under Protected Cultivation Worldwide

According to FAO and industry estimates, the total global area under protected cultivation exceeds 500,000 hectares, with steady growth each year. The adoption rate varies significantly across countries, depending on technological advancement, economic capacity, and market demand. Developed countries lead in sophisticated climate control systems, while developing nations focus more on low-cost, naturally ventilated structures. Asia currently accounts for the largest share of the global area, followed by Europe and North America.

1.4.2. Technological Advancement in Developed Nations

Countries like the Netherlands, Spain, Israel, and the USA have pioneered advanced greenhouse technologies with fully automated systems for climate control, fertigation, artificial lighting, and hydroponics. These high-tech facilities are designed for precision agriculture, optimizing resource use while maximizing yields and quality. Netherlands, in particular, is known for producing vegetables and flowers in multi-span glasshouses equipped with computerized environmental control systems.

1.4.3. Asia – The Global Leader in Area Coverage

Asia dominates in terms of area under protected cultivation, with China alone accounting for over 75 per cent of the world's total greenhouse area. Most of these are low-cost, naturally ventilated plastic-covered structures aimed at small and medium-scale vegetable production. Other leading Asian countries include Japan, South Korea, and India, where the technology is being adapted to local conditions and market requirements. Asia's focus is on increasing productivity and ensuring food security for large populations.

1.4.4. China's Massive Greenhouse Industry

China's protected cultivation sector is the largest in the world, covering millions of hectares, especially in Shandong, Hebei, and Liaoning provinces. The structures range from simple plastic tunnels to advanced solar greenhouses that require minimal heating. These systems support vegetable, fruit, and flower production throughout the year. Government policies, subsidies, and rural development programs have played a major role in this large-scale adoption.

1.4.5. Japan and South Korea – Precision and Efficiency

Japan and South Korea have integrated modern greenhouse systems with automation, hydroponics, and vertical farming techniques to cope with limited agricultural land. These countries produce high-value crops like tomatoes, strawberries, and flowers with exceptional quality standards. Energy efficiency and smart farming practices are major focuses, with AI and IoT applications becoming increasingly common in commercial greenhouses.

1.4.6. India – Growing Adoption with Government Support

India's protected cultivation sector is expanding rapidly due to government schemes like the National Horticulture Mission (NHM) and state-level subsidies. Polyhouses, shade nets, and walk-in tunnels are being established for crops like capsicum, cucumber, gerbera, and orchids. Although the area is smaller compared to global leaders, the focus is on affordable, climate-resilient designs suitable for diverse agro-climatic zones.

1.4.7. Europe – High-Tech and Market-Driven

Europe is home to some of the most technologically advanced greenhouse operations in the world. Countries like the Netherlands, Spain, Italy, and France specialize in large-scale, market-oriented production of vegetables, fruits, and ornamental plants. The emphasis is on maximizing yields per unit area while minimizing environmental impact through energy-efficient designs, renewable energy use, and waste recycling.

1.4.8. Netherlands - The Greenhouse Capital of the World

The Netherlands has become a global benchmark for greenhouse farming, despite having limited land resources. The country exports large quantities of tomatoes, cucumbers, peppers, and cut flowers to other parts of Europe. Dutch greenhouses use advanced glasshouse designs, artificial lighting, and climate control systems to maintain optimal growing conditions year-round. The country also exports greenhouse technology worldwide.

1.4.9. Spain - Greenhouses in Arid Regions

Spain's Almería region is famous for its "sea of plastic," a vast expanse of plastic-covered greenhouses producing vegetables for European markets. These structures help conserve water and protect crops from harsh Mediterranean climates. Most Spanish greenhouses are naturally ventilated, making them cost-effective for large-scale production, although modern upgrades are increasingly common.

1.4.10. North America - Commercial and Technological Growth

In North America, protected cultivation is expanding in both the USA and Canada, driven by consumer demand for fresh, locally grown produce. High-tech greenhouses produce crops like lettuce, tomatoes, and herbs for supermarkets year-round. Canada's greenhouse industry is concentrated in Ontario and British Columbia, where controlled environments help overcome cold winters and short growing seasons.

1.4.11. USA - From Flowers to Food Crops

Initially focused on ornamental plants and flowers, U.S. greenhouse operations have shifted toward high-value vegetable production in response to health-conscious consumers. States like California, Arizona, and Texas are leaders in commercial greenhouse vegetable farming. Hydroponics and aquaponics systems are becoming popular for sustainable urban agriculture.

1.4.12. Latin America - Emerging Opportunities

Countries like Mexico, Chile, and Brazil are adopting protected cultivation to increase productivity and meet export demands. Mexico, in particular, has become a major exporter of greenhouse-grown tomatoes, peppers, and cucumbers to the USA and Canada. The region's warm climate allows for cost-effective production with minimal heating requirements.

1.4.13. Middle East - Overcoming Harsh Climates

Protected cultivation has become essential in arid and desert countries such as the UAE, Saudi Arabia, and Qatar. Greenhouses here are designed to cope with extreme heat and water scarcity, often integrating cooling systems and hydroponics. Israel is a leader in this region, known for drip irrigation, climate-adapted greenhouse designs, and high-value crop production.

1.4.14. Africa – Gradual but Promising Growth

In Africa, protected cultivation is still in its early stages, but countries like Kenya, Ethiopia, and South Africa are expanding their greenhouse production, mainly for export-oriented flowers and vegetables. Low-cost structures are common, but there is increasing interest in climate-resilient and resource-efficient designs to cope with erratic weather patterns.

1.4.15. Russia and Eastern Europe – Cold Climate Adaptations

Russia, Poland, and other Eastern European countries use heated greenhouses to grow vegetables during long, cold winters. Modernization programs are replacing outdated Soviet-era facilities with energy-efficient designs. Demand for year-round fresh produce is driving investments in LED lighting, thermal screens, and geothermal heating.

1.4.16. Australia and New Zealand – Quality over Quantity

Protected cultivation in Australia and New Zealand is relatively small in area but focused on producing high-quality, premium crops. These countries integrate modern climate control, hydroponics, and organic-certified systems to cater to domestic and export markets. Sustainability and water efficiency are major priorities.

1.4.17. Role of Protected Cultivation in Global Food Security

Globally, protected cultivation plays a critical role in ensuring food security by enabling stable production despite climatic uncertainties. It reduces dependence on seasonal imports, ensures continuous supply, and supports urban agriculture. By producing more food per unit of land, it helps address the challenge of feeding a growing population without expanding agricultural land.

1.4.18. Challenges in Global Adoption

Despite its benefits, the expansion of protected cultivation faces challenges such as high initial investment, energy costs, and technical know-how requirements. In developing nations, affordability and access to quality materials remain constraints, while in developed countries, sustainability and energy efficiency are the primary focus. Addressing these issues will determine the pace of global adoption.

Challenges and Limitations

- **High Initial Investment:** Setting up protected cultivation structures like greenhouses can involve significant capital costs.
- **Technical Knowledge Gaps:** Lack of technical expertise and skilled labour can hinder adoption, especially in developing regions.
- **Energy Consumption:** Operating greenhouses can be energy-intensive, requiring careful consideration of energy-efficient systems and renewable energy sources.

- **Pest and Disease Management:** While generally reduced, managing certain pests and soil-borne pathogens within controlled environments can still be challenging.
- **Assured Marketing:** Requires assured markets for high-value produce to ensure profitability due to higher production costs.

1.4.19. Future Prospects

The future of protected cultivation is promising, with innovations in renewable energy integration, automation, and AI-driven farming expected to revolutionize the sector. Vertical farming, smart greenhouses, and closed-loop hydroponic systems are likely to dominate future developments.

- **Increased Adoption:** As global population and climate change pressures continue to mount, protected cultivation offers a sustainable solution to meet the growing demand for horticultural products.
- **Technological Integration:** Integration of precision agriculture, AI, and IoT solutions for crop monitoring, automated harvesting, and data analysis will further optimize protected cultivation systems.
- **Research and Innovation:** Continued research and development in areas like energy-efficient systems, renewable energy adoption, and integrated pest management will further enhance the sustainability and economic viability of protected cultivation, according to ResearchGate.

Types of Protected Structures

2.0 Introduction

Protected structures in horticulture refer to any structure used to create a controlled or modified environment for plant growth, thereby extending the growing season, protecting crops from harsh weather and pests, and enabling year-round production of specific crops. Protected structures in horticultural crop production are specialized facilities designed to create a controlled or partially controlled microclimate, enabling the cultivation of crops beyond normal seasonal and climatic limitations. They protect plants from adverse weather conditions such as excessive rainfall, strong winds, frost, hailstorms, and extreme heat, while also reducing the incidence of pests and diseases. These structures help in optimizing temperature, humidity, light intensity, carbon dioxide levels, and ventilation to ensure better crop growth and higher yields. Based on the degree of environmental control and construction design, protected structures can be classified into various types such as greenhouses, shade net houses, polyhouses, plastic tunnels, insect-proof net houses, and low-cost structures. Greenhouses are fully covered frameworks with transparent materials like glass or polyethylene, allowing maximum light penetration and enabling precise climate regulation through advanced systems like heating, cooling, and irrigation automation. Polyhouses are similar to greenhouses but are covered primarily with UV-stabilized polyethylene sheets, making them cost-effective for small and medium farmers. Shade net houses are constructed with synthetic nets of varying shade percentages, suitable for crops requiring diffused light and protection from scorching sunlight. Plastic tunnels, either low tunnels or walk-in tunnels, provide a low-cost solution for early-season vegetable production and frost protection. Insect-proof net houses are designed with fine mesh netting to prevent the entry of insect vectors, thereby reducing the need for chemical pesticides and supporting organic farming practices. There are also naturally ventilated polyhouses, which rely on passive ventilation without expensive cooling systems, and fan-and-pad greenhouses that ensure precise temperature and humidity control through mechanical systems. Low-cost bamboo-based or wooden-frame polyhouses are suitable for small-scale farmers in rural areas, while high-tech greenhouses with hydroponic or aeroponic systems cater to export-oriented, high-value crop production. The choice of protected structure depends on factors like crop type, climatic conditions, investment capacity, and intended market. Overall, the adoption of different types of protected structures in horticulture enables off-season production, higher productivity, better quality produce, and increased profitability for farmers, making them an essential component of modern agricultural systems.

The various types of protected structures can be summarised as follows:

2.0.1. Greenhouses

Greenhouses are fully enclosed structures made with transparent or translucent coverings such as glass, polycarbonate, or UV-stabilized polyethylene film. They allow maximum light penetration while offering precise control over environmental parameters like temperature, humidity, and ventilation. Equipped with systems such as heating, cooling, shading, misting, and CO_2 enrichment, greenhouses create ideal growing conditions for high-value horticultural crops throughout the year. They are classified into naturally ventilated and climate-controlled (fan-and-pad or evaporative cooling) types. Greenhouses enable off-season production, quality improvement, and high yield of crops like tomatoes, capsicum, cucumber, gerbera, and orchids. The structure's framework can be made from galvanized iron (GI), aluminium, or wood, depending on cost and durability requirements. Glass greenhouses are more durable but costlier, while polyethylene-covered ones are more economical but require periodic replacement. In export-oriented floriculture, greenhouses ensure uniform quality that meets international standards. They are essential in regions with extreme climates, enabling consistent production unaffected by external weather fluctuations. Automation in modern greenhouses minimizes labour requirements and enhances efficiency. Crop growth is faster, with reduced pest and disease incidence compared to open-field cultivation. They also support advanced cultivation methods like hydroponics, aeroponics, and NFT (Nutrient Film Technique). Greenhouses are a backbone of protected cultivation in developed countries and are gaining popularity in India due to government subsidies and increasing market demand.

- ☆ **Low-cost polyhouse/greenhouse:** A low-cost polyhouse is constructed using a 700-gauge UV-stabilized polythene sheet, which is tightly secured with the help of bamboo, ropes, and nails. This simple yet effective structure creates a protective covering over crops, reducing damage from wind, rain, and pests. The enclosed environment helps in conserving soil moisture and maintaining a favourable microclimate. Inside the polyhouse, the temperature generally remains 6–10°C higher than the outside environment, which significantly promotes faster germination, better crop growth, and higher yields. This economical option is especially suitable for small and marginal farmers, as it requires minimal investment while offering considerable benefits in off-season and quality crop production (Figure 2.1).

Figure 2.1: Low-Cost Greenhouse.
(*Source*: www.vigyanvarta.com)

☆ **High-tech greenhouse:** A high-tech greenhouse is a scientifically designed structure, usually constructed using strong and durable materials such as galvanized iron or aluminium, ensuring greater strength and longer life compared to low-cost alternatives. These structures are often dome-shaped or cone-shaped, allowing efficient distribution of sunlight and better wind resistance. Unlike traditional soil-based cultivation, high-tech greenhouses rely on specialized growing media such as peat, perlite, solarite, vermiculite, and rock wool, which ensure optimum aeration, water retention, and root development. Although these media are 5–6 times more expensive, they are highly durable and promote uniform crop growth under controlled conditions. In the Indian context, more affordable and locally available substitutes like coco peat, coconut fibres, and rice husks are widely used to make the system cost-effective while still maintaining efficiency. High-tech greenhouses are integrated with modern technologies such as automated climate control, fertigation systems, and advanced irrigation, enabling year-round production of high-value crops, exotic vegetables, and flowers, thus making them highly profitable despite higher initial investment (Figure 2.2).

Figure 2.2: Hi-Tech Greenhouse
(*Source*: www.vigyanvarta.com)

- ☆ **Medium cost greenhouse**: A medium-cost greenhouse serves as a balanced option between low-tech and high-tech structures, offering improved durability and climate regulation at a moderate investment. The framework is generally built using Galvanised Iron (GI) pipes, providing higher strength and longer service life compared to bamboo-based structures. These greenhouses are typically quonset-shaped polyhouses, designed to withstand moderate wind loads and environmental fluctuations. The cladding material used is a single-layered Ultra Violet (UV) stabilized polyethylene sheet of 800-gauge thickness, which effectively protects crops from harmful solar radiation while allowing sufficient light transmission for photosynthesis. To maintain optimal internal conditions, exhaust fans regulated by thermostats are incorporated, ensuring automatic ventilation and temperature control.

Figure 2.3: Medium Cost Greenhouse.
(*Source*: www.vigyanvarta.com)

The structural frame has a lifespan of nearly 20 years, whereas the polyethylene glazing material lasts around 2 years before requiring replacement. This type of greenhouse enables farmers to cultivate vegetables, flowers, and seedlings with greater yield stability, reduced pest incidence, and improved quality, making it a practical option for medium-scale enterprises seeking year-round production with moderate investment (Figure 2.3).

- ☆ **Plastic Tunnels:** Plastic tunnels, also known as low tunnels or walk-in tunnels depending on their height, are one of the most economical forms of protected cultivation used widely for vegetable and flower production. They are constructed by bending galvanised iron (GI) pipes, PVC pipes, or bamboo sticks into semi-circular arches that are covered with transparent UV-stabilised polyethylene film. The covering acts as a protective barrier against frost, heavy rainfall, and strong winds, while at the same time creating a greenhouse effect by trapping solar radiation and raising the temperature inside the tunnel. Depending on their size, plastic tunnels may allow farmers to simply walk through them or may only cover the crop canopy close to the ground.

Figure 2.4: Plastic Tunnels.
(*Source*: www.vigyanvarta.com)

In low tunnels, the internal temperature can be maintained 5–7°C higher than the outside, making them especially useful for early-season planting of cucurbits, solanaceous crops, and leafy vegetables. In walk-in tunnels, farmers can perform intercultural operations such as weeding, spraying, and harvesting conveniently. The tunnels help in extending the growing season, protecting crops from biotic and abiotic stresses, and ensuring higher yields with better quality. They also reduce water requirement by minimising evaporation losses and allow for efficient fertigation. Plastic tunnels are relatively easy to construct, dismantle, and shift from one field to another, thus making them highly suitable for small and marginal farmers who cannot afford permanent greenhouses. In India, they are particularly popular for raising off-season vegetables, hybrid seedlings, flowers like marigold and chrysanthemum, and even strawberries in cooler regions. Their low initial investment, coupled with quick returns, makes them an effective transitional technology between open-field cultivation and fully controlled high-tech greenhouses (Figure 2.4).

The comparative features of low-cost, medium-cost, high-tech greenhouses, and plastic tunnels is summarised in Table 2.1.

Table 2.1: Comparative Features of Low-Cost, Medium-Cost, High-Tech Greenhouses, and Plastic Tunnels

Feature/ Aspect	*Low-Cost Polyhouse*	*Medium-Cost Greenhouse*	*High-Tech Greenhouse*	*Plastic Tunnels*
Framework Material	Bamboo, ropes, nails	Galvanised Iron (GI) pipes	Galvanised Iron/ Aluminium	Bamboo, PVC pipes, or GI hoops
Covering Material	700-gauge UV-stabilised polythene	800-gauge UV-stabilised polyethylene	Glass, polycarbonate sheets, or durable UV-stabilised polyethylene	UV-stabilised polyethylene film
Shape/Design	Simple arch or shed type	Quonset-shaped polyhouse	Dome-shaped/ Cone-shaped	Semi-circular tunnel shape
Temperature Rise	6–10°C higher than outside	6–8°C higher with exhaust fan control	Precisely regulated with sensors	5–7°C higher than outside
Ventilation System	Natural (manual opening)	Exhaust fans with thermostat	Fully automated climate control (fans, cooling pads, foggers)	Natural ventilation (open ends/ vents)
Growing Media	Mostly soil-based	Soil or soil-less media	Peat, perlite, vermiculite, rock wool (coco peat/rice husk in India)	Mainly soil, sometimes mulched with plastic
Durability	2–3 years	Frame ~20 years, film ~2 years	Frame 25–30 years, cladding 10–15 years	1–2 seasons (short lifespan)

Feature/ Aspect	*Low-Cost Polyhouse*	*Medium-Cost Greenhouse*	*High-Tech Greenhouse*	*Plastic Tunnels*
Cost	Very low	Moderate	5–6 times costlier than medium-cost	Very low (cheapest option)
Suitable For	Small/marginal farmers, off-season crops	Medium-scale farmers, vegetables, flowers	High-value crops, exotic vegetables, export-oriented farming	Early-season vegetables, flowers, strawberries
Advantages	Low investment, simple to build, good for off-season crops	Moderate control of climate, durable frame, year-round production	Precise control of all parameters, high yields, export quality	Very low cost, portable, season extension, easy to install
Limitations	Less durable, limited climate control	Film needs replacement every 2 years, higher cost than low-tech	High investment, requires technical knowledge	Short lifespan, less protection against extreme conditions

2.0.2. Polyhouses

Polyhouses are greenhouse-like structures covered with polyethylene sheets, generally UV-stabilized for durability against solar degradation. They are cost-effective alternatives to glass greenhouses and are widely adopted by small and medium farmers. Polyhouses offer moderate climate control by maintaining optimum temperature and humidity, protecting crops from adverse weather such as frost, heavy rains, and scorching sunlight. They are suitable for vegetables, flowers, and nursery raising. The frame can be made from bamboo, GI pipes, or wood, with galvanized iron being preferred for longer life. Polyhouses can be naturally ventilated, relying on side roll-up curtains for airflow, or climate-controlled with fans, pads, and foggers. They extend the growing season and allow farmers to grow crops like capsicum, tomato, gerbera, chrysanthemum, and strawberry during off-season periods. Polyhouses also improve water-use efficiency through drip irrigation and fertigation systems. They reduce pesticide usage by providing a barrier against pests, thus supporting organic production. Construction and operational costs are relatively lower than high-tech greenhouses, making them more accessible in rural and peri-urban areas. They are easy to repair and maintain, with plastic covers replaced every 3–5 years depending on quality. Polyhouses contribute to higher yields, better quality produce, and improved profitability. They are an ideal entry point for farmers transitioning from open-field cultivation to protected farming systems.

2.0.3. Shade Houses

Shade houses are structures covered with shade nets made from HDPE or synthetic fibres to reduce light intensity and protect crops from harsh sunlight. They regulate temperature, lower evapotranspiration rates, and provide a favourable microclimate for shade-loving or partial-shade plants like orchids, anthuriums, ferns, and leafy greens. Shade percentages vary (35 per cent, 50 per cent, 75 per cent, 90 per cent) depending on crop requirements. The frames are constructed from bamboo, GI pipes, or wood, with nets fixed over the top and sometimes sides. Shade houses are particularly useful in tropical and subtropical climates where excessive solar radiation can damage crops. They protect plants from wind, hail, and bird damage while allowing adequate air circulation. By reducing temperature and conserving moisture, they enhance seed germination, rooting, and vegetative growth. In horticulture, shade houses are used extensively for nursery raising, hardening tissue culture plants, and floriculture production. They also help in producing quality grafts and seedlings for fruit crops. Some shade houses are equipped with misting systems to increase humidity for propagation purposes. The structure's design can be fixed or retractable to adjust shading levels as per crop stage. Low maintenance and long life of nets make shade houses a cost-effective solution. They also prevent leaf scorching and improve aesthetic quality of ornamental plants, which is vital for market acceptance.

2.0.4. Net Houses (Insect-Proof Net Houses)

Net houses are protective structures enclosed with fine mesh nets to physically block insect pests, birds, and excessive sunlight from reaching crops. The mesh size, often 40–60 mesh, is chosen to exclude common insect vectors like whiteflies, thrips, and aphids, which spread viral diseases in crops. Net houses create a barrier that reduces the need for chemical pesticides, supporting integrated pest management (IPM) and organic farming. The structure is usually made with GI pipes or bamboo frames, with nets stretched tightly over all sides and the roof. They provide partial shade, moderate temperature control, and protection against physical damage from wind or hail. Net houses are widely used for high-value crops such as capsicum, tomato, cucumber, and flowers like marigold and chrysanthemum. They are also suitable for seed production programs where genetic purity and disease-free conditions are essential. The ventilation in net houses ensures good air exchange, preventing fungal diseases due to excess humidity. In some designs, net houses are combined with plastic covers on the roof for rain protection. Maintenance involves periodic inspection and repairing torn nets to maintain pest exclusion efficiency. Net houses can be permanent or seasonal, depending on the crop cycle. They are a sustainable solution for producing residue-free vegetables and flowers for urban markets.

2.0.5. Low Tunnels/Row Covers

Low tunnels, also called row covers, are miniature greenhouse-like structures made of transparent plastic film or netting stretched over low wire hoops along crop rows. They create a warm, humid microclimate around plants, protecting them from frost, wind, rain, and insect attack. Low tunnels are inexpensive and easy to construct using locally available materials like bamboo sticks or mild steel rods. They are mainly used for early-season vegetable cultivation such as cucumbers, melons, beans, and leafy greens, enabling farmers to advance planting dates and harvest earlier. By trapping solar heat during the day, low tunnels protect crops from cold nights, extending the growing season. The plastic covering is perforated or lifted periodically for ventilation to prevent overheating. In insect-proof versions, fine mesh fabric is used instead of plastic film. Low tunnels are particularly beneficial in temperate and subtropical regions where frost and low temperatures limit crop growth in winter. They also conserve soil moisture and reduce weed growth by creating a covered environment. The simplicity of design allows rapid installation and dismantling, making them a flexible option for smallholders. They require minimal investment and labour, offering high returns when used for high-value crops during market shortage periods.

2.0.6. Walk-in Tunnels

Walk-in tunnels are temporary, semicircular or hoop-shaped structures tall enough for a person to enter and manage crops comfortably. They are generally made from galvanized iron (GI) pipes or bamboo frames, covered with transparent or semi-transparent polyethylene film. The height typically ranges from 2 to 3 meters, providing ample space for crop growth and cultural operations. Walk-in tunnels offer partial climate control, protecting crops from frost, heavy rainfall, wind, and insect infestation. They are commonly used for off-season vegetable production, raising flower seedlings, and nursery crops. Ventilation is provided by roll-up side curtains or open ends to avoid overheating. The design allows easy dismantling and relocation as per cropping needs, making them highly flexible for farmers with limited land. Walk-in tunnels help in advancing planting schedules, thereby fetching higher market prices for early-season produce. They also improve crop quality by preventing direct contact with rainwater, which reduces disease incidence. In addition, the warm and humid microclimate inside accelerates growth and increases yield. Their construction is relatively low-cost compared to permanent greenhouses, yet they offer similar protective benefits. Many small and medium-scale farmers prefer walk-in tunnels for their affordability, portability, and high return on investment.

2.0.7. Plastic Tunnels

Plastic tunnels are protective structures similar to low tunnels but larger in size and designed for extended crop coverage. They are semicircular or gothic-shaped, constructed using steel, aluminium, or bamboo frames covered with UV-stabilized polyethylene sheets. Plastic tunnels create a warm, humid microclimate ideal for early or late production of vegetables, flowers, and certain fruit crops. The trapped heat helps in frost protection during winter and accelerates crop growth.

These structures can be either low-height for quick seasonal use or high tunnels for longer crop cycles. Ventilation is managed through roll-up sides or removable covers to prevent overheating during sunny days. Plastic tunnels also conserve soil moisture, reduce weed growth, and shield crops from wind and heavy rainfall. They are widely used in strawberry, capsicum, cucumber, and melon cultivation. Being cost-effective and easy to install, plastic tunnels are a popular choice among smallholders and commercial growers alike. Their modular design allows easy extension or relocation. Maintenance mainly involves periodic replacement of the plastic film, typically every 3–5 years. By enabling out-of-season cultivation and improving produce quality, plastic tunnels help farmers secure better market prices.

2.0.8. Cold Frames

Cold frames are low, box-like structures with transparent or translucent covers made from glass, acrylic, or polyethylene. They trap solar heat during the day, creating a slightly warmer environment than the outside air, thus protecting plants from frost and low temperatures. Cold frames are typically used for hardening seedlings, rooting cuttings, or extending the growing season in cool climates. They are inexpensive and can be built using wood, bricks, or concrete for the base, with a hinged lid for easy access and ventilation. The slope of the cover allows maximum sunlight penetration while facilitating water runoff. In horticulture, cold frames are particularly useful for leafy vegetables, herbs, and early-season flower crops. They also serve as an intermediate stage between greenhouse propagation and open-field planting, allowing plants to acclimatize to outdoor conditions gradually. By maintaining a warmer environment at night, cold frames prevent cold injury to young plants. In addition to protecting from frost, they reduce wind stress and conserve soil moisture. Cold frames require minimal maintenance, making them a cost-effective option for small-scale growers and hobby gardeners. They can be combined with heating mats or insulation for additional warmth in extremely cold regions.

2.0.9. Mist Chambers

Mist chambers are specialized enclosed structures designed to maintain high humidity levels through fine water misting, promoting rapid rooting and plant propagation. They are commonly used for vegetative propagation of cuttings, nursery production, and acclimatization of tissue culture plants. The chamber frame is made from GI pipes or wood, covered with polythene or fiberglass sheets, with misting nozzles installed overhead. The system is connected to a timer or humidity sensor to regulate mist frequency, ensuring the air remains saturated without waterlogging the plants. High humidity reduces transpiration stress, preventing wilting of cuttings before root formation. Mist chambers are widely used in horticulture for crops like roses, carnations, gerbera, and fruit tree grafts. They can also be used for raising delicate seedlings during hot, dry seasons. Temperature inside is moderated by evaporative cooling from misting, creating a favourable environment for rooting. Proper drainage is essential to prevent fungal diseases. Maintenance includes cleaning nozzles, ensuring leak-free water supply, and controlling algae growth on surfaces. Mist chambers can be small for home nurseries or large for commercial propagation units. They significantly improve rooting success rates, shorten propagation time, and increase overall nursery productivity.

2.0.10. Rain Shelters

Rain shelters are simple protective structures designed primarily to shield crops from excessive rainfall, preventing waterlogging and disease spread. They typically consist of bamboo, wooden, or GI pipe frames covered with UV-stabilized polyethylene or transparent plastic sheets on the roof, while sides may remain open for ventilation. Rain shelters are especially useful for high-value crops like capsicum, tomato, cucumber, and flowers in areas with heavy monsoon rains. By preventing direct contact between rainwater and foliage, they reduce the incidence of fungal diseases like blight and mildew. The dry environment inside also improves pollination and fruit set in certain crops. Rain shelters are cost-effective and can be easily erected before the rainy season and dismantled afterward. In addition to rain protection, they offer partial shade, wind reduction, and microclimate moderation. They also help in minimizing soil erosion and nutrient leaching due to heavy rainfall. Farmers in hilly and high-rainfall regions particularly benefit from rain shelters as they ensure continuous production even during wet months. The durability of rain shelters depends on the quality of covering material and frame construction. Proper design ensures water runoff from the roof without leakage onto crops. Overall, rain shelters are a practical, low-cost solution for protecting crops during monsoon seasons.

2.1 Classification Based on Site

Protected structures can be classified according to the site or location where they are established, as the surrounding environmental conditions play a decisive role in determining their design, materials, and operational efficiency. Site-related factors directly influence the internal microclimate, structural stability, and the cost-effectiveness of the system. Important considerations include topography, which affects construction feasibility and drainage; climate, which determines the type of covering material and ventilation requirements; and soil type, which impacts root development and irrigation design. Wind direction and speed are critical for structural safety, as strong winds can cause damage if not accounted for in orientation and bracing. Proximity to water sources ensures reliable irrigation and fertigation, while access to markets is vital for timely sale of perishable produce. A well-selected site minimizes energy costs for heating, cooling, or pumping water, while a poorly chosen location can lead to operational inefficiencies and crop losses. The site also influences pest and disease incidence, as certain locations may harbour higher insect populations or pathogens. In regions prone to natural hazards like cyclones, floods, or frost, the protected structure must be adapted accordingly for resilience. Overall, proper site selection is not just a preliminary step but a long-term investment that ensures optimal crop growth, efficient resource utilization, reduced maintenance costs, and sustained profitability in protected cultivation.

2.1.1. Plain Area Structures

Plain area structures are designed for flat or gently sloping lands, making construction simpler, faster, and more economical. These locations require minimal land levelling, reducing the overall cost of site preparation. Foundations in such areas can be shallow yet stable, allowing for the easy installation of polyhouses, net houses, and greenhouses. They are ideal for a wide range of horticultural crops due to moderate climatic conditions and easy access to irrigation. Orientation and design can be optimized for maximum sunlight capture without major adjustments for slope or elevation. Transportation of materials and market access are generally easier in plain areas. Since wind speeds tend to be moderate, less reinforcement is needed compared to hilly or coastal structures. Drainage systems are simpler to install, and mechanized operations can be performed more efficiently. The relative ease of construction in plains also makes expansion more feasible. Overall, plain area structures offer flexibility, lower costs, and suitability for diverse crop cultivation.

2.1.2. Hilly or Sloping Area Structures

Hilly or sloping area structures are adapted for uneven terrains, where flat construction space is limited. These structures are often built on stepped or terraced bases to create level surfaces for cultivation. Additional engineering measures, such as reinforced foundations, windbreaks, and anchoring systems, are required to withstand strong winds and heavy rainfall common in hilly regions. Orientation is carefully planned to maximize sunlight exposure despite varying slopes. Materials need to be lightweight yet strong, making GI pipes and aluminium preferred over heavy steel. Drainage is a major consideration, as rapid water runoff can erode soil and damage foundations. Crops grown in such areas benefit from cooler temperatures and high sunlight intensity, making them ideal for floriculture and temperate vegetables. However, transportation of construction materials and produce may be more challenging due to steep and narrow roads. Skilled labour is often required for precise installation. Well-designed hilly area structures can make effective use of otherwise difficult terrains.

2.1.3. Coastal Area Structures

Coastal area structures are specifically designed to handle high humidity, saline winds, and occasional cyclonic storms. The salty environment accelerates corrosion, so materials like galvanized iron, aluminium, or PVC-coated frames are preferred to ensure durability. The design must incorporate strong wind-resistant features such as bracing, deep anchoring, and aerodynamic roof shapes. Covering materials should be UV-stabilized and resistant to salt damage. Adequate ventilation is critical to prevent fungal diseases promoted by high humidity. Coastal structures may also include cyclone nets or protective barriers to reduce wind damage. Roof slopes are designed to withstand heavy rainfall while ensuring rapid water drainage. Crops like orchids, anthuriums, cucumbers, and certain exotic vegetables thrive well when humidity is controlled. Access to abundant water is an advantage, but it often requires desalination or treatment to reduce salinity for irrigation. With proper design, coastal structures can operate successfully despite challenging environmental conditions.

2.1.4. Arid or Semi-Arid Zone Structures

Arid and semi-arid zone structures are located in regions with high temperatures, intense solar radiation, low humidity, and scarce water resources. These structures prioritize shading, evaporative cooling, and water conservation measures to create favourable microclimates for crop growth. Shade nets, misting systems, and foggers are commonly integrated to reduce heat stress. Materials must withstand extreme temperature fluctuations and strong winds typical of desert areas. Drip irrigation and fertigation systems are essential for efficient water use. Roofs are often designed with reflective coatings to reduce heat absorption. Crops grown here include melons, tomatoes, capsicum, and leafy greens, often targeted for high-value markets. Solar-powered ventilation and pumping systems are also gaining popularity to reduce energy costs. Sand and dust storms are a concern, so protective barriers and sealed joints are incorporated into the design. These structures enable successful horticulture in regions where open-field cultivation would be highly risky and unproductive.

2.1.5. High Rainfall Zone Structures

High rainfall zone structures are built in tropical and monsoon-prone regions where heavy and prolonged rains can damage crops. The primary design feature is a steeply sloped roof to allow quick water runoff and prevent accumulation that can collapse the structure. Roof and side coverings are often made from UV-stabilized polyethylene or polycarbonate sheets to protect against rain splash. Adequate side covers and gutters prevent soil erosion and waterlogging. Frames are reinforced to handle the weight of water during downpours. Ventilation is carefully managed to avoid excessive humidity, which can lead to fungal diseases. Crops like capsicum, cucumber, tomato, and cut flowers benefit greatly from such structures as they remain dry and disease-free. The design also takes into account strong winds often accompanying heavy rains. Proper drainage channels around the structure ensure excess water is quickly diverted. These structures provide reliable protection during long rainy seasons, enabling year-round production. The classification based on site is crucial because it determines not only the design and materials used but also the crop selection, irrigation systems, and environmental control measures needed for successful protected cultivation.

2.2 Classification Based on Climate

Protected structures can also be classified according to the prevailing climatic conditions of the region in which they are constructed. Climate plays a decisive role in determining the type of structure, its covering material, ventilation arrangements, and environmental control mechanisms. Each climatic zone presents distinct challenges such as extreme temperatures, high humidity, low moisture levels, heavy snowfall, or intense rainfall, all of which require specific design adaptations to ensure optimal crop growth and productivity. The design must carefully account for temperature range, humidity levels, light intensity, wind speed, and seasonal rainfall patterns. Choosing a structure suited to local climate conditions not only enhances yield and produce quality but also minimizes operational costs by reducing the need for intensive heating, cooling, or dehumidification systems. Based on climatic factors, protected structures may be classified into the following types:

2.2.1. Hot and Arid Climate Structures

Regions with hot and arid climates are characterized by extremely high temperatures, intense solar radiation, low relative humidity, and scarce water resources. The primary goal of protected structures in these areas is to reduce internal heat load, prevent excessive water loss, and create a suitable microclimate for crop growth. Shade nets, aluminized reflective films, or double-layer polyethylene covers are often used to lower temperature and light intensity. Whitewashing or reflective paint may be applied to roofs for additional heat reduction. Evaporative cooling systems such as fan-and-pad arrangements, foggers, or misting devices are crucial for maintaining optimum temperatures. Ventilation openings are kept large to ensure rapid air circulation and prevent heat buildup. Dust-proof insect nets are installed to block sand particles and pests in desert regions. Drip irrigation combined with fertigation ensures precise water and nutrient supply with minimal wastage. Crops like capsicum, tomatoes, cucumbers, leafy greens, and melons thrive well under such structures in these regions.

2.2.2. Hot and Humid Climate Structures

These structures are suited for tropical and coastal regions where high humidity, moderate to high temperatures, and heavy rainfall are common. The main challenge here is to prevent fungal and bacterial diseases caused by excessive moisture while ensuring adequate ventilation. Naturally ventilated polyhouses with wide side openings, ridge vents, and roll-up side curtains are preferred to promote cross-ventilation and humidity control. Insect-proof nets with large mesh sizes allow air movement while keeping pests out. Roof designs often include overhangs or gutters to direct rainwater away from crops. Rain shelters with UV-stabilized polyethylene or shade nets protect plants from heavy downpours while maintaining light diffusion. Corrosion-resistant materials like galvanized iron or aluminium are essential to withstand saline winds in coastal areas. Dehumidifiers or simple forced ventilation systems may be used during prolonged wet conditions. Crops like orchids, anthuriums, heliconias, leafy vegetables, and shade-loving flowers perform well in these structures.

2.2.3. Temperate Climate Structures

Temperate regions experience moderate climatic conditions but often have distinct seasons with cold winters and warm summers. Protected structures in such zones are designed for year-round production by balancing heating needs in winter and cooling requirements in summer. Polyhouses with roll-up sides, retractable shade nets, and thermal screens are commonly used to regulate temperature and light. In winter, heating systems such as hot air blowers, thermal blankets, or geothermal heating help maintain suitable conditions for crops. During summer, shading and natural ventilation reduce heat buildup. The flexibility of these structures allows for cultivating a wide range of crops such as tomatoes, capsicum, cucumbers, cut flowers, and fruit seedlings throughout the year. Structural designs often incorporate insulation layers and adjustable openings for precise environmental control. Rainwater harvesting systems may be added to supplement irrigation in dry months, enhancing resource efficiency.

2.2.4. Cold and Snow-Prone Climate Structures

In high-altitude and snow-prone areas, the primary objective is to retain heat and withstand heavy snow loads. Such structures require robust frames made of galvanized steel or aluminium and steep roof slopes to facilitate snow shedding. Covering materials like double-glazed glass or double-layer polyethylene films improve insulation and reduce heat loss. Thermal curtains or insulated panels may be installed for extra warmth retention during extreme cold. Heating systems such as hot water pipes, biomass heaters, or geothermal heat pumps are often used. Snow load calculations are essential during design to ensure structural safety. Sidewalls are insulated, and openings are minimized to conserve heat. These structures allow the production of leafy greens, herbs, strawberries, and seedlings during extended cold seasons when open-field cultivation is impossible. Adequate light transmission is ensured despite the insulating materials to maintain healthy plant growth.

2.2.5. High Rainfall and Monsoon Climate Structures

These structures are found in tropical and subtropical zones with prolonged heavy rainfall and high humidity during monsoon seasons. The key design consideration is efficient rainwater management to prevent waterlogging and crop damage. Steeply sloped roofs, extended overhangs, and durable gutter systems ensure rapid water drainage. Sidewalls may be partially covered with transparent or semi-transparent materials to allow light while protecting crops from rain splash. Floors are elevated or provided with well-drained beds to avoid root zone saturation. Materials are chosen for their resistance to prolonged wet conditions and fungal growth. Natural ventilation through ridge vents or roll-up sides helps maintain air movement and reduces humidity-related disease risks. These structures are well-suited for crops like capsicum, cucumbers, tomatoes, cut flowers, and medicinal plants that require protection from excess rain while still benefiting from natural light. Rainwater harvesting systems can also be integrated to store water for use in drier months, improving sustainability.

2.3 Types of Structures: Greenhouse, Polyhouse, Shade Net House

Protected cultivation employs various types of structures designed to create favourable microclimatic conditions for crop growth, irrespective of external weather fluctuations. Among the most widely used are greenhouses, polyhouses, and shade net houses, each serving distinct purposes and offering unique advantages. A greenhouse is a framed structure covered with transparent or translucent material, allowing maximum light penetration while enabling precise control over temperature, humidity, and ventilation. It is ideal for high-value crops requiring year-round production under stable environmental conditions. A polyhouse is a type of greenhouse covered with UV-stabilized polyethylene film, offering cost-effective protection and climate modification for vegetables, flowers, and nursery plants. Its flexibility in design and affordability make it popular among medium and small-scale farmers. A shade net house, on the other hand, uses knitted or woven shade nets to reduce light intensity, protect crops from harsh sunlight, wind, and pests, and maintain moderate temperatures–making it suitable for shade-loving plants and nursery raising. The selection among these structures depends on factors like crop type, climate, investment capacity, and intended level of environmental control. While greenhouses allow advanced automation and year-round cultivation, polyhouses strike a balance between cost and control, and shade net houses provide an economical solution for partial protection needs. Understanding the design, functionality, and application of each type is essential for maximizing productivity and profitability in protected cultivation.

2.3.1 Greenhouse

A greenhouse is a purpose-built structure designed to create and maintain an optimal environment for plant growth by providing protection from unfavourable external weather conditions and enabling precise control over internal climatic factors. The structure is typically covered with transparent or translucent cladding materials such as glass, polycarbonate sheets, or UV-stabilized polyethylene films, which allow maximum transmission of sunlight while diffusing the light to prevent scorching and ensure uniform distribution. The enclosed design traps solar radiation, creating a "greenhouse effect" that raises the internal temperature, making it possible to cultivate crops even during colder seasons. Inside a greenhouse, environmental parameters such as temperature, relative humidity, light intensity, CO_2 concentration, and air movement can be adjusted through a range of systems, including natural or forced ventilation, heating units, evaporative cooling pads, shade curtains, and artificial lighting. This level of control helps in extending the growing season, enabling off-season production, and achieving consistently high-quality yields. Greenhouses are constructed using various frame materials like galvanized iron, aluminium, bamboo, or wood, chosen based on cost, durability, and intended lifespan. The design may vary from simple tunnel types to advanced multi-span structures integrated with fully automated climate control systems. The advantages of greenhouse cultivation include protection from heavy rainfall, strong winds, hailstorms, frost, and extreme heat, along with reduced pest and disease incidence due to the physical barrier. These structures are ideal for growing high-value crops such as roses, gerberas, capsicum, cucumber, tomato, and strawberries, as well as for nursery raising and tissue culture plant acclimatization. In commercial horticulture, greenhouses are extensively used for export-oriented production, as the controlled environment ensures produce that meets stringent international quality standards.

Importance of Greenhouse

Greenhouses are crucial for creating a controlled environment that allows cultivation of crops throughout the year, regardless of adverse external weather conditions. By regulating temperature, humidity, CO_2 concentration, and light intensity, they enable consistent, high-quality yields and extend the growing season. Their ability to protect crops from heavy rainfall, frost, hailstorms, strong winds, and extreme heat reduces crop losses and ensures better productivity. Greenhouses are particularly valuable for high-value and export-oriented crops such as roses, gerberas, capsicum, cucumbers, tomatoes, and strawberries, as they meet strict quality standards required in global markets. They also play a key role in nursery raising, tissue culture acclimatization, and producing disease-free planting material.

2.3.2 Polyhouse

A polyhouse is a specialized type of greenhouse where the structural frame is covered with UV-stabilized polyethylene (PE) film instead of glass or polycarbonate sheets. This covering material is flexible, lightweight, and cost-effective, yet durable enough to withstand various climatic conditions for a specified lifespan (usually 3–5 years for quality films). The UV stabilization prevents degradation of the plastic due to prolonged exposure to sunlight, while the transparency of the film allows sufficient sunlight for photosynthesis. Polyhouses utilize the same principle of the greenhouse effect, trapping solar radiation to maintain warmer internal temperatures compared to the outside environment. The internal microclimate in a polyhouse can be manipulated through systems for natural ventilation, forced air circulation, evaporative cooling, shading, and drip or sprinkler irrigation. In more advanced designs, polyhouses are equipped with automated climate control systems that regulate temperature, humidity, and CO_2 levels to optimize plant growth. The structure can be single-span or multi-span, depending on the scale of production, and is generally supported by galvanized iron (GI) pipes for corrosion resistance and strength. Polyhouses offer several advantages, including extended growing seasons, higher yields, better crop quality, and reduced incidence of pests and diseases due to the protective barrier. They also minimize the effects of heavy rains, strong winds, hailstorms, and excessive sunlight. By moderating extreme temperature fluctuations, polyhouses make it possible to cultivate crops in regions or seasons where open-field cultivation would not be feasible. This type of protected structure is widely used for producing high-value horticultural crops such as tomatoes, capsicum, cucumbers, roses, gerberas, chrysanthemums, and exotic vegetables like lettuce and broccoli. Due to their relatively lower construction cost compared to glass greenhouses, polyhouses are especially popular among small and medium-scale farmers. Additionally, they play an important role in export-oriented horticulture, as they allow growers to produce uniform, high-quality produce that meets international market standards throughout the year.

Importance of Polyhouse

Polyhouses provide an affordable yet effective alternative to glass greenhouses, making climate-controlled cultivation more accessible to small and medium-scale farmers. By maintaining a favourable microclimate using UV-stabilized polyethylene film, polyhouses allow the production of crops in off-season or in unsuitable climates, increasing farmer income and market competitiveness. Their lower construction cost, combined with the ability to protect against heavy rains, strong winds, and excessive sunlight, makes them ideal for diversified horticulture production. Polyhouses contribute to higher yields, superior crop quality, and reduced pest and disease incidence, which benefits both domestic markets and export-oriented horticulture. They are especially useful for cultivating vegetables, flowers, and exotic crops that command premium prices year-round.

2.3.3 Shade Net House

A shade net house is a framed structure covered with specially manufactured shade nets made from high-density polyethylene (HDPE) or other durable synthetic fibres, designed to filter and diffuse sunlight reaching the crops. These nets are treated with UV stabilizers to prevent degradation from prolonged sun exposure and to maintain their strength over several years. Shade nets are available in different shade percentages–commonly 25 per cent, 35 per cent, 50 per cent, 75 per cent, and 90 per cent –which indicate the proportion of sunlight blocked. The choice of shading percentage depends on the light requirements of specific crops, with lower percentages suited for sun-loving plants and higher percentages for shade-loving species. The primary function of a shade net house is to reduce excessive solar radiation and protect crops from harmful effects such as leaf scorching, wilting, and sunburn, while still allowing adequate light for photosynthesis. Additionally, it provides protection from heavy rain, strong winds, hailstorms, and bird damage. The microclimate inside a shade net house is generally cooler than the outside, which helps in lowering plant stress, reducing evapotranspiration, and saving irrigation water. Shade net houses can be naturally ventilated since the porous netting allows free airflow, preventing excessive humidity build-up and reducing the risk of fungal diseases. They can be constructed using galvanized iron pipes, bamboo, or wood for the framework, making them adaptable for both low-cost temporary setups and more permanent installations. These structures are particularly useful for raising nurseries, hardening tissue-cultured plants, and growing shade-loving or semi-shade crops such as orchids, ferns, anthuriums, lettuce, spinach, and certain medicinal herbs. They are also widely used in horticulture for protected seedling production before transplanting in open fields or other protected structures. Due to their lower construction cost and simpler design compared to greenhouses or polyhouses, shade net houses are an affordable option for small and marginal farmers. They are easy to install, require minimal maintenance, and can be dismantled or relocated when needed. By offering an intermediate level of protection, shade net houses enable farmers to improve crop quality, reduce losses from extreme weather, and achieve better market prices with relatively low investment.

Importance of Shade Net House

Shade net houses are essential for protecting crops from excessive sunlight, leaf scorching, and other forms of heat stress, while still allowing adequate light for photosynthesis. They are particularly beneficial for raising seedlings, hardening tissue culture plants, and growing shade-loving or semi-shade crops like orchids, lettuce, and certain medicinal herbs. By reducing temperature and evapotranspiration, they save irrigation water and improve plant health. Shade net houses also protect against heavy rain, hail, strong winds, and bird damage, helping to maintain crop quality and minimize losses. Their low cost, easy installation, and portability make them an attractive option for small and marginal farmers, offering a cost-effective solution to improve crop production and secure better prices in the market.

Comparison between Greenhouse, Polyhouse and Shade Net House is depicted in Table 3.1.

Table 3.1: Comparison between Greenhouse, Polyhouse and Shade Net House

Feature	*Greenhouse*	*Polyhouse*	*Shade Net House*
Cladding Material	Glass, polycarbonate sheets, or polyethylene film	UV-stabilized polyethylene (PE) film	UV-stabilized HDPE shade nets
Light Transmission	Very high (glass ~90 per cent, polycarbonate ~80 per cent)	High (~85–90 per cent for clear PE film)	Moderate (depends on shading per cent : 25 90 per cent)
Temperature Control	Excellent – can be fully climate-controlled	Good – can be natural or climate-controlled	Limited – mainly reduces temperature through shading
Humidity Control	Excellent with automated systems	Good with ventilation and fogging systems	Limited – natural airflow only
Protection from Rain/Wind	Very high	High	Moderate
Protection from Sunlight	High with shading devices	High with shading nets inside	Moderate to high depending on shade percentage

Feature	*Greenhouse*	*Polyhouse*	*Shade Net House*
Durability	10–20 years (glass), 5–10 years (polycarbonate)	3–5 years for film; frame lasts 15+ years	3–5 years for net; frame lasts 5–10 years
Construction Cost	High (most expensive)	Medium (cheaper than glass greenhouse)	Low (most affordable)
Maintenance Requirement	High	Moderate	Low
Suitable Crops	High-value crops like roses, gerberas, capsicum, cucumber, strawberries, exotic vegetables	Vegetables (tomato, capsicum, cucumber), flowers, exotic leafy greens	Shade-loving crops (orchids, ferns, lettuce), nurseries, hardening plants
Season Extension	Excellent – year-round cultivation possible	Very good – extended growing season	Limited – mainly for hot/sunny seasons
Best Use	Precision farming, export-quality production	Affordable commercial protected cultivation	Low-cost crop protection and nursery raising

Cladding Materials for Protected Structures

3.0 Introduction: Cladding Materials for Protected Structures

Cladding material refers to the transparent or translucent covering mounted on the walls and roof of a greenhouse to allow sunlight to enter while providing protection from adverse weather conditions. It plays a crucial role in controlling light transmission, temperature, humidity, and overall microclimate inside the structure. Common cladding materials include glass, polyethylene films, polycarbonate sheets, acrylic, and fiberglass, each with distinct properties and lifespan. The choice of material depends on crop requirements, climatic conditions, cost, and durability. Proper selection and maintenance of cladding ensure optimal plant growth and higher productivity.

Cladding materials form one of the most crucial components of any protected cultivation structure, as they act as the physical barrier between the internal controlled environment and the external natural conditions. The choice of cladding material determines the efficiency, durability, light transmission, thermal performance, and overall cost-effectiveness of the structure. In horticultural protected cultivation, cladding materials are carefully selected to optimize crop growth by ensuring adequate sunlight penetration, maintaining suitable temperatures, reducing excessive evaporation, and protecting plants from wind, rain, hail, frost, and pests. The materials used vary according to the type of structure, climatic zone, crop requirement, budget, and lifespan expectations. Cladding may be rigid or flexible, transparent or translucent, single-layered or multi-layered, and may also include UV stabilization or thermal modification to improve performance. Historically, glass was the earliest cladding material used for greenhouses, offering high transparency and durability, but it is heavy, brittle, and requires a strong supporting frame, making it costly for large-scale use in developing regions. The development of modern polymers revolutionized protected cultivation by introducing lightweight, affordable, and versatile cladding options such as polyethylene films, polycarbonate sheets, and acrylic sheets. Polyethylene films, especially UV-stabilized low-density polyethylene (LDPE), are now the most common in polyhouses due to their low cost, easy installation, and good light diffusion properties. However, they require periodic replacement every 2–4 years depending on thickness and quality. Polycarbonate sheets, available in single or multiwall designs, combine good light transmission with excellent impact resistance and better thermal insulation, making them suitable for high-tech greenhouses, though their cost is higher than polyethylene. Acrylic and fiberglass-reinforced plastic (FRP) sheets are also used in specific applications where durability and higher light transmission are required.

Apart from polymer-based materials, shade nets and insect-proof nets are employed in shade net houses and net houses, where the primary objective is to reduce light intensity, prevent insect entry, or control microclimate without complete environmental isolation. These nets come in different mesh sizes, colours, and shading percentages to suit different crops and seasons. Shade nets, typically made from high-density polyethylene (HDPE), are UV-stabilized for longer life and available in green, black, white, or aluminized forms. Insect-proof nets with fine mesh sizes protect crops from pests such as whiteflies, thrips, and aphids while still allowing sufficient ventilation.

The physical properties of cladding materials–such as thickness, tensile strength, elongation, tear resistance, light transmittance, thermal conductivity, and resistance to weathering–directly influence their suitability for a given location and crop type. For example, in cooler climates, multiwall polycarbonate sheets provide better insulation to reduce heating costs, while in hot tropical regions, diffused light polyethylene films or aluminized shade nets can prevent heat stress and improve crop quality. Additionally, advanced cladding technologies now include anti-drip, anti-dust, and infrared (IR)-blocking coatings to further improve greenhouse efficiency. Anti-drip films prevent condensation droplets from forming and dripping on plants, reducing fungal disease risk, while IR-blocking materials help retain heat during cold nights or reduce excess heat during the day, depending on the requirement.

Durability and maintenance are equally important considerations. While glass may last decades if properly maintained, plastic films are prone to degradation from UV exposure, wind abrasion, and handling damage, requiring regular inspection and timely replacement. Cost analysis often involves balancing the initial investment with operational benefits, crop yield improvement, and lifespan of the material. Environmental concerns have also gained importance, leading to a focus on recyclable plastics and sustainable materials in modern protected cultivation systems.

Ultimately, cladding materials are not merely coverings–they are active components of the protected cultivation system that directly influence plant physiological processes, microclimate control, and economic viability. An informed selection, based on scientific understanding of material properties and local agro-climatic conditions, is essential for maximizing productivity and ensuring long-term sustainability of protected structures. This chapter explores in detail the various cladding materials used in different types of protected cultivation structures, their properties, advantages, limitations, and suitability for specific horticultural applications, providing growers, researchers, and policymakers with a comprehensive knowledge base for informed decision-making.

3.1 Types of Cladding Materials

Cladding materials for protected cultivation can be broadly classified based on their physical form, composition, light transmission capacity, thermal properties, and suitability for specific structural applications. They serve as the primary barrier between the external environment and the controlled conditions inside the structure, directly influencing crop growth, yield, and quality. These materials may be rigid or flexible, transparent or diffused, single-layered or multi-layered, and may include functional additives like UV stabilizers, anti-drip agents, or infrared modifiers. Common categories include traditional glass, modern polymer films such as polyethylene, rigid thermoplastics like polycarbonate and acrylic, fiberglass-reinforced sheets, as well as shade and insect-proof nets. Each material type offers unique advantages and limitations in terms of durability, cost, light transmission, insulation, and maintenance requirements. Selection depends on factors such as climatic zone, crop type, structure design, and economic feasibility. Understanding these classifications helps growers choose the most suitable cladding for efficient and sustainable protected cultivation. The major types are explained below.

3.1.1 Glass

Glass is the oldest and most traditional cladding material used in greenhouses, valued for its exceptional clarity and ability to transmit 88–92 per cent of sunlight, which is ideal for photosynthesis. It is chemically inert, UV-stable, and resistant to scratching, ensuring long-term performance without significant degradation in light quality. Its durability allows it to last several decades with proper maintenance, making it suitable for permanent, high-end greenhouse structures. However, its heavy weight requires strong supporting frames, and its brittleness makes it prone to shattering under impact. Despite these drawbacks, glass remains a preferred choice for research facilities, botanical gardens, and premium commercial greenhouses where maximum light penetration and longevity are priorities.

Properties

- ☆ ***Light transmission- 88–92 per cent for clear glass:*** Clear horticultural glass allows very high levels of sunlight to pass through, ensuring optimal photosynthetically active radiation (PAR) for plant growth. This high clarity promotes healthy, uniform crop development throughout the year. Unlike some plastics, glass maintains consistent light transmission over its entire lifespan.
- ☆ ***High durability (lifespan 20–30 years):*** Glass is a long-lasting cladding material that can serve effectively for two to three decades with minimal degradation. It withstands harsh weather conditions such as heavy rains, winds, and snow loads without warping. This extended service life offsets its higher initial investment cost.
- ☆ ***Chemically inert and non-reactive:*** Glass does not react with fertilizers, pesticides, or cleaning chemicals used in greenhouse operations. Its inert nature prevents the release of harmful substances and ensures that no chemical residues affect crop quality. This makes it ideal for sensitive horticultural and research applications.
- ☆ ***Resistant to UV degradation:*** Unlike most plastics, glass does not break down or lose transparency under prolonged UV exposure. It provides consistent protection and performance in regions with intense sunlight. This UV resistance ensures steady light quality for plants over many years of use.

Advantages

- ☆ ***Excellent transparency for photosynthetically active radiation (PAR):*** Glass transmits a high percentage of PAR, which is essential for efficient photosynthesis and robust plant growth. This ensures healthy, vigorous crops with uniform development. Its optical clarity remains consistent over time, providing stable light quality.
- ☆ ***Long lifespan with minimal loss of light transmission:*** With proper care, glass can last 20–30 years without significant reduction in light penetration. Unlike plastics, it does not yellow, become brittle, or lose transparency with age. This long-term performance makes it a cost-effective choice for permanent structures.
- ☆ ***Good scratch resistance and ease of cleaning:*** The hard, smooth surface of glass resists scratches from dust, tools, or maintenance activities. Dirt and algae can be easily washed away with water or mild cleaning agents. This helps maintain maximum light transmission and greenhouse aesthetics.
- ☆ ***Fire-resistant and recyclable:*** Glass is a non-combustible material that does not burn or emit toxic fumes in case of fire, enhancing safety. At the end of its service life, it can be fully recycled into new glass products. This makes it an environmentally sustainable cladding choice.

Limitations

- ☆ ***Heavy weight requires strong frame structure:*** Glass is significantly heavier than plastic cladding materials, which means the greenhouse frame must be built with stronger, more expensive materials to support the load. This increases the overall construction cost. It also limits its use in low-cost or lightweight structures.
- ☆ ***Brittle and prone to shattering under impact:*** Although durable, glass can break easily if struck by hailstones, falling branches, or accidental impacts. Shattered glass poses safety risks to workers and crops. Special tempered or laminated glass can reduce this risk but adds to the cost.
- ☆ ***Higher initial cost compared to plastics:*** The purchase and installation cost of glass is considerably higher than that of polyethylene films or polycarbonate sheets. Additional expenses arise from the need for a strong support structure. This makes glass less feasible for low-budget or small-scale projects.
- ☆ ***Poor thermal insulation unless double-glazed:*** Single-layer glass allows significant heat loss in winter and heat gain in summer, increasing the need for heating or cooling systems. Double-glazed units improve insulation but are much more expensive. This can limit energy efficiency in extreme climates.

Uses

Glass is widely used in high-technology greenhouses where precise environmental control and maximum light penetration are essential for year-round cultivation. Its high clarity and stability make it particularly suitable for temperate and cool climates, where maximizing sunlight is critical to maintain optimal plant growth without excessive heating costs. In research facilities, glass is preferred due to its consistent optical properties, chemical inertness, and ability to maintain uniform light conditions for experimental accuracy. Botanical gardens often use glass structures not only for plant cultivation but also for aesthetic appeal, visitor visibility, and long-term durability. Its transparency allows clear viewing of plants from outside, enhancing educational and tourism value. Glass-clad conservatories are ideal for growing exotic and ornamental species that require controlled climates. It is also used in orchid houses, heritage greenhouses, and specialized crop production systems such as seed banks and germplasm conservation units. In regions prone to snow, glass structures can support heavy loads without sagging, provided the frame is strong enough. These characteristics make glass the material of choice for projects where longevity, clarity, and structural integrity outweigh initial costs.

3.1.2 Polyethylene Films (PE Films)

Polyethylene is the most widely used flexible cladding material for polyhouses, particularly in tropical and subtropical regions where cost-effectiveness and ease of installation are major priorities. Available mainly as low-density polyethylene (LDPE) and linear low-density polyethylene (LLDPE), it provides good light transmission (80–90 per cent) and can be manufactured as single-layer or co-extruded multilayer films with UV stabilizers, anti-drip coatings, and infrared (IR) modifiers for enhanced performance. Its lightweight nature allows for quick installation over simple frameworks, making it suitable for both smallholder farmers and large commercial operations. The films are available in various thicknesses, typically ranging from 100 to 200 microns, and in widths that can cover large spans without joints, reducing light obstruction. Diffused polyethylene films are particularly valued for minimizing plant scorching and promoting uniform light distribution inside the structure. Although economical and versatile, polyethylene films have a shorter lifespan (2–4 years for UV-stabilized variants) compared to rigid materials, requiring periodic replacement. Despite this, their affordability, adaptability to different designs, and ability to integrate functional additives make them the most popular cladding choice for low to medium-cost protected cultivation systems worldwide.

Types

- ☆ ***Low-Density Polyethylene (LDPE) – Most Common:*** LDPE is the standard choice for greenhouse films due to its affordability, flexibility, and good light transmission. It is easy to handle, cut, and install over various frame designs. LDPE films can be transparent or diffused, depending on crop needs. UV-stabilized LDPE offers a service life of 2–3 years under tropical sunlight. Its popularity comes from the balance between cost and performance for small to medium-scale growers.
- ☆ ***Linear Low-Density Polyethylene (LLDPE)–better Mechanical Strength:*** LLDPE films are tougher and more resistant to tearing and punctures compared to LDPE, making them suitable for areas with high wind loads or rough handling during installation. They offer similar light transmission but with improved tensile and impact strength. LLDPE's flexibility allows for tight tensioning over frames, reducing flapping and wear. It has better resistance to stress cracking, prolonging its useful life. This makes it ideal for commercial greenhouses requiring higher durability.
- ☆ ***Co-extruded multilayer films with UV, anti-drip, and IR additives:*** Modern greenhouse films often use co-extrusion technology to combine multiple functional layers in one sheet. The UV-stabilized outer layer protects against solar degradation, extending film life. Anti-drip properties prevent condensation droplets from falling on plants, reducing disease risk. IR additives help retain heat during cold nights or reduce excessive heat buildup in hot days, depending on the formulation. These films provide enhanced microclimate control for better yield and quality.

Properties

- ***Light transmission 80–90 per cent (diffused or clear):*** Polyethylene films allow a high percentage of sunlight to pass through, ensuring adequate photosynthetically active radiation (PAR) for plant growth. Clear films provide maximum penetration, while diffused films scatter light to reduce shadowing and prevent leaf scorching. This flexibility helps growers match film type to crop requirements and seasonal conditions.
- ***Thickness 100–200 microns (400–800 gauge):*** The film thickness directly affects strength, durability, and resistance to tearing. Thicker films (150–200 microns) are more durable and withstand extreme weather better, but are slightly costlier. The standard range ensures sufficient protection while keeping the structure lightweight and economical.
- ***Lifespan 2–4 years (UV-stabilized):*** Without UV stabilization, polyethylene films degrade rapidly under sunlight, becoming brittle within months. UV-stabilized variants can last 2–4 years depending on climatic conditions, quality, and maintenance. Proper tensioning and cleaning can further extend service life.
- ***Flexible, lightweight, and easy to install:*** Polyethylene films can be easily cut, stretched, and fitted over simple or complex frame designs without heavy machinery. Their low weight reduces structural load and allows quick installation or replacement. This makes them highly adaptable for both smallholder farms and commercial operations.

Advantages

- ***Low cost and easily replaceable:*** Polyethylene films are among the most affordable cladding materials, making them accessible for small and medium-scale farmers. Even when damaged or degraded, they can be replaced quickly without heavy investment. This cost-effectiveness encourages wider adoption of protected cultivation in resource-limited areas.
- ***High light diffusion reduces plant scorching:*** Diffused polyethylene films scatter incoming sunlight, distributing it more evenly inside the structure. This minimizes hot spots and leaf burn, especially in high-radiation tropical and subtropical climates. Even light distribution promotes uniform plant growth and higher yields.
- ***Available in various widths for large spans without joints:*** Greenhouse-grade polyethylene films are manufactured in wide sheets, allowing them to cover large structural spans without seams. Fewer joints mean reduced light obstruction and lower risk of tearing or leakage. This also improves the overall aesthetics and structural integrity of the cladding.
- ***Can incorporate additives for specific functions (e.g., IR-blocking, anti-dust):*** Modern polyethylene films can be engineered with functional

additives to improve performance. IR-blocking additives help manage temperature by retaining heat in winter or reducing heat gain in summer. Anti-dust and anti-drip coatings maintain optical clarity and reduce disease risk, enhancing crop quality.

Limitations

- Degrades under UV exposure without stabilizers
- Prone to tearing in high winds if not properly tensioned
- Shorter lifespan compared to rigid materials

Uses

Commercial polyhouses, tunnels, low-cost protected cultivation structures.

3.1.3 Polycarbonate Sheets

Polycarbonate is a strong, rigid thermoplastic widely used as a premium cladding material for greenhouses and other protected structures. It is available in single-layer, twin-wall, and multiwall configurations, each offering varying levels of light transmission, thermal insulation, and strength. Known for its exceptional impact resistance–about 250 times stronger than glass–it can withstand hail, windstorms, and mechanical damage without shattering, making it highly durable in extreme weather conditions. Polycarbonate sheets provide good light transmission (80–85 per cent for clear and 60–75 per cent for diffused types), allowing for optimal photosynthesis while reducing the risk of scorching. Multiwall designs trap air between layers, significantly improving thermal insulation, which helps maintain stable temperatures and reduce heating costs in colder climates. UV-protected grades resist yellowing and degradation, extending their service life to 10–15 years with minimal maintenance. They are lightweight compared to glass, reducing the load on the supporting structure and enabling easier installation. While more expensive than polyethylene films, their long lifespan, durability, and energy-saving properties make them cost-effective over time, especially for high-technology or commercial greenhouses.

Properties

- ***Light transmission 80–85 per cent (clear), 60–75 per cent (diffused):*** Clear polycarbonate sheets allow a high percentage of sunlight to pass through, ensuring ample photosynthetically active radiation (PAR) for plant growth. Diffused sheets scatter light uniformly, preventing shadows and reducing heat stress. This helps maintain consistent growth rates across all plants. The choice between clear and diffused depends on crop type and climatic conditions.
- ***High impact resistance (250× stronger than glass):*** Polycarbonate is renowned for its exceptional toughness, making it highly resistant to hail, windborne debris, and accidental impact. This strength significantly

reduces the risk of damage and crop loss in extreme weather. Unlike glass, it does not shatter into dangerous shards. This makes it a safer option for workers and plants alike.

- ☆ ***Thermal conductivity: lower than glass → better insulation:*** Polycarbonate has a lower thermal conductivity than glass, meaning it retains heat more effectively in cooler conditions. Multiwall designs create insulating air pockets that help stabilize internal temperatures. This can reduce heating costs in winter and minimize temperature fluctuations. Better insulation also supports year-round cultivation in cold climates.
- ☆ ***Lifespan 10–15 years (UV-protected):*** UV-protected polycarbonate sheets resist yellowing and surface degradation from prolonged sun exposure. With proper care, they can last over a decade, maintaining both structural integrity and light transmission. Periodic cleaning helps maximize performance throughout their lifespan. This longevity makes them a cost-effective investment despite higher initial costs.

Advantages

- ☆ ***Excellent durability and resistance to hail, wind, and mechanical damage:*** Polycarbonate can withstand extreme weather events, including heavy hailstorms and strong winds, without cracking or breaking. Its high impact strength ensures long-term protection for crops. This durability minimizes repair costs and reduces downtime in greenhouse operations.
- ☆ ***Lightweight compared to glass, reducing frame cost:*** Polycarbonate sheets weigh far less than glass, allowing for lighter, less expensive supporting structures. This reduces construction costs while still maintaining strength and stability. The lightweight nature also makes transportation and installation easier.
- ☆ ***Available in diffused light variants for uniform crop growth:*** Diffused polycarbonate evenly scatters incoming light, preventing hot spots and ensuring all plants receive balanced illumination. This improves photosynthesis efficiency and crop uniformity. It is especially beneficial for tall or densely planted crops.
- ☆ ***Provides good thermal insulation in cold climates:*** Multiwall polycarbonate traps insulating air between layers, reducing heat loss during cold weather. This helps maintain optimal internal temperatures with less energy input. Better insulation supports year-round production in regions with harsh winters.

Limitations

- ☆ ***Higher cost than polyethylene films:*** Polycarbonate requires a higher initial investment, making it less affordable for small-scale or low-budget growers. While its long lifespan can offset the cost over time, the upfront expense can be a barrier. This makes it more common in commercial or high-tech greenhouse projects.

- ☆ ***Light transmission slightly less than glass:*** Even in clear form, polycarbonate transmits slightly less light (80–85 per cent) compared to glass (88–92 per cent). This minor reduction can impact light-loving crops in low-sunlight regions. Diffused variants further lower transmission but improve light distribution.
- ☆ ***May yellow over time if UV layer wears out:*** Without proper UV protection, polycarbonate can gradually yellow and lose transparency due to prolonged sun exposure. This reduces light penetration and can affect crop productivity. Regular cleaning and using high-quality UV-stabilized sheets help maintain performance.

Uses

Permanent and semi-permanent high-tech greenhouses, cold climate protected cultivation, research stations.

3.1.4 Acrylic Sheets (Polymethyl Methacrylate – PMMA)

Acrylic is a transparent thermoplastic material that closely resembles glass in clarity, offering light transmission levels of around 88–92 per cent, making it ideal for maximizing photosynthetically active radiation (PAR). It is lighter than glass, which reduces structural load and allows for easier handling and installation, while still being more impact-resistant than glass, providing better protection against accidental damage. Acrylic has excellent weathering properties and remains clear for many years, resisting yellowing under UV exposure when properly stabilized. It offers better thermal insulation than glass, helping to moderate internal temperatures and reduce heating costs in cooler climates. Available in various thicknesses and finishes, including clear and diffused types, acrylic sheets can be customized to meet specific crop and climate needs. Although more expensive than polycarbonate or polyethylene films, their optical quality, durability, and aesthetic appeal make them suitable for premium greenhouses, botanical displays, and research facilities. With a lifespan of 15–20 years when UV-protected, acrylic is a long-term investment for growers seeking high transparency and lasting performance.

Properties

- ☆ **Light transmission 88–92 per cent (comparable to glass)**: Acrylic sheets allow 88–92 per cent of visible light to pass through, offering clarity close to that of premium glass. This high light transmission ensures bright and well-lit interiors, making them ideal for applications where visibility is crucial. The material also diffuses light effectively, reducing glare and creating a pleasant visual environment.

- ☆ **Lifespan 15–20 years (UV-stabilized)**: UV-stabilized acrylic can maintain its clarity and strength for 15–20 years, even under continuous sunlight exposure. The special UV treatment prevents yellowing, fading, and brittleness over time. This long service life reduces replacement costs and ensures reliable performance for decades.
- ☆ **Good resistance to weathering**: Acrylic sheets can withstand harsh environmental conditions such as rain, wind, and temperature fluctuations without losing structural integrity. They resist cracking, warping, or discoloration caused by prolonged outdoor exposure. This durability makes them suitable for both indoor and outdoor installations in varied climates.

Advantages

- ☆ ***High optical clarity for maximum photosynthesis:*** Acrylic sheets provide excellent transparency, ensuring plants receive maximum sunlight for efficient photosynthesis. This clarity supports healthy growth, higher yields, and better-quality produce. The material also maintains its optical quality over time, ensuring consistent light availability.
- ☆ ***Lighter than glass, reducing structural load:*** Acrylic is significantly lighter than glass, making it easier to handle, transport, and install. Its reduced weight lowers the stress on supporting structures, allowing for more economical and flexible designs. This also improves safety by reducing risks during installation or replacement.
- ☆ ***Better thermal insulation than glass:*** Acrylic offers superior thermal insulation, helping maintain a stable temperature inside protected cultivation structures. This reduces heat loss in cooler climates and minimizes overheating in warmer conditions. As a result, it lowers energy costs and creates a more controlled growing environment for crops.

Limitations

- ☆ More expensive than polycarbonate and polyethylene: Acrylic sheets generally cost more than alternative materials like polycarbonate and polyethylene. The higher price can increase the initial investment for large-scale installations. However, this cost is often offset by their superior clarity, durability, and long service life.
- ☆ Brittle under extreme cold conditions: In extremely low temperatures, acrylic can become less flexible and more prone to cracking or shattering under impact. This brittleness limits its use in regions with severe winter conditions unless special grades are chosen. Proper handling and installation are essential to avoid damage in such climates.

Uses

Acrylic sheets are widely used in specialty greenhouses where high light transmission and durability are essential for optimal plant growth. In orchid houses, they help create the precise light and temperature conditions needed for delicate, high-value flowers. Display houses, such as those in botanical gardens or exhibitions, benefit from acrylic's clarity, allowing visitors an unobstructed view of plants while protecting them from harsh weather. Their lightweight nature makes them suitable for intricate structural designs without heavy frameworks. Additionally, their long-lasting performance ensures minimal maintenance and consistent aesthetics over many years.

3.1.5 Fiberglass Reinforced Plastic (FRP) Sheets

Fiberglass Reinforced Plastic (FRP) sheets are composite materials made by embedding fine glass fibres within a durable plastic matrix. This combination offers exceptional strength, rigidity, and resistance to impact, making FRP an excellent choice for long-term structural applications. The embedded glass fibres provide reinforcement, while the plastic component ensures flexibility and resistance to corrosion. FRP sheets are lightweight compared to many traditional building materials, which simplifies handling and installation. They also possess good thermal insulation properties, helping regulate internal temperatures in protected cultivation structures. Their resistance to UV rays and weathering allows them to perform reliably in harsh outdoor environments. Overall, FRP sheets combine strength, durability, and low maintenance, making them ideal for agricultural and industrial uses.

Properties

- ✰ ***Light transmission 75–85 per cent initially, decreases over time:*** FRP sheets allow 75–85 per cent of visible light to pass through when new, providing adequate illumination for plant growth. However, due to surface weathering and fiber exposure, light transmission gradually decreases over time. Regular cleaning and maintenance can help slow this decline.
- ✰ ***Lifespan 8–10 years with UV protection:*** When manufactured with UV-protective coatings, FRP sheets can last 8–10 years under outdoor conditions. The UV layer helps prevent discoloration, brittleness, and loss of mechanical strength. This ensures consistent performance over their intended service life.
- ✰ ***Resistant to corrosion and weather extremes:*** FRP sheets are highly resistant to rust, chemical corrosion, and degradation from moisture. They can withstand extreme weather conditions such as heavy rain, strong winds, and fluctuating temperatures without losing structural integrity. This makes them suitable for diverse climatic zones.

Advantages

- ***Strong yet lightweight:*** FRP sheets combine high tensile strength with low weight, making them easy to handle and install without heavy support structures. This strength ensures durability and resistance to mechanical stress. Their lightness also reduces transportation costs and installation effort.
- ***Can be moulded into curved shapes for aesthetic designs:*** FRP can be easily moulded into various shapes, including attractive curves and custom profiles. This flexibility allows designers to create visually appealing structures without compromising functionality. Such designs are especially useful in display houses and modern greenhouse architecture.
- ***Suitable for windy and coastal areas due to high strength:*** The high structural strength of FRP sheets enables them to withstand strong winds, making them ideal for storm-prone regions. Their corrosion resistance also makes them well-suited for coastal environments with salty, humid air. This ensures long-term stability and minimal maintenance in challenging climates.

Limitations

- ***Loses transparency with age due to fibre exposure:*** Over time, the surface of FRP sheets weathers, causing the glass fibres to become exposed. This exposure scatters light, reducing overall transparency and affecting illumination inside the structure. While this is a natural aging process, protective coatings can help slow it down.
- ***Requires proper cleaning to maintain light transmission:*** Dust, algae, and environmental pollutants can accumulate on the surface of FRP sheets, further reducing light penetration. Regular cleaning with mild detergents and non-abrasive tools is essential to maintain optimal clarity. Neglecting maintenance can accelerate light loss and reduce crop productivity.

Uses

FRP sheets are widely used in semi-permanent structures where durability and strength are essential but full-scale permanent construction is not required. In research houses, they provide a controlled environment while withstanding frequent modifications and experimental setups. Their high wind resistance makes them particularly valuable in areas prone to storms and cyclones, ensuring structural stability and safety. FRP's corrosion resistance also makes it ideal for coastal regions where salty air can damage other materials. The ability to mould FRP into various shapes adds design flexibility for both functional and aesthetic purposes. Overall, FRP sheets offer a balance of strength, weather resistance, and adaptability for specialized agricultural and industrial applications.

3.1.6 Shade Nets

Shade nets are protective fabrics made from UV-stabilized High-Density Polyethylene (HDPE) designed to regulate sunlight exposure and control temperature in agricultural and horticultural settings. These nets work by filtering and diffusing incoming solar radiation, thereby reducing light intensity to optimal levels for plant growth. They also help lower ambient temperatures, minimizing heat stress and improving crop quality. Available in different shade percentages, they can be tailored to suit the specific light requirements of various crops and growth stages. Their lightweight, flexible, and durable nature makes them easy to install, relocate, and maintain. In addition to agricultural use, shade nets are also employed in nurseries, animal shelters, parking areas, and recreational spaces for sun protection.

Properties

- ☆ ***Available shading percentages 30 per cent, 50 per cent, 75 per cent, 90 per cent :*** Shade nets come in different shading percentages to match the light requirements of specific crops. Lower shading percentages (30–50 per cent) are suitable for plants that need more sunlight, while higher percentages (75–90 per cent) are ideal for shade-loving species. This flexibility allows growers to optimize light intensity for maximum yield and quality.
- ☆ ***Colours- Green, black, white, aluminized:*** Shade nets are available in various colours, each serving a specific purpose. Green and black nets are commonly used for general shading and temperature reduction, while white nets offer diffused light for uniform plant growth. Aluminized nets help reflect excess sunlight and heat, keeping the environment cooler.
- ☆ ***Lifespan 3–5 years:*** Made from UV-stabilized HDPE, shade nets are designed to withstand prolonged sun exposure without rapid degradation. Their typical lifespan ranges from 3 to 5 years under normal use. Proper installation and occasional cleaning can help extend their service life.

Advantages

- ☆ ***Reduces heat stress in plants:*** Shade nets help lower ambient and leaf temperatures by filtering intense sunlight. This reduction in heat stress allows plants to maintain normal physiological functions and prevents wilting. As a result, crop growth, flowering, and yield improve even during hot weather.
- ☆ ***Protects from excess radiation, hail, and wind damage:*** By diffusing and blocking excess solar radiation, shade nets prevent sunburn and leaf scorch. They also act as a barrier against hail, reducing physical damage to plants, and provide a windbreak that minimizes mechanical stress. This combined protection ensures better crop survival and quality.

- ***Lightweight and easy to install:*** Shade nets are made of lightweight HDPE material, making them easy to transport, handle, and set up without heavy machinery. They can be quickly installed over structures or directly over crops. Their flexibility also allows for seasonal use and easy relocation when needed.

Limitations

- ***Not suitable for crops requiring full environmental control:*** Shade nets primarily regulate light and temperature but cannot offer complete control over humidity, CO_2 levels, or other microclimatic factors. This makes them unsuitable for high-value crops that demand precise environmental regulation. Such crops are better grown in greenhouses or polyhouses.
- ***Does not retain heat during cold nights:*** Because shade nets are made of open-weave material, they allow heat to escape freely after sunset. This inability to retain warmth can lead to significant temperature drops at night, especially in cooler regions. As a result, they provide little protection against frost or chilling injury.

Uses

Shade net houses for nursery raising, ornamental plant production, and vegetables in hot climates. Shade nets are widely used in nursery raising to protect young seedlings from harsh sunlight, ensuring healthy and uniform growth. In ornamental plant production, they provide the diffused light necessary to enhance leaf colour, flower quality, and overall aesthetic appeal. For vegetable cultivation in hot climates, shade nets help lower field temperatures and reduce heat stress, improving yield and quality. They are also beneficial for extending the growing season in regions with intense summer heat. Additionally, their portability allows for flexible seasonal use across different crops and locations.

3.1.7 Insect-Proof Nets

Insect-proof nets are protective coverings made from fine mesh fabric, specifically designed to physically block insect pests from entering crop-growing areas. They are typically made from UV-stabilized materials that can withstand prolonged outdoor exposure without degrading. The mesh size is carefully selected to exclude common pests like whiteflies, aphids, thrips, and leaf miners, while still allowing adequate airflow for ventilation. By preventing pest entry, these nets reduce the need for chemical pesticides, promoting safer and more sustainable crop production. They are widely used in greenhouses, net houses, and open-field structures to protect high-value crops. In addition to pest control, insect-proof nets also act as partial barriers against wind, dust, and excessive sunlight.

Properties

- ☆ Mesh size: 25–50 mesh depending on target pest
- ☆ Lifespan: 3–5 years

Advantages

- ☆ ***Reduces pesticide use by preventing insect entry:*** Insect-proof nets create a physical barrier that stops pests from reaching crops, significantly lowering the incidence of pest-borne diseases. This reduces the reliance on chemical pesticides, leading to safer produce and lower production costs. It also supports eco-friendly and sustainable farming practices.
- ☆ ***Maintains natural airflow and light:*** The fine mesh design allows sufficient ventilation, ensuring that temperature and humidity remain within acceptable ranges for plant growth. It also permits natural light penetration, enabling photosynthesis without compromising pest protection. This balance supports healthy crop development while minimizing stress.

Limitations

- ☆ ***May slightly reduce light intensity:*** Although insect-proof nets are designed to allow good light penetration, the fine mesh can still cause a minor reduction in light intensity. This can be more noticeable in low-light seasons or for light-demanding crops. Choosing the correct mesh size and colour can help minimize this effect.
- ☆ ***Requires regular cleaning to prevent clogging:*** Dust, pollen, and debris can accumulate in the net's mesh over time, restricting airflow and further reducing light transmission. Regular cleaning is necessary to maintain optimal ventilation and light conditions. Neglecting maintenance can lead to poor crop growth and increased humidity-related issues.

Uses

Net houses for vegetables, seed production, and organic farming: Insect-proof nets are widely used in net houses for vegetable cultivation to protect crops from insect damage and disease transmission. In seed production, they help maintain genetic purity by preventing cross-pollination from insect vectors. For organic farming, they provide an effective non-chemical method of pest control, ensuring compliance with organic certification standards. Their use supports higher-quality produce, reduced crop losses, and more sustainable agricultural practices. Additionally, they are adaptable for use in both open-field and protected cultivation systems.

3.1.8 Specialty Films and Advanced Cladding Materials

Specialty films and advanced cladding materials represent the next generation of protective coverings for agriculture, designed to offer multiple benefits beyond basic shelter. These films are engineered with advanced additives and multi-layer structures to regulate temperature, humidity, and light quality within cultivation spaces. Some variants incorporate anti-drip properties to prevent water droplets from forming on the surface, reducing the risk of fungal diseases. Others feature light-diffusing capabilities that scatter sunlight evenly, minimizing plant stress and promoting uniform growth. Certain specialty films include infrared-blocking layers to reduce heat buildup during the day and retain warmth at night. Anti-dust and anti-static coatings help maintain transparency by preventing dirt accumulation. UV-stabilized materials ensure long-term durability and resistance to degradation from sunlight. In addition, pest-repellent or photo-selective films can be used to deter specific insects or enhance flowering in certain crops. These materials are suitable for high-value horticultural production, research facilities, and climate-sensitive cultivation. Overall, they enable precise microclimate management, improving productivity and quality while reducing environmental impact.

Examples

- ☆ Anti-drip films – prevent condensation droplets
- ☆ IR-blocking films – reduce heat loss in winter or heat gain in summer
- ☆ Thermal screens – used internally to conserve energy at night
- ☆ Aluminized reflective films – improve light distribution and reduce heat

Advantages

- ☆ **Increases energy efficiency and yield:** Specialty films help regulate internal temperatures by reducing excessive heat loss at night and minimizing overheating during the day. This improved thermal balance lowers energy consumption for heating or cooling systems. By maintaining optimal growing conditions, plants experience faster growth, better nutrient uptake, and higher yields. Consistent environmental control also allows for extended growing seasons and multiple harvests.
- ☆ ***Enhances crop quality and reduces disease risk:*** Light-diffusing and anti-drip properties of specialty films ensure even light distribution and prevent water droplet formation on plant surfaces. This reduces the occurrence of fungal and bacterial diseases caused by excess moisture. The controlled microclimate supports uniform flowering, better fruit set, and improved colour and size of produce. As a result, farmers can achieve higher market value for their crops with reduced chemical intervention.

Limitations

- ☆ ***Higher cost than standard films:*** Specialty films are more expensive than conventional polyethylene covers due to their advanced manufacturing processes and added functionalities. This higher initial investment can be a concern for small-scale or budget-conscious farmers. However, the long-term benefits in yield, quality, and energy savings often offset the upfront cost.

- ☆ ***Requires precise installation for best performance:*** To achieve the intended benefits, specialty films must be installed with proper tension, alignment, and sealing. Incorrect installation can reduce their effectiveness in regulating temperature, light, and humidity. Regular inspection and maintenance are also necessary to ensure consistent performance over time.

3.2 Properties and Selection Criteria

Selecting the right cladding material or covering for protected cultivation structures is a crucial decision that directly impacts crop growth, yield, and overall economic returns. Various physical, optical, thermal, and mechanical properties must be evaluated to ensure that the chosen material aligns with the climatic conditions, crop requirements, and budget considerations. The following key properties should be examined in detail before making a selection.

3.2.1. Light Transmission

Light transmission is the percentage of sunlight that penetrates the covering material and reaches the plants inside. This factor is fundamental because plants depend on light for photosynthesis, the process that drives growth and yield. Materials like glass and acrylic offer excellent transparency, transmitting up to 88–92 per cent of visible light. Polycarbonate sheets and polyethylene films transmit slightly less but may still provide sufficient illumination depending on the crop. In high-radiation regions, materials with lower light transmission or with shading properties help prevent overheating and sunburn in plants. The choice of transmission level should match the crop's light requirements–tomatoes and capsicum thrive in higher light, while leafy greens prefer moderate levels. Seasonal variations in natural light must also be considered when selecting the material. Over time, factors like dust accumulation, weathering, or surface degradation can reduce light transmission, so maintenance is essential. Specialized films with anti-dust coatings or light-diffusing additives can help sustain optimal light penetration. In short, proper selection ensures balanced illumination, promoting healthy, uniform plant growth and maximizing production efficiency.

3.2.2. Light Diffusion

Light diffusion refers to the scattering of incoming sunlight, dispersing it evenly across the growing area. Instead of producing harsh shadows and hotspots, diffused light penetrates deeper into the plant canopy, reaching lower leaves that might otherwise be shaded. This balanced light distribution improves photosynthetic efficiency, particularly in densely planted crops or multi-tier growing systems. Materials such as polycarbonate sheets, diffused polyethylene films, and certain specialty claddings are designed with additives or surface textures to enhance diffusion. High light diffusion is especially important in tropical or subtropical regions, where intense direct sunlight can cause leaf burn or excessive heat load. By softening light intensity, diffusion reduces plant stress, promotes uniform growth, and improves flowering and fruit set. In ornamental cultivation, diffused light enhances colour uniformity and overall aesthetic appeal. For crops sensitive to photoinhibition, such as orchids or lettuce, light diffusion can be a key factor in maintaining quality. The optimal diffusion percentage depends on the crop and season, but in general, a balance between high transmission and adequate diffusion is ideal. Properly selected diffused coverings can increase both yield and product quality.

3.2.3. UV Stability and Protection

UV stability is the ability of a cladding material to resist damage caused by prolonged exposure to ultraviolet (UV) radiation. Without UV stabilizers, most plastics degrade quickly, becoming brittle, discoloured, and structurally weak. UV-stabilized materials retain their clarity and strength for a longer period, making them more cost-effective in the long run. In addition to protecting the material itself, certain coverings can filter out harmful UV wavelengths that are detrimental to plant health or can encourage pest activity. Some crops benefit from partial UV blocking, which reduces insect populations such as whiteflies and thrips, while others may require full-spectrum light including UV for optimal growth and quality. Specialty films can be engineered to selectively transmit or block UV light depending on the intended purpose. For example, in seed production, UV-blocking films help maintain genetic purity by discouraging insect vectors. The correct level of UV protection extends the lifespan of the cladding, reduces replacement costs, and enhances crop protection. Farmers should consider both climatic intensity and crop type when choosing UV-stabilized coverings to ensure maximum benefits.

3.2.4. Thermal Insulation

Thermal insulation measures how effectively a covering material prevents heat exchange between the inside and outside environment. High insulation capacity reduces heat loss on cold nights and prevents overheating during hot days. This temperature stability is essential for sensitive crops, especially in extreme climates or during off-season production. Double-layer polycarbonate sheets, twin-wall acrylic, and air-inflated polyethylene films are examples of materials with good insulation properties. Better insulation not only improves plant growth conditions but also reduces energy costs associated with heating or cooling systems. For example, in cold climates, improved insulation can prevent frost damage, allowing for earlier planting or year-round production. In hot climates, reflective or infrared-blocking coatings can minimize heat buildup. The insulation value is often expressed as an "R-value" or "U-value," with higher R-values indicating better insulation. Choosing the right insulation helps maintain optimal growth temperatures, improves resource efficiency, and extends the productive season.

However, high insulation should be balanced with adequate light transmission to avoid compromising plant photosynthesis.

3.2.5. Mechanical Strength

Mechanical strength refers to a material's ability to withstand physical forces such as wind pressure, hail impact, snow loads, and accidental damage during handling. Strong coverings prevent costly crop losses due to structural failure or material tearing. FRP sheets, polycarbonate panels, and toughened glass are examples of materials with high mechanical strength. In coastal regions or cyclone-prone areas, strength against high wind loads is a critical selection criterion. In hail-prone zones, impact resistance can mean the difference between minor surface marks and complete structural replacement. Materials with higher tensile strength and flexibility can better absorb impact without breaking. Reinforced edges, proper anchoring, and strong framing further enhance overall structural resilience. Mechanical strength also affects long-term durability, as weaker materials may fail prematurely under repetitive stress. Farmers must consider their local weather conditions, seasonal hazards, and crop value when selecting cladding to ensure both safety and longevity.

3.2.6. Weight of the Material

The weight of the cladding material influences both the ease of installation and the type of support structure required. Lightweight materials such as polyethylene films and shade nets reduce the load on the frame, allowing for simpler, less expensive structural designs. This can be particularly advantageous for temporary or portable structures. However, very lightweight coverings may be more prone to wind uplift and require extra anchoring or tensioning systems. Heavier materials like glass or multi-wall polycarbonate offer superior durability and stability but demand stronger, more costly support frames. In high-wind regions, weight can be beneficial for preventing displacement, whereas in earthquake-prone areas, lighter materials reduce the risk of structural collapse. Transport and handling costs are also affected by material weight, with lighter options being more economical and easier to work with. Ultimately, the ideal weight balances durability, stability, and installation practicality.

3.2.7. Durability and Lifespan

Durability describes how well a material maintains its properties under prolonged exposure to environmental stressors such as UV radiation, temperature fluctuations, wind, and precipitation. Lifespan is the expected number of years the material can function effectively before requiring replacement. While polyethylene films typically last 2–5 years, materials like acrylic sheets or glass can serve for 15–20 years or more. Factors affecting durability include UV stabilization, thickness, resistance to mechanical damage, and maintenance practices. In harsher climates, more durable materials may be cost-effective despite their higher initial cost. For example, polycarbonate sheets with UV coating may last twice as long as standard films, reducing long-term replacement expenses. The lifespan should be evaluated alongside economic returns from the crop to determine the best investment.

3.2.8. Maintenance Requirements

Maintenance plays a major role in ensuring that cladding materials perform optimally throughout their lifespan. Some materials, like smooth-surfaced glass or acrylic, require minimal upkeep, while others–such as FRP or certain films–may accumulate dust, algae, or mineral deposits more quickly. Regular cleaning restores light transmission and prevents pest or disease buildup. In regions with heavy dust or pollution, anti-dust or anti-static coatings can significantly reduce cleaning frequency. Maintenance costs and labour availability should be factored into material selection, as high-maintenance options may not be practical for all growers. Proper care not only maintains performance but also extends material life, reducing long-term costs.

3.2.9. Cost and Economic Feasibility

The cost of a cladding material should be evaluated not only in terms of initial purchase price but also its lifespan, maintenance requirements, and potential yield benefits. High-performance materials like specialty greenhouse films or polycarbonate may require higher upfront investment but can result in greater long-term profitability through improved yields and lower energy consumption. Economic feasibility also depends on the value of the crop being grown–premium crops may justify higher investments in superior materials. A cost-benefit analysis helps compare options by factoring in all associated expenses and returns over the material's lifespan. Subsidy availability or government support schemes can also influence feasibility.

3.2.10. Crop and Climate Compatibility

The choice of cladding must align with both the crop's specific growth requirements and the local climate conditions. Shade nets may suit tropical vegetables and ornamentals but would be ineffective in cold climates where heat retention is critical. Similarly, crops like roses or tomatoes that demand high light levels require high-transmission materials, while lettuce or ferns may benefit from diffused or lower-intensity light. Local climatic factors such as wind patterns, rainfall intensity, snowfall, and seasonal temperature variations influence the optimal choice. Selecting the right combination of properties ensures that the microclimate inside the structure supports optimal plant growth, maximizes yield, and minimizes environmental stress.

Greenhouse Design and Environment Control

4.0 Introduction

A greenhouse is a specially designed structure that provides a controlled environment for plant growth, protecting crops from unfavourable weather conditions and allowing year-round production. It creates an artificial microclimate where temperature, humidity, light, and ventilation can be regulated to suit specific crop requirements. The concept of greenhouse cultivation originated in temperate regions to extend the growing season, but modern designs have evolved to suit tropical, subtropical, and arid climates as well. The success of greenhouse cultivation depends largely on appropriate design, which should harmonize structural strength, cost-effectiveness, and environmental control efficiency. Greenhouse design involves careful consideration of factors such as location, orientation, structural materials, cladding type, ventilation systems, and climate control equipment. An effective design must also withstand local climatic stresses, including wind loads, rainfall, snow, and high temperatures. The integration of environmental control systems–such as heating, cooling, shading, and irrigation–ensures stable growing conditions, minimizes plant stress, and optimizes productivity. Automation and smart control technologies are increasingly being incorporated to monitor and adjust parameters in real time, improving efficiency and reducing labour requirements. The choice of design also depends on the type of crop being grown, the production scale, and economic feasibility. High-value crops like cut flowers, exotic vegetables, and nursery plants benefit most from precision-controlled environments. In addition, sustainability considerations, such as energy efficiency, water conservation, and renewable energy integration, are becoming essential aspects of modern greenhouse planning. Overall, greenhouse design and environmental control work together to transform traditional agriculture into a high-tech, resource-efficient, and profitable system. A well-designed greenhouse not only enhances yield and quality but also enables cultivation in regions and seasons where open-field farming is not viable. This chapter explores the principles, components, and technologies that define effective greenhouse systems, providing a foundation for successful protected cultivation. Smart greenhouse system is depicted in Figure 4.1.

Figure 4.1: Smart Greenhouse System.
(*Source*: https://superlicious.pk)

4.1 Principles of Greenhouse Design

- ☆ **Solar Energy Utilization:** The greenhouse is designed to capture maximum sunlight through transparent or translucent cladding materials. Incoming solar radiation passes through the covering, is absorbed by internal surfaces, and converted into heat energy that supports plant growth.
- ☆ **Heat Retention Mechanism:** Once solar energy is converted into heat, the enclosed structure minimizes heat loss through convection, conduction, and radiation, ensuring a warm microclimate even during cooler periods.
- ☆ **Controlled Ventilation:** Proper ventilation systems regulate air exchange, preventing overheating, controlling humidity levels, and maintaining optimal CO_2 concentration for photosynthesis.
- ☆ **Thermal Mass and Heat Distribution:** Materials with high thermal mass, such as water barrels or stone floors, store excess heat during the day and release it gradually at night, stabilizing temperature fluctuations.

- ✰ **Light Diffusion and Distribution:** Diffused light ensures even distribution of sunlight, preventing leaf scorching and enhancing uniform photosynthetic activity across the crop canopy.
- ✰ **Climate Regulation Technologies:** Integration of heating, cooling, shading, and misting systems enables precise control over temperature, humidity, and light, adapting to seasonal and daily changes.
- ✰ **Structural Orientation and Design Efficiency:** The greenhouse orientation is optimized for maximum sunlight exposure throughout the year, often with an east–west alignment in temperate zones and north–south in tropical regions. Principles of greenhouse design and their functions is summarised in Table 4.1.

Table 4.1: Principles of Greenhouse Design and their Functions

Principle	*Description*
Solar Energy Utilization	Transparent cladding captures sunlight, converts it into heat for plant growth.
Heat Retention Mechanism	Enclosure reduces heat loss via convection, conduction, and radiation.
Controlled Ventilation	Regulates air exchange, prevents overheating, controls humidity and CO_2.
Thermal Mass and Heat Distribution	High-mass materials store heat by day, release it at night to stabilize temperature.
Light Diffusion and Distribution	Diffused light prevents scorching and ensures uniform photosynthesis.
Climate Regulation Technologies	Heating, cooling, shading, and misting enable precise climate control.
Structural Orientation and Design Efficiency	Orientation maximizes sunlight–east–west in temperate, north–south in tropics.

4.2 Structural Components

Protected structures in horticulture are purpose-built systems designed to create a favourable and controlled environment for plant growth, regardless of external climatic variations. These structures physically enclose plants, providing a barrier against adverse weather conditions such as heavy rainfall, frost, hail, strong winds, excessive heat, or prolonged drought. By moderating temperature, humidity, and light intensity, they allow for year-round cultivation and improved crop quality. The primary function of these structures is to protect plants from biotic stresses like pests, insects, and diseases, while simultaneously minimizing abiotic stresses caused by unpredictable weather. They help maintain an optimal microclimate that promotes efficient photosynthesis, better nutrient uptake, and enhanced growth rates. Structural components vary depending on the type of protected cultivation system–greenhouse, polyhouse, shade net house, low tunnels, or walk-in tunnels–but they generally consist of a rigid or semi-rigid frame, a suitable cladding material, and accessories for ventilation, shading, irrigation, and climate control. The frame serves as the backbone of the structure, providing stability and resistance to wind loads, snow loads, and other mechanical stresses. Common materials used for frames include galvanized iron (GI) pipes, aluminium, steel, or even bamboo for low-cost versions. The cladding material–such as glass, polycarbonate sheets, or UV-stabilized polyethylene films–determines the degree of light transmission, insulation, and durability of the structure.

Ventilation systems, whether natural or forced, are integrated to regulate air movement, prevent excessive heat buildup, and maintain ideal humidity levels. Additional components like insect-proof nets, shading screens, and thermal curtains further enhance environmental control. Drip or micro-irrigation systems are often installed to ensure efficient water and nutrient delivery directly to the plant root zone. Advanced protected structures also incorporate automated climate control technologies, including temperature and humidity sensors, exhaust fans, foggers, misting systems, and CO_2 enrichment setups. These systems help optimize growing conditions and reduce manual intervention, improving production efficiency. By combining structural strength, functional design, and environmental adaptability, protected horticulture systems maximize productivity, reduce crop losses, extend growing seasons, and enable the cultivation of high-value crops in regions where open-field farming would be uneconomical or impractical (Figure 4.2). The protected cultivation of vegetables, flowers, and fruits is depicted in Figure 4.3.

Figure 4.2: Benefits of Protected Cultivation: Maximizing Yield with Polyhouse.
(*Source*: https://agrodome.in)

They enable growers to:

- ✰ Extend the growing season well beyond what is possible in open-field cultivation, allowing off-season production and better market opportunities.

Figure 4.3: Protected Cultivation of Vegetables, Flowers, and Fruits.
(*Source*: https://s3.ap-southeast-1.amazonaws.com)

- ✰ Enhance crop quality and yield by maintaining optimal temperature, humidity, and light conditions, thereby reducing environmental stress on plants.
- ✰ Lower pesticide and fertilizer usage through physical barriers against insects, birds, and diseases, coupled with precise nutrient and water delivery systems.
- ✰ Diversify production by enabling the cultivation of exotic, high-value, or climate-sensitive crops that would not normally thrive in the region's natural conditions.
- ✰ Conserve water resources through advanced irrigation techniques like drip or micro-sprinklers, ensuring efficient water use and minimal wastage.

4.2.1. Greenhouses

A greenhouse is a purpose-built, enclosed structure designed to create an optimal and controlled environment for plant growth, independent of prevailing outdoor weather conditions. It is typically permanent or semi-permanent in nature and is covered with transparent or translucent cladding materials that allow sunlight to enter while providing protection from adverse environmental factors. Common cladding options include glass, which offers excellent light transmission and durability; polycarbonate sheets, which are lightweight, impact-resistant, and thermally efficient; and UV-stabilized polyethylene films, which are cost-effective and suitable for short- to medium-term use. The primary function of a greenhouse is to harness and regulate solar radiation. Sunlight passes through the covering material, where it is absorbed by the internal surfaces (soil, plants, benches) and converted into heat. This heat is retained inside by minimizing losses through convection, conduction, and radiation, creating a warmer microclimate that supports plant growth even during cooler periods. This principle is known as the greenhouse effect. Greenhouses enable growers to manipulate key environmental parameters such as temperature, humidity, ventilation, and carbon dioxide (CO_2) concentration to match the physiological needs of specific crops. Maintaining optimum CO_2 levels inside the greenhouse can significantly enhance photosynthesis and increase yields. Temperature control is achieved through a combination of passive methods (orientation, natural ventilation) and active systems (heating units, cooling pads, exhaust fans). Humidity is managed through misting, fogging, or dehumidification systems, depending on the crop and climate.

The degree of environmental control varies with the design:

- ✰ **Naturally Ventilated Greenhouses:** Naturally ventilated greenhouses are the simplest and most economical structures widely used by farmers across different agro-climatic zones (Figure 4.4). These greenhouses rely primarily on natural forces such as wind movement, pressure differences, and thermal buoyancy for maintaining an internal environment suitable for crop growth. They are designed with roof vents, side vents, and mesh-covered openings that allow air circulation, which helps in removing excess heat and humidity. The absence of complicated mechanical systems makes them cost-effective, easy to maintain, and suitable for small and medium-scale farmers. Shading techniques such as shade nets, whitewash paints, or UV-stabilized films are often used to regulate solar radiation and protect crops from scorching sunlight.

Figure 4.4: Naturally Ventilated Greenhouse.
(*Source*: https://5.imimg.com)

During the summer season, cross-ventilation helps in bringing fresh air inside while warm air escapes through the roof vents. In winter, the side vents can be kept partially closed to retain warmth while still allowing moderate air exchange. These structures are highly beneficial for cultivating flowers, vegetables, and nurseries where moderate climatic regulation is sufficient. Since they do not depend heavily on electricity or advanced technology, operational costs remain low, making them sustainable for rural areas. Their design is flexible, allowing construction with locally available materials such as bamboo, wood, or galvanized iron pipes, thus reducing investment costs. Farmers practicing organic cultivation also prefer naturally ventilated greenhouses as they provide a near-natural environment. However, the control over temperature, humidity, and CO_2 is limited compared to hi-tech systems. Despite these limitations, they are ideal in regions with favourable weather conditions where only partial protection is needed. The simplicity of operation, lesser dependence on skilled labour, and low maintenance make naturally ventilated greenhouses a practical choice for resource-poor farmers. Moreover, they can be integrated with drip irrigation or fertigation systems to enhance water-use efficiency. These greenhouses are widely used in India for crops like tomato, capsicum, cucumber, and cut flowers. Their affordability and adaptability make them one of the most popular types of protected cultivation structures across the world.

☆ **Climate-Controlled or Hi-Tech Greenhouses:** Climate-controlled or hi-tech greenhouses represent the advanced stage of protected cultivation technology (Figure 4.5). These structures are equipped with fully automated systems that continuously monitor and adjust the internal environment to create ideal growing conditions throughout the year. Sensors installed inside the greenhouse record temperature, humidity, light intensity, and carbon dioxide levels in real time. Based on these readings, controllers automatically activate heating units, cooling pads, exhaust fans, foggers, shading screens, and irrigation systems.

Figure 4.5: Climate-Controlled or Hi-Tech Greenhouse.
(*Source*: https://www.growspan.com)

This ensures that crops are never exposed to stress conditions, leading to higher yields and better-quality produce. Hi-tech greenhouses are especially useful in regions with extreme weather conditions, where natural ventilation alone cannot provide stable conditions. The high precision of climate control allows farmers to grow exotic vegetables, high-value flowers, and off-season crops that fetch premium prices in the market. In addition to environmental regulation, fertigation and hydroponic systems can be integrated to deliver nutrients and water in exact proportions, thereby increasing input-use efficiency. With the use of artificial lighting systems, crop cycles can be extended, and production can continue even during low-light seasons. Although the initial investment is high, the returns from year-round cultivation of high-value crops justify the costs in the long run. These structures require skilled operation and regular maintenance, but modern digital interfaces and mobile applications have made monitoring and control much easier. Hi-tech greenhouses also support vertical farming, automation of harvesting, and integration with Internet of Things (IoT) technologies. The consistent environment reduces pest and disease incidence, minimizing the need for chemical pesticides and improving food safety. Farmers can maintain a continuous supply to supermarkets, exporters, and processing industries, ensuring steady income. In countries like the Netherlands, Israel, and now India, hi-tech greenhouses are central to modern agriculture and export-oriented farming. Their ability to conserve resources, maximize production, and provide precision agriculture solutions makes them the future of protected farming. Despite the challenges of high cost and dependence on uninterrupted power supply, they remain the most efficient and reliable systems for sustainable, commercial-scale cultivation.

Structural design is another critical aspect. The frame must be strong enough to withstand wind loads, snow loads, and other mechanical stresses. Materials commonly used include galvanized iron (GI) pipes, aluminium, or steel, chosen for their strength, durability, and resistance to corrosion. The shape of the greenhouse–such as even-span, uneven-span, Quonset, or gothic arch–affects light penetration, airflow, and structural stability. Greenhouses also support the integration of irrigation and fertigation systems, ensuring efficient water and nutrient delivery. Drip irrigation is most commonly used to conserve water while providing uniform moisture at the root zone. Automated fertigation units mix and supply nutrients in precise quantities, enhancing plant health and reducing waste. For crops requiring specific photoperiods, supplementary lighting can be installed to extend daylight hours or enhance light intensity during cloudy days. Conversely, shade screens or thermal curtains can be deployed to reduce heat load during intense sunlight. Greenhouses are widely used for cultivating high-value crops such as cut flowers (roses, carnations, gerberas), exotic vegetables (capsicum, cucumber, lettuce), nursery plants, and herbs. They are particularly beneficial in regions with extreme climates–cold deserts, high altitudes, or arid zones–where open-field cultivation would be unfeasible or economically unviable. By enabling year-round production, reducing dependency on external weather, and producing uniform, high-quality crops, greenhouses have become a cornerstone of protected cultivation. They also open opportunities for export-oriented farming, as crops grown under controlled conditions can meet strict international quality standards. In modern agriculture, greenhouses represent a fusion of engineering, plant science, and environmental control technologies–a system where every design element, from the covering material to the ventilation system, plays a critical role in ensuring sustainable and profitable crop production.

4.2.2. Polyhouses

A polyhouse is a type of protected cultivation structure that operates on the same basic principles as a greenhouse but is covered with polyethylene (PE) sheets instead of glass or polycarbonate panels. This choice of cladding material significantly reduces construction costs, making polyhouses an affordable option for small and medium-scale farmers while still offering adequate environmental control. Polyethylene sheets used in polyhouses are typically UV-stabilized, meaning they are treated to resist degradation from ultraviolet rays. This increases their lifespan–generally ranging from 3 to 5 years–and helps maintain consistent light transmission. The thickness of the film (usually between 120 to 200 microns) influences both durability and thermal efficiency. Polyhouses protect crops from adverse weather conditions such as strong winds, heavy rainfall, hailstorms, frost, and excessive sunlight. By enclosing the plants, they also act as a physical barrier against insect pests, birds, and some airborne diseases, thereby reducing the need for pesticides and promoting healthier crops. The microclimate inside a polyhouse can be modified to suit crop needs, though the level of control is generally moderate compared to high-tech glass greenhouses. Temperature and humidity regulation are often achieved through natural ventilation–via roll-up side curtains or roof vents–supplemented with shade nets, foggers, or evaporative cooling pads in warmer regions.

Polyhouses are well-suited for cultivating a wide variety of crops, including high-value vegetables (capsicum, tomato, cucumber, lettuce), cut flowers (gerbera, rose, carnation), herbs, and nursery seedlings. They are especially beneficial for off-season production, enabling farmers to target high-demand market windows and fetch premium prices. Structural components of a polyhouse include a framework–commonly made from galvanized iron (GI) pipes or steel–to support the PE film and withstand wind and load stress. Depending on budget and design, frames can be single-span, multi-span, Quonset-type, or gothic-arch shaped. The design influences both durability and internal light distribution. The relatively low construction cost of polyhouses makes them a popular choice for government subsidy programs under schemes such as the National Horticulture Mission (NHM) and the National Horticulture Board (NHB). These programs encourage farmers to adopt protected cultivation practices to increase productivity and profitability. While polyhouses are less durable than glass or polycarbonate greenhouses and require periodic replacement of the film, their lower initial investment means they offer a quicker return on investment for many growers. They are also easier to construct and repair, allowing farmers to undertake partial replacements without dismantling the entire structure. Polyhouses can be fitted with drip irrigation and fertigation systems to deliver water and nutrients directly to the plant roots, improving resource-use efficiency. This not only conserves water but also minimizes nutrient loss, resulting in healthier plants and reduced production costs.

For crops that require specific light intensities, shade nets can be installed inside the polyhouse to filter sunlight and prevent leaf scorching. Conversely, transparent films ensure high light penetration during low-light conditions, especially in winter months. Overall, polyhouses provide a cost-effective solution for farmers seeking to improve yield, quality, and cropping intensity without the high capital investment of fully climate-controlled greenhouses. They represent a balanced approach–offering sufficient environmental protection and yield benefits–while remaining economically accessible to a wide range of agricultural producers.

4.2.3. Shade Houses

A shade house is a type of protected cultivation structure designed to moderate the intensity of sunlight reaching plants by using shade nets as the primary covering material. These nets are typically manufactured from polyethylene (PE) or other durable synthetic fibres and are woven or knitted to achieve specific levels of shading. The main purpose of a shade house is to create a cooler, less stressful environment for plants, reducing solar radiation, temperature, and wind speed, while still allowing adequate air circulation. Shade houses are particularly beneficial for shade-loving or partial-shade crops, which can be damaged by excessive direct sunlight. Examples include orchids, ferns, anthuriums, leafy greens (lettuce, spinach), foliage plants, and ornamental flowers. They are also widely used for nursery management, where young seedlings and saplings require protection during the initial growth stages before being transplanted into open fields or greenhouses. One of the key advantages of shade houses is their ability to protect plants from multiple environmental stresses–not just intense sunlight, but also heavy rainfall, strong winds, and occasional hailstorms. By reducing light intensity, they also help lower the rate of evaporation from both soil and plant surfaces, conserving moisture and improving water-use efficiency. Shade nets are available in different shading percentages, typically ranging from 35 per cent to 75 per cent, depending on crop requirements. Lower shading (35–50 per cent) is used for crops that need moderate light, such as certain vegetables and flowers, while higher shading (60–75 per cent) is reserved for delicate ornamental species and tropical foliage plants. The choice of shading level is critical for achieving optimal photosynthesis without heat stress.

Structurally, a shade house consists of a supporting frame made from materials such as galvanized iron (GI) pipes, steel, bamboo, or wood, depending on budget and expected lifespan. The frame must be strong enough to withstand wind loads and the weight of wet shade nets during rainy conditions. Nets are usually fixed to the frame with ropes, clips, or specialized fastening systems to ensure they remain taut and functional. In addition to light control, shade houses can be fitted with side nets or windbreaks to minimize insect entry, particularly for nursery plants. This reduces the reliance on chemical pesticides and helps in producing healthier, pest-free seedlings. Shade houses also play a significant role in hardening nursery plants–a process where seedlings are gradually exposed to more sunlight and harsher conditions before field planting. This transition period helps plants develop stronger stems, better root systems, and higher survival rates in open-field conditions. Irrigation in shade houses is typically provided through micro-irrigation systems such as drip lines, micro-sprinklers, or misters. These systems ensure precise water application while maintaining appropriate humidity levels, which is particularly important for tropical ornamentals and leafy greens. From an economic perspective, shade houses are more affordable to construct than polyhouses or glass greenhouses, making them accessible to small-scale farmers, floriculturists, and nursery operators. Maintenance costs are relatively low, although shade nets may need replacement every 3–5 years depending on exposure and quality. In regions with hot, dry climates, shade houses are an invaluable tool for reducing heat stress and improving crop quality, whereas in humid tropical regions, they help filter intense midday sun and shield plants from torrential rains. By balancing light, temperature, and airflow, shade houses enable consistent, high-quality production while reducing environmental stress on crops.

4.2.4. Net Houses

A net house is a type of protected cultivation structure primarily designed to provide a physical barrier against insect pests, birds, and sometimes excessive sunlight, while still allowing free movement of air and adequate penetration of natural light. Unlike greenhouses or polyhouses, which rely heavily on environmental modification, net houses focus on pest exclusion and crop protection, making them ideal for regions where insect pressure is high but climate regulation requirements are moderate. The covering material used in net houses is insect-proof netting–typically made from high-density polyethylene (HDPE) or similar synthetic fibres–that is UV-stabilized to withstand prolonged sun exposure. The net mesh size is selected based on the target pest:

- ✰ 40–50 mesh for protection against whiteflies, aphids, and thrips.
- ✰ 25–30 mesh for protection against larger insects and birds.

The frame structure can be built from galvanized iron (GI) pipes, steel, aluminium, or bamboo, depending on budget, durability needs, and intended lifespan. The design should be strong enough to withstand wind pressure, as net houses are often used in open, pest-prone areas. Net houses create a semi-controlled environment by reducing pest infestation levels, preventing physical damage from birds, and slightly lowering wind speed around the crops. This results in healthier plants with better growth rates and higher yields without the need for frequent pesticide applications, which not only reduces production costs but also minimizes chemical residues on produce. They are widely used for producing healthy seedlings for transplantation into open fields or other protected structures. By raising seedlings in pest-free conditions, farmers can ensure vigorous, disease-free plants that establish quickly after transplanting. Net houses are also used for growing vegetables (tomato, cucumber, capsicum), flowers (marigold, chrysanthemum), and even certain fruit crops in areas with high pest incidence. In addition to pest protection, net houses can be fitted with shade nets on top to provide partial protection from intense sunlight, which helps reduce heat stress during peak summer. This makes them suitable for cultivating crops in both tropical and subtropical climates. Irrigation within net houses is commonly managed through micro-irrigation systems such as drip lines or micro-sprinklers, ensuring precise water application and maintaining ideal soil moisture. The enclosed structure also helps reduce evaporation losses, making water use more efficient. From an economic perspective, net houses are cost-effective compared to polyhouses or glass greenhouses, with lower construction and maintenance costs. The lifespan of the nets generally ranges from 3 to 5 years, after which they may need replacement due to wear, dust accumulation, or UV damage. Net houses are particularly important in integrated pest management (IPM) programs, as they reduce the initial pest population pressure and help maintain lower pest levels throughout the cropping season. This makes them valuable for organic farming and export-oriented production, where chemical pesticide use must be minimized. By combining low-cost construction, effective pest exclusion, and minimal environmental modification, net houses offer a practical, sustainable solution for improving crop health, quality, and productivity in pest-prone regions.

4.2.5. Low Tunnels

Low tunnels are small, temporary protective structures designed to cover individual crop rows or beds, providing localized protection against adverse environmental conditions and pests. They consist of hoops or frames made from bamboo, galvanized iron (GI) rods, PVC pipes, or other lightweight materials, over which transparent polyethylene films or insect-proof nets are stretched. The primary function of low tunnels is to create a favourable microclimate around the plants by trapping heat, reducing wind speed, and preventing direct contact with rain or frost. This localized environmental modification accelerates plant growth, protects young seedlings, and reduces stress from fluctuating weather conditions. Low tunnels are especially effective in early-season crop production, allowing farmers to sow and transplant crops earlier than in open fields. The resulting earlier harvests help growers enter the market ahead of peak supply, often securing premium prices. These structures are commonly used for crops such as melons, cucumbers, squash, tomatoes, peppers, spinach, and other leafy greens. They are also valuable for frost-sensitive crops during late winter or early spring. By raising temperatures inside the tunnel by 3–5°C compared to ambient levels, they extend the growing season and enhance crop uniformity. Covering materials are selected based on the protection required:

- Transparent polyethylene films are used for heat retention and frost protection.
- Insect-proof nets are used to reduce pest infestations while allowing good ventilation.

Low tunnels are manually operated–the covers can be lifted or removed during warm days to prevent overheating and replaced at night or during bad weather. This flexibility makes them easy to manage without sophisticated equipment. Structurally, they are simple to construct, requiring minimal investment and materials that are locally available. This makes them accessible to small and marginal farmers, particularly in resource-limited regions. In addition to weather protection, low tunnels reduce evaporation losses from soil, conserving water and maintaining more stable moisture conditions around the root zone. When combined with drip irrigation, they ensure efficient use of water and nutrients. Another advantage of low tunnels is their mobility–once the crop is harvested, the materials can be dismantled, stored, and reused for subsequent seasons, lowering long-term costs. While low tunnels offer many benefits, they do have limitations: they provide only partial climate control, require regular monitoring to prevent overheating, and may be damaged by strong winds or heavy snowfall if not properly anchored. In modern farming systems, low tunnels are often integrated into protected cultivation programs as a low-cost, entry-level option for growers. They are particularly valuable for off-season vegetable production, seedling protection, and organic farming, where reduced pesticide use and season extension are priorities. By combining simplicity, affordability, and proven effectiveness, low tunnels offer farmers a practical way to protect crops, improve yields, and achieve better market timing without large capital investments.

4.2.6. High Tunnels (Walk-in Tunnels)

High tunnels, also known as walk-in tunnels, are elongated protective structures designed to cover entire crop beds or fields while providing enough interior height for workers to move freely and perform crop management tasks such as pruning, trellising, irrigation maintenance, pest monitoring, and harvesting. They are essentially an enlarged version of low tunnels but are more durable, versatile, and convenient for large-scale production. Constructed with arched or hoop-shaped frames made from galvanized iron (GI) pipes, steel, or PVC, high tunnels are covered with transparent polyethylene films that allow ample sunlight penetration while offering protection from adverse weather conditions. The covering material is usually UV-stabilized to prolong its lifespan, typically lasting 3–5 years before replacement is needed. High tunnels offer better protection against environmental stresses compared to low tunnels, shielding crops from wind, frost, heavy rainfall, hail, and excessive sunlight. They help maintain a warmer microclimate inside, often 2–4°C higher than ambient temperatures, which promotes early crop establishment and extended harvest periods. One of their key advantages is cost-effectiveness. While they are larger and more functional than low tunnels, they remain less expensive than fully climate-controlled greenhouses, as they rely primarily on passive solar heating and natural ventilation rather than advanced automated systems. This makes them an attractive choice for medium-scale farmers and market gardeners.

Ventilation in high tunnels is managed through roll-up or drop-down sidewalls, end-wall vents, or manually opened flaps. This simple yet effective system allows growers to regulate temperature, humidity, and airflow without expensive equipment. In warm weather, sidewalls are kept open to prevent overheating, while in cooler conditions, they are closed to retain heat. High tunnels are widely used for seasonal vegetable and flower production, especially crops like tomatoes, peppers, cucumbers, strawberries, lettuce, cut flowers, and herbs. They enable earlier planting in spring and prolonged harvesting into late autumn, thus increasing the marketing window and profitability. They are also suitable for organic farming, as the protective barrier reduces pest pressure and minimizes disease incidence, allowing growers to reduce chemical pesticide use. Additionally, high tunnels protect crops from soil splash during heavy rains, which helps prevent the spread of certain soil-borne diseases. Many high tunnels are equipped with drip irrigation and fertigation systems, ensuring efficient water and nutrient delivery directly to the root zone. Mulching, trellising systems, and crop rotation plans are often integrated into high tunnel management to maximize yield and maintain soil health. While high tunnels are robust and versatile, they do require regular maintenance–cover replacement, structural inspection, and anchoring reinforcement to withstand strong winds or heavy snow loads. By combining affordability, ease of construction, and significant season-extension capabilities, high tunnels have become a vital tool in protected cultivation systems worldwide. They offer a middle ground between low-cost, temporary low tunnels and high-cost, fully automated greenhouses, making them ideal for growers seeking high productivity with moderate investment.

Factors Influencing Structure Selection

The choice of a protected structure depends on several factors:

- **Climate conditions:** The local climate is the primary determinant in selecting a suitable protected structure. Areas with high temperature fluctuations require structures that can maintain stable internal conditions. In regions with heavy rainfall or hailstorms, durable roofing materials and strong frames are necessary. Similarly, high wind zones demand wind-resistant structures to prevent damage. Humidity levels also influence the choice, as excessive moisture can encourage disease. Therefore, understanding microclimatic variations is essential. The structure should be customized to withstand the prevailing environmental stresses.

- **Type of crop:** Different crops have unique requirements of temperature, light intensity, and humidity. For instance, leafy greens thrive in moderate light, while fruiting vegetables often need higher solar radiation. Flowers may require specific photoperiod management for proper blooming. High-value crops like capsicum or strawberries often demand sophisticated climate-controlled structures. Similarly, nurseries for seedlings need uniform temperature and humidity. Thus, the crop's biological needs largely dictate the design and features of the protected structure.
- **Level of environmental control needed:** Protected structures vary from simple low-cost tunnels to fully automated climate-controlled greenhouses. If the goal is only basic protection from rain or frost, a naturally ventilated polyhouse may suffice. However, for precise control of temperature, humidity, and CO_2 levels, high-tech greenhouses are essential. The level of control depends on whether year-round production is desired. Crops requiring exact conditions like orchids or exotic vegetables need advanced systems. Hence, farmers must match their crop's sensitivity with the degree of control offered.
- **Budget and resources:** The financial capacity of a grower greatly influences structure selection. Low-cost options such as shade nets and plastic tunnels are affordable but provide limited protection. Medium-cost polyhouses balance affordability and functionality for small-scale farmers. High-tech greenhouses require significant capital investment but ensure higher productivity and premium quality. Operational costs, including electricity, water, and maintenance, also need consideration. The availability of skilled labour for handling sophisticated systems further impacts feasibility. Thus, budget and resource planning guide the final choice.
- **Market opportunities:** Market demand strongly shapes decisions about protected cultivation. Structures that allow off-season production give farmers a competitive advantage by fetching higher prices. Specialty crops like exotic vegetables, cut flowers, or herbs often require advanced structures but yield better returns. If local markets support only seasonal produce, simpler structures may suffice. Export-oriented production generally demands higher quality standards, necessitating modern facilities. Access to transportation and storage infrastructure also matters. Hence, aligning structure selection with market prospects ensures profitability.

By carefully considering these factors, growers can select the most appropriate protected structure to maximize crop yield and profitability while promoting sustainable agricultural practices.

4.3 Environmental Factors Affecting Growth

Plant growth and development are strongly influenced by environmental conditions, which play a decisive role in shaping physiological and biochemical processes. Factors such as light, temperature, water, air, soil, and nutrients regulate photosynthesis, respiration, transpiration, and overall metabolism. Adequate sunlight ensures energy for food production, while temperature governs enzymatic activities and flowering. Water availability determines nutrient absorption, turgidity, and survival under stress. Similarly, soil quality provides anchorage and essential minerals, while air composition, especially CO_2 and oxygen, supports photosynthesis and respiration. Nutrient balance is critical, as both deficiencies and excesses can limit productivity. Humidity, wind, and altitude further modify crop performance by altering microclimatic conditions. Biotic interactions, including pests, diseases, and beneficial microbes, also shape growth patterns. Thus, plant performance is a combined outcome of multiple interacting environmental factors. Below are the major environmental factors explained in detail:

4.3.1. Light (Solar Radiation)

- ☆ Light is the primary energy source for photosynthesis, directly affecting plant growth and productivity.
- ☆ Intensity, quality (wavelength), and duration (photoperiod) are important components.
- ☆ High light intensity generally increases photosynthesis, but excessive light can cause scorching.
- ☆ Different crops have different light requirements–shade-loving crops like ferns need diffused light, while sun-loving crops like tomato and cotton require strong light.
- ☆ Photoperiod influences flowering–long-day plants (wheat, spinach) require longer daylight, whereas short-day plants (rice, chrysanthemum) require shorter daylight hours.
- ☆ Seasonal variations in light duration affect crop cycles and yield patterns.

4.3.2. Temperature

- ☆ Temperature regulates biochemical and physiological processes such as germination, enzyme activity, and flowering.
- ☆ Each plant species has a minimum, optimum, and maximum temperature range for growth.
- ☆ Low temperatures slow down metabolic activity, delay germination, and may cause chilling or frost injury.
- ☆ High temperatures can increase transpiration, cause wilting, or lead to heat stress.

- Some plants are adapted to cool climates (peas, cabbage), while others thrive in warm climates (cotton, maize).
- Temperature fluctuations also influence pest and disease prevalence.

4.3.3. Water (Moisture Availability)

- Water is vital for photosynthesis, nutrient transport, and cell turgidity.
- Inadequate water supply leads to drought stress, wilting, and reduced productivity.
- Excess water or waterlogging reduces soil aeration, causing root rot and poor growth.
- Crops have different water requirements–rice needs standing water, while pulses prefer well-drained soils.
- Irrigation management and rainfall distribution play critical roles in ensuring adequate moisture.
- Relative humidity in the air also affects transpiration and water balance in plants.

4.3.4. Air (CO_2 and Oxygen)

- Carbon dioxide is the raw material for photosynthesis, and its concentration directly affects productivity.
- A higher concentration of CO_2 under controlled environments (CO_2 enrichment in greenhouses) enhances yields.
- Oxygen is crucial for respiration, particularly in root zones.
- Poor soil aeration due to compaction or waterlogging reduces oxygen availability, affecting root growth.
- Wind is also an atmospheric factor–mild winds help pollination and gas exchange, but strong winds damage crops.

4.3.5. Soil Conditions

- Soil provides anchorage, nutrients, and water to plants.
- Soil texture (sand, loam, clay) influences water retention and drainage.
- Soil structure affects root penetration and aeration.
- pH level determines nutrient availability–acidic soils may lock up phosphorus, while alkaline soils may reduce iron availability.
- Organic matter enhances soil fertility, microbial activity, and moisture retention.
- Different crops demand specific soil conditions–wheat prefers loamy soils, while groundnut grows well in sandy loam.

4.3.6. Nutrients

- ✰ Nutrients are essential elements required for physiological and structural functions.
- ✰ Macronutrients (N, P, K, Ca, Mg, S) are needed in large amounts, while micronutrients (Zn, Fe, Mn, B, Cu, Mo, Cl) are required in small amounts.
- ✰ Deficiency of any nutrient leads to characteristic symptoms–nitrogen deficiency causes yellowing, phosphorus deficiency stunts growth, potassium deficiency causes leaf scorching.
- ✰ Excess nutrients can also cause toxicity and imbalance.
- ✰ Balanced fertilization and soil testing are necessary for optimum plant health and growth.

4.3.7. Biotic Factors

- ✰ Apart from physical factors, living organisms also influence plant growth.
- ✰ Pests, diseases, and weeds compete with or damage crops, reducing yields.
- ✰ Symbiotic organisms such as Rhizobium in legumes fix atmospheric nitrogen and enhance fertility.
- ✰ Mycorrhizal fungi improve water and nutrient absorption.
- ✰ Intercropping and biodiversity in cropping systems reduce the risk of pest outbreaks.
- ✰ Thus, both beneficial and harmful organisms play roles in determining plant growth.

Concise summary of environmental factors affecting growth is depicted in Table 4.2.

Table 4.2: Concise Summary of Environmental Factors Affecting Growth

Factor	*Role in Plant Growth*
Light	Provides energy for photosynthesis; intensity, duration, and quality affect growth and flowering.
Temperature	Regulates enzymatic activity, germination, respiration, and flowering; extremes cause stress.
Water	Essential for photosynthesis, nutrient transport, and turgidity; scarcity causes drought stress, excess leads to waterlogging.
Air (CO_2 and O_2)	CO_2 drives photosynthesis, oxygen is vital for respiration; wind influences pollination and transpiration.
Soil	Provides anchorage, nutrients, and water; texture, structure, and pH determine fertility.
Nutrients	Macro and micronutrients are essential; deficiencies or excesses limit plant health and yield.

Factor	*Role in Plant Growth*
Humidity	Affects transpiration, water balance, and disease incidence in crops.
Wind	Assists pollination and gas exchange; strong winds can damage plants.
Altitude	Influences temperature, oxygen levels, and growing season length.
Biotic Factors	Pests, diseases, weeds hinder growth; beneficial microbes like Rhizobium and mycorrhiza enhance nutrition.

4.4 Methods of Environmental Control

Environmental control in crop production refers to the scientific and practical regulation of external conditions such as light, temperature, humidity, water, soil, and air in order to create an environment most suitable for plant growth and development. Plants are highly sensitive to changes in their surroundings, and even slight variations in temperature, moisture, or radiation can influence physiological processes like photosynthesis, respiration, transpiration, and nutrient uptake. By ensuring stable and favourable environmental conditions, farmers can improve productivity, enhance crop quality, and ensure uniformity in growth. One of the major advantages of environmental control is the ability to extend the growing season and make crop cultivation possible even in regions with unfavourable climates. For example, off-season vegetable production, floriculture, and high-value horticulture are made possible by creating controlled environments such as greenhouses and polyhouses. Environmental regulation also helps in protecting plants from extreme climatic stresses such as frost, drought, heavy rainfall, and strong winds, which otherwise limit agricultural potential. It plays an important role in minimizing crop losses due to sudden weather changes and allows farmers to maintain consistent supply for markets. Controlled environments are also highly beneficial for producing high-value specialty crops, export-oriented produce, and nursery plants where quality standards are very strict. Furthermore, by manipulating factors like photoperiod, CO_2 concentration, and soil nutrition, crops can be made to flower or fruit earlier, improving both economic returns and food security. Modern agriculture increasingly relies on a combination of simple cultural practices and advanced technologies for environmental control. Traditional methods include mulching, shading, irrigation scheduling, and windbreaks, while modern protected cultivation employs sensors, automation, fertigation, hydroponics, fogging systems, and climate control software. These approaches not only increase yield but also ensure resource efficiency by saving water, fertilizers, and energy. In the face of global climate change, environmental control has emerged as a vital strategy to maintain agricultural stability and sustainability. Thus, the integration of environmental control measures into crop production offers farmers an opportunity to overcome natural limitations, reduce risks, and achieve year-round, profitable, and high-quality agricultural production.

The major methods are described below:

4.4.1. Temperature Control

- **Ventilation:** Ventilation is one of the most effective methods to regulate temperature inside protected structures such as polyhouses and greenhouses. Natural ventilation involves side vents, roof vents, or roll-up walls that allow hot air to escape and cooler air to enter. Forced ventilation uses exhaust fans, circulation fans, or blowers to actively move air and maintain uniform temperature. It also helps in controlling humidity levels, preventing diseases caused by excess moisture. Adequate airflow supplies plants with fresh carbon dioxide needed for photosynthesis. Proper ventilation reduces heat stress and maintains a stable microclimate for optimal crop growth (Figure 4.6).

Figure 4.6: Naturally Ventilated Polyhouse.
(*Source*: https://www.indiamart.com)

- **Shading:** Shading techniques are applied to reduce excess solar radiation during hot seasons, which otherwise causes leaf burn and wilting. Shade nets of varying percentages (25–75 per cent) are used depending on crop light requirements. Whitewash or shading paints applied on greenhouse covers also help in reflecting part of the incoming sunlight. Shading not only lowers temperature but also reduces evapotranspiration losses, saving water. Crops sensitive to high light intensity, such as leafy vegetables and nursery seedlings, benefit greatly from shading. It helps in maintaining a cooler and balanced internal climate during peak summer (Figure 4.7).
- **Heating Systems:** In cold regions or during winters, heating systems are necessary to protect crops from frost and chilling injury. Electric heaters, hot air blowers, and gas-fired heating units are commonly used to raise air temperature in greenhouses. For large-scale setups, hot water pipes or geothermal heating systems may also be employed.

Figure 4.7: Shade Nets to Protect Plants from Sun Damage.
(*Source*: https://www.agriplast.co.in)

Heating ensures that the optimum temperature range required for seed germination, flowering, and fruit set is maintained. Sensitive crops like cucumber, tomato, and capsicum require stable warmth to avoid delayed growth. Proper heating safeguards crop quality and ensures continuous production even in adverse cold climates (Figure 4.8).

Figure 4.8: Heat Greenhouse in Winter (14 Proven Methods)
(*Source*: https://help.gardeningexpress.co.uk)

- ☆ **Thermal Screens and Mulching:** Thermal screens are special insulating sheets or cloths drawn over crops at night to reduce heat loss and radiation cooling. They help in retaining warmth and protecting plants from frost damage. Mulching with organic or plastic materials around the root zone also conserves soil heat and prevents temperature fluctuations. This technique additionally suppresses weed growth and reduces soil moisture loss. By moderating the root environment, mulching promotes better nutrient uptake. Together, thermal screens and mulches create a favourable microclimate that supports healthy and uniform plant growth (Figure 4.9).

Figure 4.9: Thermal Screens.
(*Source*: https://fertri.com)

☆ **Insulation Materials:** Insulation is a key strategy in maintaining consistent temperatures within protected structures.

Figure 4.10: Polycarbonate Solid Sheet
(*Source*: https://www.polycarbonateglobal.com)

Polyethylene sheets, double-layer poly films, or polycarbonate panels are commonly used as covering materials in greenhouses. These materials trap solar energy during the day and prevent rapid heat loss at night. The air gap between double-layer films acts as an insulator, improving energy efficiency. Insulation also helps in reducing heating costs during winters. In high-tech setups, thermal blankets or reflective foils further improve insulation. Well-insulated structures enable year-round cultivation of high-value crops by stabilizing the internal climate (Figure 4.10).

- ☆ **Importance of Temperature Control:** Temperature regulation is central to successful protected cultivation, as it influences all major physiological processes in plants. Maintaining an optimal range ensures faster germination, balanced vegetative growth, proper flowering, and higher fruit set. Crops like tomato, cucumber, and capsicum, which are sensitive to extremes, can be cultivated year-round under controlled temperatures. This enables off-season production and supply to markets when demand and prices are high. Proper temperature management also minimizes risks of heat stress, chilling injury, and diseases. Hence, efficient temperature control directly translates into higher productivity and profitability for farmers.

4.4.2. Light Control

- ☆ **Shading Materials:** Shading materials are essential in regulating the intensity of light inside greenhouses and polyhouses. Shade nets of varying densities, such as 25 per cent, 50 per cent, or 75 per cent, are widely used depending on crop requirements. These nets help reduce excessive solar radiation, which can otherwise lead to leaf scorching and poor photosynthetic efficiency. By diffusing sunlight, they create a favourable microclimate for tender crops. Additionally, shading reduces water loss through transpiration and lowers the risk of heat stress. In hot summer months, shade nets ensure healthy growth and better flower and fruit quality. Thus, they act as a protective shield for sensitive horticultural crops.
- ☆ **Artificial Lighting:** Artificial lighting plays a crucial role in supplementing natural sunlight, especially in regions with short-day conditions or during winter. Greenhouses often use LED, fluorescent, or high-pressure sodium lamps to provide consistent illumination. These light sources extend the duration of photosynthesis and promote vegetative as well as reproductive growth. LEDs are preferred due to their energy efficiency, low heat emission, and ability to emit specific wavelengths. By using artificial lights, growers can increase yields and maintain year-round production. This technique is particularly useful for off-season cultivation of vegetables and flowers. Artificial lighting, therefore, ensures stable productivity under varying climatic conditions.

- **Photoperiod Manipulation:** Photoperiod manipulation refers to controlling the length of day and night to regulate plant flowering and growth. Different crops respond differently–long-day plants like spinach need extended light, while short-day crops like chrysanthemum require longer darkness. Greenhouse operators use artificial lighting or blackout curtains to alter the natural photoperiod. This technique helps synchronize flowering, improve uniformity, and meet market demand for flowers and fruits. It is widely practiced in floriculture and commercial horticulture industries. By managing photoperiods, growers can produce crops out of season and achieve higher profitability. Hence, it is a powerful tool for precision crop management.
- **Reflective Surfaces:** Reflective surfaces are an innovative way to enhance light distribution inside greenhouses and polyhouses. Aluminium foils, white plastic sheets, and lime-coated walls are commonly used to redirect sunlight evenly across the crop canopy. This reduces shading effects caused by plant density and structural components. Better light penetration ensures uniform photosynthesis, resulting in healthy and vigorous plants. Reflective surfaces also help in reducing temperature fluctuations by bouncing back excessive heat. They are cost-effective, easy to install, and highly efficient in boosting productivity. Farmers cultivating crops in dense arrangements benefit greatly from this method.
- **Light Quality Management:** Light quality management focuses on providing crops with the most beneficial wavelengths of light for their growth. Research shows that red and blue wavelengths are most effective for photosynthesis and morphogenesis. Red LEDs promote flowering and fruiting, while blue LEDs enhance leaf expansion and stem elongation control. By using LED panels with specific spectra, growers can fine-tune plant growth according to their needs. This method improves crop yield, nutritional quality, and uniformity. Light quality management also helps reduce energy wastage compared to broad-spectrum lighting. It represents a scientific and sustainable approach to modern protected cultivation.
- **Importance of Light Regulation in Crops:** Light regulation is a vital factor in the successful cultivation of high-value crops like chrysanthemum, strawberry, and lettuce. These crops have specific light requirements that influence flowering, fruiting, and leaf development. For instance, chrysanthemums are highly sensitive to photoperiod, while strawberries require adequate light for fruit set and sweetness. Lettuce, being a leafy crop, responds strongly to light intensity and quality for head formation. Proper regulation ensures enhanced yield, uniform growth, and better marketability. Farmers practicing protected cultivation rely heavily on shading, artificial lights, and reflective surfaces to meet crop-specific needs. Hence, light management forms the backbone of greenhouse technology.

4.4.3. Humidity Control

- ✰ **Ventilation Systems:** Ventilation is one of the most effective and widely adopted methods of controlling excess humidity in greenhouses. Natural ventilation relies on roof vents, side vents, or louvered windows that allow humid air to escape, while cooler and drier outside air enters. Forced ventilation uses exhaust fans and blowers that rapidly push moist air out, maintaining better control even under extreme weather conditions. Proper ventilation minimizes condensation on leaves, stems, and greenhouse structures, thereby reducing the risk of fungal spores germinating on wet surfaces. It is particularly useful in preventing diseases such as downy mildew, powdery mildew, and botrytis, which thrive in humid environments. Ventilation also ensures that carbon dioxide levels remain sufficient for photosynthesis, avoiding stagnation of air. Well-ventilated houses show fewer pest and disease outbreaks due to reduced favourable microclimates. Additionally, temperature regulation is achieved as ventilation allows heat escape during hot days. By combining roof vents, side vents, and fans, farmers can maintain an ideal balance of humidity and temperature. Thus, ventilation is considered the first line of Défense against excessive humidity and a cornerstone of greenhouse climate management.
- ✰ **Fogging and Misting:** Fogging and misting systems are essential tools for raising humidity during hot and dry weather conditions inside greenhouses. These systems release ultra-fine droplets of water into the air, which quickly evaporate and increase relative humidity without leaving excess water on plant surfaces. This is highly beneficial for nurseries, young seedlings, and cuttings that require a consistently humid environment for rooting and establishment. By lowering leaf surface temperature, fogging helps in preventing wilting, sunburn, and heat stress. The micro-droplets also reduce transpiration rates, allowing plants to conserve water and maintain turgidity. Advanced automated fogging systems are equipped with sensors that monitor humidity levels and activate misting only when required, making them energy- and water-efficient. In crops like cucumber, tomato, and ornamentals, misting ensures uniform vegetative growth and higher survival rates of transplants. Moreover, fogging indirectly assists in nutrient absorption by maintaining favourable stomatal activity. It also prevents leaf-tip burn in leafy vegetables like lettuce under dry air conditions. Overall, fogging and misting systems are vital technologies for maintaining balanced humidity and stable growth in sensitive crops.
- ✰ **Drip Irrigation and Mulching:** Drip irrigation combined with mulching is an efficient method to maintain soil moisture while avoiding excessive atmospheric humidity in greenhouses. Drip systems deliver water directly to the root zone in controlled quantities, reducing evaporation

and minimizing leaf wetting, which otherwise contributes to high humidity and disease outbreaks. Mulching with plastic films, straw, or organic matter further helps in conserving soil moisture, moderating soil temperature, and reducing evaporation losses. This synergy ensures that crops receive adequate water for optimal growth without adding unnecessary moisture to the greenhouse air. The system also improves water-use efficiency, making it an eco-friendly choice for farmers. By reducing free water on leaf surfaces, the risks of fungal diseases like damping-off, fusarium wilt, and bacterial leaf spots are minimized. Mulching additionally suppresses weed growth, improves soil structure, and enhances microbial activity, leading to healthier root systems. Crops like capsicum, cucumber, and strawberry benefit greatly from drip with mulching as it supports uniform growth and better-quality produce. In regions facing water scarcity, this method is a sustainable solution. Thus, drip irrigation with mulching is both a disease-preventive and resource-efficient strategy for humidity management.

☆ **Dehumidifiers and Absorbers:** In advanced greenhouse setups, mechanical dehumidifiers and chemical absorbers are employed for precise humidity management. Mechanical dehumidifiers work by condensing excess water vapor from the air, collecting it as water, and releasing drier air back into the environment. These are highly effective in closed systems where natural or forced ventilation is limited, such as during winter or in regions with high outside humidity. On the other hand, chemical absorbers like silica gel, calcium chloride, or activated carbon are used in smaller greenhouses to trap excess moisture passively. Such systems help maintain optimal relative humidity levels, generally between 60–80 per cent, depending on crop needs. By doing so, they significantly reduce the risk of fungal and bacterial infections such as blight, leaf spots, and mildew. Although dehumidifiers involve higher installation and operational costs, they ensure consistency in crop quality and yield, making them suitable for high-value crops like orchids, roses, and strawberries. Moreover, in hydroponic and vertical farming systems, dehumidifiers are indispensable for precise climate control. Farmers engaged in export-oriented or pharmaceutical crop cultivation find them particularly beneficial. Hence, dehumidifiers and absorbers represent modern solutions for scientific humidity regulation in greenhouses.

☆ **Crop Spacing and Pruning:** Crop spacing and pruning are simple, low-cost, but highly effective cultural practices to regulate humidity within greenhouse environments. When crops are planted too closely, air circulation is restricted, creating localized humid zones around plant canopies. Such microclimates become breeding grounds for fungal pathogens, causing rapid disease spread. By ensuring adequate spacing between plants, growers allow better airflow that naturally lowers humidity and keeps foliage drier. Pruning involves the removal of excess leaves, side shoots, or branches that block air movement and trap moisture. This not only reduces transpiration load but also improves light penetration, leading to healthier photosynthesis. Well-spaced and pruned crops exhibit improved flowering and fruiting due to better nutrient distribution. Additionally, pruning helps in early detection and removal of diseased plant parts, preventing pathogen spread. For crops like tomato, cucumber, and grape, systematic pruning is an integral part of greenhouse management. When combined with proper trellising, spacing and pruning maximize yield potential while minimizing fungal risks. Overall, these cultural methods enhance both air circulation and crop quality while keeping production costs low, making them indispensable for greenhouse growers.

- **Importance of Proper Humidity Management:** Proper humidity management is central to achieving healthy plant growth, disease resistance, and high productivity in greenhouse cultivation. Excess humidity creates conditions conducive for fungal pathogens such as powdery mildew, downy mildew, and late blight, which can devastate crops if uncontrolled. Conversely, very low humidity can result in dehydration, poor seed germination, stunted growth, and increased vulnerability to pests like spider mites. The goal is to maintain an optimum humidity range that balances transpiration, nutrient uptake, and photosynthesis. This ensures active stomatal functioning, which directly impacts crop vigour and yield. Crops such as chrysanthemum, rose, lettuce, and strawberry are particularly sensitive to humidity fluctuations and require precise control. Farmers adopt integrated strategies–ventilation, drip irrigation, fogging, pruning, and dehumidifiers–to achieve this balance. Advanced sensor-based systems are now available to monitor and regulate humidity in real time, reducing guesswork. Proper management also lowers the need for chemical fungicides, promoting safer and more sustainable farming. Ultimately, scientific humidity control not only prevents crop losses but also improves quality and shelf life of produce. Thus, it is a fundamental pillar of modern protected cultivation systems.

4.4.4. Water and Irrigation Management

- **Drip Irrigation:** Drip irrigation is one of the most efficient water management techniques used in protected cultivation. It delivers water directly to the root zone of plants through a network of pipes, emitters, and tubes, ensuring precise application of moisture where it is needed most. This method significantly reduces water wastage compared to traditional surface irrigation, as there is minimal evaporation and runoff. By keeping the plant canopy dry, drip irrigation reduces the chances of fungal diseases that thrive under wet leaf conditions. It also facilitates fertigation–the application of fertilizers through irrigation–which enhances nutrient uptake and plant growth. The system allows farmers to regulate the frequency and quantity of water, thereby maintaining a constant soil moisture level. Crops such as tomato, cucumber, capsicum, and strawberry show higher yields and better quality when cultivated under drip systems. Moreover, drip irrigation is suitable for areas with limited water availability, making it a sustainable practice. By saving up to 40–60 per cent of water compared to conventional methods, it has become an essential part of greenhouse farming. Thus, drip irrigation ensures precision, efficiency, and sustainability in crop production (Figure 4411).

Figure 4.11: Drip Irrigation in Protected Cultivation.
(*Source*: https://pixabay.com)

- **Sprinkler Irrigation:** Sprinkler irrigation is another widely used method for greenhouse and open-field cultivation, providing water in the form of fine sprays that mimic natural rainfall. This system distributes water uniformly across the crop canopy and soil, ensuring all plants receive adequate moisture. Apart from irrigation, sprinklers also help in moderating canopy temperature during hot periods, reducing heat stress. They wash off dust from leaves and improve photosynthetic efficiency. Portable sprinkler sets and micro-sprinklers are available for small-scale farms, while automated systems are preferred in larger greenhouses. The technique is highly effective for crops with shallow roots, nurseries, and leafy vegetables where even water distribution is crucial. Sprinklers are also used for frost protection by maintaining a thin film of water that prevents freezing injuries. However, care must be taken to avoid excessive wetting of foliage, which can encourage fungal diseases. Modern sprinkler systems are designed with adjustable nozzles to control droplet size and water pressure, thus preventing soil erosion. Overall, sprinkler irrigation not only ensures uniform watering but also creates a favourable microclimate for sensitive crops (Figure 4.12).

Figure 4.12: Sprinkler Irrigation
(*Source*: https://irrigationmontroyal.com)

- ☆ **Subsurface Irrigation:** Subsurface irrigation is an advanced method that delivers water below the soil surface directly to the root zone. Unlike surface irrigation, this technique reduces evaporation losses, making it highly efficient in terms of water use. Pipes or drip lines are buried underground, allowing constant soil moisture to be maintained without frequent manual intervention.

Figure 4.13: Subsurface Irrigation
(*Source*: https://media.geeksforgeeks.org)

This prevents waterlogging on the surface and keeps the plant canopy dry, thereby lowering the risk of fungal and bacterial infections. Subsurface irrigation is particularly useful in sandy soils where water percolation is high and in regions where water scarcity is a major issue. It ensures deep root development, which strengthens plants and enhances their tolerance to drought conditions. Automated subsurface systems can also integrate fertigation, ensuring nutrients reach the root zone without wastage. While initial installation costs are relatively high, long-term savings in water, fertilizer, and labour make it cost-effective. High-value crops like grapes, capsicum, and cut flowers benefit greatly from this technique. Thus, subsurface irrigation represents a modern, resource-efficient, and sustainable approach to greenhouse farming (Figure 4.13).

- **Rainwater Harvesting:** Rainwater harvesting is an eco-friendly water management practice increasingly adopted in greenhouse and polyhouse farming. It involves the collection and storage of rainwater from rooftops, greenhouse structures, or catchment areas in tanks or reservoirs. The stored water can later be used during dry spells, ensuring a reliable supply for irrigation throughout the year. This method reduces dependency on groundwater and external water sources, which are often costly or unreliable. Collected rainwater is usually of good quality, free from harmful salts and chemicals, making it suitable for sensitive crops. By integrating rainwater harvesting with drip or sprinkler systems, farmers can practice sustainable water management. In regions with erratic rainfall, it acts as a buffer against drought, ensuring uninterrupted production. Additionally, it helps in groundwater recharge when excess water is diverted into soak pits.

Figure 4.14: Rainwater Harvesting.
(*Source*: https://www.freepik.com)

Governments and agricultural institutions encourage farmers to adopt rainwater harvesting by offering subsidies and technical support. When combined with filtration systems, harvested water can even be used for fertigation. Thus, rainwater harvesting is a sustainable, low-cost, and long-term solution for water security in protected farming (Figure 4.14).

- ☆ **Soil Moisture Sensors:** Soil moisture sensors are advanced tools that provide real-time data on soil water content, helping farmers optimize irrigation practices. These sensors measure moisture at different soil depths and send signals to automated irrigation systems, which then supply water only when needed. By preventing over-irrigation, they reduce waterlogging, nutrient leaching, and energy wastage. Conversely, they also prevent under-irrigation, ensuring crops never face drought stress. Farmers can use portable handheld sensors or install wireless sensor networks for large-scale greenhouse operations. Data from sensors can be integrated with mobile apps or computer dashboards, making decision-making easier. Such systems promote precision farming, saving up to 30–40 per cent of irrigation water. Soil moisture sensors are particularly valuable for high-value crops like lettuce, herbs, strawberries, and ornamental plants, where water balance directly impacts yield and quality. They also reduce labour requirements by automating irrigation schedules. With the growing use of Internet of Things (IoT) in agriculture, these devices are becoming more affordable and accessible. Hence, soil moisture sensors are revolutionizing water management in modern protected farming systems.
- ☆ **Importance of Effective Water Management:** Effective water management is a cornerstone of successful greenhouse cultivation, as it ensures plants neither suffer from drought stress nor waterlogging. Both extremes can severely affect plant physiology–drought stress reduces photosynthesis and stunts growth, while waterlogging deprives roots of oxygen and encourages root rot. By adopting scientific irrigation methods like drip, sprinkler, and subsurface systems, growers can maintain optimal soil moisture levels for consistent plant performance. Integration of technologies such as soil moisture sensors, automated irrigation, and fertigation further enhances water-use efficiency. Practices like rainwater harvesting and recycling wastewater make greenhouse farming more sustainable and eco-friendlier. Effective water management not only conserves precious water resources but also reduces input costs and improves crop yield and quality. It ensures uniform growth, timely flowering, and better fruit setting in high-value crops. Moreover, it minimizes the occurrence of soil-borne and fungal diseases linked to excess moisture. As water scarcity intensifies globally, smart irrigation strategies will become indispensable. Thus, efficient water management is not just an agronomic necessity but also an environmental responsibility in protected cultivation systems.

4.4.5. Soil and Nutrient Control

- **Soil Sterilization:** Soil sterilization is a crucial step in greenhouse farming to eliminate harmful pathogens, nematodes, weed seeds, and soil-borne diseases before cultivation begins. Traditional soil fumigation using chemicals like formaldehyde and methyl bromide was common, but due to environmental concerns, safer alternatives are now preferred. Steam sterilization is widely used in modern greenhouses as it kills pathogens effectively without leaving toxic residues. In this method, hot steam is passed through the soil, raising its temperature to 70–80°C, which destroys disease-causing organisms. Solarization, using transparent polythene sheets to trap solar heat, is another eco-friendly technique for pathogen control. Sterilized soil provides a healthy environment for seedlings, ensuring higher survival and better growth. This practice is particularly beneficial in crops like tomato, cucumber, and capsicum, which are prone to soil-borne diseases. By starting with clean soil, farmers reduce dependency on fungicides and pesticides. Although initial costs may be high, the long-term benefits of reduced disease pressure and improved yields outweigh the investment. Thus, soil sterilization acts as a preventive measure, ensuring a disease-free foundation for protected cultivation (Figure 4.15).

Figure 4.15: Soil Sterilization.
(*Source*: https://www.freepik.com)

- **Soil pH Management:** Maintaining the right soil pH is essential for optimal nutrient availability and plant growth. Most greenhouse crops perform best in a slightly acidic to neutral pH range (5.5–7.0). When soils are too acidic, essential nutrients like phosphorus, calcium, and magnesium become unavailable, while toxic elements like aluminium may become soluble. To correct acidity, agricultural lime or dolomite is added to raise the pH. Conversely, when soils are too alkaline, micronutrients such as iron, manganese, and zinc become deficient, leading to chlorosis and poor growth. In such cases, elemental sulphur or acidifying fertilizers like ammonium sulphate are applied to lower the pH. Regular soil testing is essential to monitor pH levels and adjust amendments accordingly. In soilless media, pH monitoring of nutrient solutions is equally important, as fluctuations can rapidly affect crop health. pH adjustment ensures proper nutrient absorption, strong root development, and balanced plant metabolism. Crops like rose, gerbera, lettuce, and strawberry are highly sensitive to pH variations and require precise management. Therefore, soil pH regulation is a cornerstone of scientific soil fertility management in controlled cultivation.

- **Soilless Culture:** Soilless culture techniques such as hydroponics, aeroponics, and cocopeat-based systems are gaining popularity in greenhouse farming. These methods completely eliminate the need for soil, allowing growers to deliver precise quantities of water and nutrients directly to plant roots. Hydroponics involves growing plants in nutrient-enriched water solutions, ensuring rapid growth and higher yields. Aeroponics, where roots are suspended in air and periodically misted with nutrient solutions, enhances oxygen availability and nutrient absorption. Cocopeat, derived from coconut husk, is widely used as a growing medium due to its high water-holding capacity, good aeration, and eco-friendly nature. Soilless culture minimizes the risk of soil-borne diseases, improves nutrient efficiency, and supports year-round production. It also allows vertical and indoor farming, optimizing space utilization. Crops like lettuce, basil, tomato, cucumber, and cut flowers thrive in soilless systems, producing uniform quality and higher productivity. Though initial setup costs are high, returns are significantly better due to reduced pesticide use and faster crop cycles. Thus, soilless culture is a futuristic approach that ensures precision nutrition, resource efficiency, and sustainable crop production.
- **Fertilizer Scheduling:** Fertilizer scheduling is a critical aspect of nutrient management in greenhouse cultivation. It involves planning the right type, quantity, and timing of fertilizer applications to match crop needs at different growth stages. Controlled-release fertilizers are often used to provide nutrients gradually, reducing losses and ensuring continuous supply. Fertigation, where fertilizers are dissolved in irrigation water and applied through drip systems, is the most efficient method as it delivers nutrients directly to the root zone. This ensures uniform distribution, higher nutrient-use efficiency, and reduced wastage. By adjusting fertilizer schedules, farmers can optimize vegetative growth, flowering, and fruiting stages for maximum yield. Nutrient solutions are tailored based on soil or substrate analysis, ensuring a balanced supply of macro (N, P, K) and micro (Fe, Mn, Zn, B) nutrients. Excessive or imbalanced fertilization can

cause nutrient toxicity, salinity buildup, and environmental pollution, hence precision scheduling is vital. Crops like cucumber, capsicum, and rose respond very well to fertigation-based nutrient plans. Modern greenhouses also integrate computer-controlled fertigation units that adjust nutrient composition in real time. Thus, systematic fertilizer scheduling enhances crop performance, reduces costs, and ensures sustainable productivity.

- **Organic Amendments:** Organic amendments play a vital role in maintaining soil health and fertility in greenhouse cultivation. Materials like farmyard manure, compost, vermicompost, biofertilizers, and green manures enrich the soil with organic matter and beneficial microbes. They improve soil structure, aeration, and water-holding capacity while also enhancing nutrient availability. Compost and biofertilizers release nutrients slowly, supporting long-term soil fertility and reducing dependence on chemical fertilizers. Organic inputs also promote microbial diversity, which suppresses soil-borne pathogens and improves plant resilience. Green manures like sun hemp and cowpea, when incorporated into the soil, add nitrogen and improve organic carbon levels. Vermicompost enhances root development by providing plant growth-promoting substances. The use of biofertilizers such as Rhizobium, Azotobacter, and mycorrhizal fungi further improves nutrient uptake, particularly nitrogen and phosphorus. Incorporating organic matter also reduces soil compaction and enhances cation exchange capacity, ensuring better nutrient retention. For greenhouse crops like cucumber, capsicum, and strawberry, organic amendments improve taste, quality, and shelf life of produce. Thus, integrating organic amendments with inorganic fertilizers ensures balanced nutrition, soil sustainability, and long-term productivity.
- **Importance of Proper Soil and Nutrient Control:** Proper soil and nutrient management is the foundation of successful greenhouse farming. Balanced nutrition enhances root growth, supports vigorous vegetative development, and ensures higher yields. Soil sterilization prevents early crop losses due to diseases, while pH management optimizes nutrient availability. Soilless culture provides precision and eliminates soil-related problems, making year-round farming possible. Fertilizer scheduling ensures that plants receive the right nutrients at the right time, reducing wastage and improving efficiency. Organic amendments enrich soil health, improve structure, and support beneficial microbial activity. Together, these practices create a balanced soil environment that promotes sustainable production. Improper soil or nutrient management can lead to nutrient deficiencies, toxicities, poor flowering, and low yields. With crops like tomato, rose, strawberry, and lettuce being highly sensitive to nutrient balance, scientific management becomes even more

important. Modern greenhouses increasingly rely on sensors, soil testing, and computer-based nutrient dosing to maintain precision. Ultimately, well-managed soils and nutrient systems lead to healthier crops, better quality produce, and maximum profitability. Hence, soil and nutrient control forms the backbone of protected cultivation systems.

4.4.6. Air and Gas Regulation

- **CO_2 enrichment:** Carbon dioxide (CO_2) enrichment is one of the most effective techniques to enhance photosynthesis and increase crop yield in greenhouses. In natural conditions, atmospheric CO_2 levels are about 400 ppm, but most greenhouse crops show higher productivity when concentrations are raised to 800–1,200 ppm. Crops like cucumber, tomato, capsicum, and rose respond particularly well to CO_2 supplementation, showing increased leaf expansion, faster growth, and higher yields. CO_2 enrichment is achieved by using liquid CO_2 cylinders, compressed gas tanks, or combustion-based CO_2 generators. The gas is released in controlled amounts through pipelines or perforated tubes to ensure even distribution. This technique is especially beneficial during mornings and cloudy days when photosynthesis is otherwise limited. Farmers often combine CO_2 enrichment with adequate lighting and temperature regulation to maximize efficiency. Studies have shown yield increases of up to 20–40 per cent in vegetables and flowers under proper CO_2 management. However, excessive enrichment beyond optimal levels can cause toxicity, so monitoring is essential. Overall, CO_2 supplementation represents a key practice in precision greenhouse farming that significantly boosts productivity and profitability.
- **Oxygen supply:** Adequate oxygen supply around plant roots is crucial for healthy growth and efficient nutrient uptake. In poorly aerated soils or waterlogged conditions, oxygen levels decline, leading to anaerobic environments that inhibit root respiration. Such conditions result in stunted growth, root rot, and reduced yields. To prevent this, soil aeration through practices like deep tillage, subsoiling, and incorporation of organic matter is carried out before greenhouse planting. In soilless culture systems, oxygen supply is maintained by circulating nutrient solutions or bubbling air into hydroponic reservoirs. Well-aerated soils encourage the growth of beneficial microbes while suppressing harmful anaerobic pathogens. Oxygen availability also enhances root metabolism, water absorption, and nitrogen fixation by symbiotic organisms. For crops like lettuce, strawberry, and cucumber, oxygen supply in the root zone is directly linked to yield quality. Technologies such as oxygen-enriched irrigation water and root-zone aeration pipes are now being explored to enhance efficiency. Thus, maintaining oxygen balance in soil and root media is an often-overlooked but vital aspect of air and gas regulation in protected farming.

- **Air circulation fans:** Air circulation fans play a central role in maintaining uniform distribution of gases, temperature, and humidity inside greenhouses. Without proper circulation, CO_2 released during enrichment may settle near the ground, leaving upper foliage deprived of sufficient carbon dioxide. Fans create a gentle airflow that ensures all crop canopies receive uniform gas exchange, supporting balanced photosynthesis. Proper air circulation also prevents the formation of hot or humid pockets that encourage fungal diseases such as botrytis and downy mildew. Additionally, moving air strengthens plant stems and reduces lodging, especially in tall crops like tomato and chrysanthemum. Ceiling-mounted horizontal airflow (HAF) fans are widely used in modern greenhouses for energy-efficient circulation. By regulating airflow, these fans also help maintain temperature uniformity, preventing stress due to localized overheating or chilling. Integration of fans with automated sensors and controllers further improves precision in climate regulation. For crops under high-density planting, air circulation is particularly critical to avoid humidity buildup. Thus, circulation fans ensure not just better air composition but also healthier microclimates for higher crop productivity.
- **Windbreaks:** Windbreaks are physical barriers, natural or artificial, designed to reduce the damaging effects of strong winds in open-field and peri-urban greenhouse farming. Rows of trees, tall shrubs, or artificial structures such as fences and nets are commonly used as windbreaks. They reduce wind velocity, protecting crops from mechanical damage, desiccation, and soil erosion. In cereals like wheat, maize, and rice, windbreaks help prevent lodging, where plants are knocked over by high winds, reducing yield losses. In horticultural crops, they protect flowers and fruits from abrasion and dropping. Windbreaks also conserve soil moisture by reducing evaporation losses caused by strong winds. Moreover, they create a microclimate with moderated temperature and humidity, which benefits crop growth. Tree-based windbreaks, such as rows of eucalyptus, neem, or casuarina, additionally provide timber, fuelwood, and biodiversity support. Artificial windbreaks using shade nets or plastic screens are often installed around polyhouses to protect fragile structures. Well-designed windbreaks not only protect crops but also enhance sustainability by integrating agroforestry benefits. Thus, they are a simple yet highly effective method of regulating airflow and reducing climate-related risks in agriculture.
- **Importance of Air Composition Regulation:** Regulating air composition is fundamental for improving crop growth, quality, and resilience in both greenhouse and open-field conditions. Adequate CO_2 enrichment enhances photosynthesis and boosts yield, while proper oxygen supply in the root zone ensures efficient nutrient uptake and root health. Uniform air

circulation prevents microclimate imbalances, reduces humidity-related diseases, and strengthens plant structure. In open fields, windbreaks minimize lodging in cereals and protect sensitive crops from wind stress. Maintaining an ideal air balance also improves fruit set, flower quality, and nutrient assimilation. Poor regulation, on the other hand, can lead to yield losses, disease outbreaks, and reduced quality of produce. Advanced greenhouses increasingly use sensors to monitor CO_2, O_2, temperature, and humidity, adjusting airflow and gas levels automatically. Integration of air and gas regulation with other climate-control systems ensures holistic crop management. For crops like cucumber, tomato, rose, and lettuce, such regulation directly influences yield potential and market value. Thus, managing air and gases scientifically is not only a productivity enhancer but also a sustainability measure in modern agriculture.

4.4.7. Pest and Disease Control (Biotic Environment)

- **Physical Barriers:** Physical barriers are the first line of Défense against insect pests and pathogens in protected cultivation. Insect-proof nets, screens, and double-door entry systems prevent the entry of harmful pests such as whiteflies, thrips, aphids, and leaf miners into greenhouses. These barriers work by creating a physical obstruction that reduces pest pressure and minimizes the need for chemical pesticides. Net houses made with UV-stabilized nylon or polyethylene mesh are commonly used to cover vents and openings, allowing ventilation while blocking insects. Proper mesh size selection is critical–for example, 40–50 mesh nets are effective against small insects like thrips. In addition to insect nets, sticky traps and reflective mulches are also deployed as physical deterrents to trap or repel pests. Barriers not only protect crops but also reduce the spread of viral diseases transmitted by insect vectors. For high-value crops such as tomato, capsicum, cucumber, and flowers, physical protection ensures quality and reduces post-harvest losses. When combined with good sanitation, these barriers provide long-term effectiveness. Thus, physical barriers are an eco-friendly and preventive measure for maintaining crop health in controlled environments.
- **Biological Control:** Biological control is an environmentally safe and sustainable approach that utilizes natural enemies of pests to keep their populations under control. Predators like ladybird beetles feed on aphids, while lacewings and predatory mites target thrips and spider mites. Parasitoids such as Trichogramma wasps lay their eggs inside pest eggs, preventing their development. Microbial biocontrol agents like Bacillus thuringiensis (Bt), Beauveria bassiana, and Trichoderma spp. are also widely used against caterpillars, soil-borne fungi, and other pathogens. These biocontrol agents are either introduced artificially or conserved naturally within the greenhouse ecosystem. For instance, *Encarsia*

formosa is effective against whiteflies in tomato and cucumber cultivation. Biological control reduces pesticide residues in produce, making it safer for consumers and export markets. It also prevents the development of pest resistance, a common problem with excessive chemical use. Farmers often combine biological agents with pheromone traps and sticky cards to enhance efficiency. With proper monitoring and release strategies, biocontrol ensures long-term pest suppression and sustainability.

- **Integrated Pest Management (IPM):** Integrated Pest Management (IPM) is a holistic approach that combines cultural, mechanical, biological, and need-based chemical practices to control pests in an eco-friendly manner. The key principle of IPM is prevention and monitoring rather than blanket pesticide use. Cultural methods such as crop rotation, resistant varieties, and sanitation reduce initial pest buildup. Mechanical approaches like sticky traps, pheromone lures, and netting physically restrict pests. Biological control agents like predators, parasitoids, and entomopathogens are integrated into the system to naturally regulate pest populations. Chemical pesticides are used only as a last resort and in minimal, targeted doses to avoid resistance buildup. IPM also relies heavily on regular scouting and monitoring to make informed decisions about pest thresholds. In greenhouse crops such as capsicum, tomato, and cut flowers, IPM reduces production costs, ensures residue-free produce, and enhances export potential. Training farmers and greenhouse managers in IPM practices further improves its success. By balancing effectiveness with environmental safety, IPM remains the backbone of modern pest and disease management.
- **Sanitation Practices:** Sanitation is a crucial aspect of pest and disease prevention in controlled environments. It involves regular cleaning and disinfection of greenhouses, tools, and equipment to minimize sources of infection. Removal and destruction of infected plants prevent the spread of pathogens like powdery mildew, blight, and viruses. Soil sterilization through solarization, steam, or chemicals like formalin reduces soil-borne diseases before planting. Tools and containers are disinfected with mild bleach or formaldehyde to prevent cross-contamination. Good sanitation also includes removing plant debris, weeds, and alternative hosts that harbor pests and pathogens. Water used in irrigation is tested and treated to ensure it is free from harmful microbes. Workers' hygiene, such as wearing clean clothing and disinfected footwear, is also enforced in high-tech greenhouses. Regular monitoring and immediate action upon disease detection are vital to effective sanitation. By maintaining a clean growing environment, farmers can reduce pesticide dependence and increase crop longevity. Thus, sanitation forms the foundation of preventive pest and disease management strategies.

- **Resistant Varieties:** The use of resistant and tolerant varieties is one of the most effective and economical methods for managing pests and diseases. Plant breeders develop cultivars that can withstand specific pathogens and insect pests through traditional breeding or modern biotechnological approaches. For example, tomato varieties resistant to Tomato Yellow Leaf Curl Virus (TYLCV) or cucumber lines resistant to downy mildew are widely adopted in protected cultivation. Resistant varieties minimize the need for chemical inputs and provide farmers with stable yields under high disease pressure. In addition, some varieties are tolerant to multiple stresses, including pests, diseases, and abiotic conditions like salinity. Marker-assisted breeding and genetic engineering further accelerate the development of resistant cultivars. However, resistance should be combined with good management practices, as pests and pathogens may evolve to overcome it. Farmers often prefer resistant hybrids in greenhouses due to their high yield potential and reduced risk of crop failure. The strategic adoption of such varieties ensures sustainable and profitable farming in controlled environments.
- **Importance of Pest and Disease Control:** Effective pest and disease management is vital for ensuring healthy crop growth, high yields, and long-term sustainability in controlled farming systems. Unlike open fields, greenhouses provide favourable conditions–stable temperature, humidity, and light–that often accelerate pest and pathogen multiplication. If not managed properly, infestations can spread rapidly, leading to severe losses. By integrating physical barriers, biological control, sanitation, resistant varieties, and IPM, farmers can minimize reliance on chemical pesticides. This not only lowers production costs but also ensures residue-free produce, essential for consumer health and international trade. Moreover, reduced chemical use promotes environmental safety by conserving pollinators and beneficial organisms. Healthy crops also produce better-quality fruits, flowers, and vegetables, fetching higher market value. Pest and disease control is therefore not just a defensive strategy but a proactive investment in productivity and profitability. In modern protected farming, managing the biotic environment effectively is as important as controlling temperature, humidity, and irrigation.

4.4.8. Use of Automation and Smart Technologies

- **Sensors and IoT Devices:** Sensors and Internet of Things (IoT) devices are revolutionizing modern farming by providing real-time monitoring of critical environmental factors such as light intensity, temperature, humidity, and soil moisture. These devices transmit data continuously to farmers or central systems through wireless networks. With accurate measurements, growers can detect even the smallest fluctuations that may affect crop growth. IoT technology enables precision farming by ensuring crops receive optimal conditions throughout their lifecycle. For example, soil moisture sensors help in applying the right amount of water at the right time. These devices reduce guesswork, thereby increasing efficiency and saving resources. Farmers can remotely monitor farm conditions using smartphones or computers. Ultimately, IoT creates a smart farm ecosystem for better decision-making.

- ☆ **Automated Climate Control:** Automated climate control systems are designed to regulate heating, cooling, ventilation, irrigation, and nutrient delivery without constant human intervention. These systems use the data collected from sensors to maintain ideal growing conditions in greenhouses or controlled environment agriculture setups. For instance, if temperature rises above the optimal range, cooling fans or foggers automatically start functioning. Similarly, irrigation systems activate when soil moisture falls below a preset level. Automation ensures uniformity in plant growth, reduces labour costs, and minimizes human errors. It also improves energy efficiency by running systems only when required. Nutrient delivery through fertigation systems can be precisely timed and measured. This smart regulation leads to higher yields with lower input costs.
- ☆ **Drones:** Drones are an advanced technological tool increasingly used in agriculture for aerial monitoring and management. Equipped with high-resolution cameras and sensors, they capture detailed images of crops to identify plant health, growth stages, and signs of disease or pest attacks. This aerial perspective allows farmers to cover large fields quickly, saving time and effort compared to manual scouting. Some drones are fitted with sprayers to apply pesticides, fertilizers, or micronutrients with pinpoint accuracy. This reduces wastage of inputs and minimizes environmental pollution. Drones also help in mapping fields and analyzing plant density or water stress zones. Farmers can use this data to make timely interventions. By improving efficiency, drones support precision agriculture and boost productivity.
- ☆ **Data Analytics and AI:** Data analytics and Artificial Intelligence (AI) are transforming agriculture by analysing huge volumes of data to predict plant needs and optimize resource use. AI systems process inputs from sensors, drones, and historical farm records to identify patterns and forecast outcomes. For example, predictive models can estimate irrigation schedules, disease outbreaks, or harvest times. This allows farmers to act before problems become severe. Machine learning algorithms also recommend the best crop varieties for specific conditions. Advanced analytics helps reduce input costs while maximizing yields. By integrating weather forecasts, AI ensures preparedness against climate risks. These technologies empower farmers with data-driven decisions. Ultimately, data analytics and AI make farming smarter, sustainable, and more profitable.

- ☆ **Robotics:** Robotics in agriculture refers to the use of automated machines for tasks such as precision planting, mechanical weeding, and harvesting. Robots are designed to perform repetitive, labour-intensive activities with high accuracy and speed. For example, robotic seeders plant seeds at uniform depth and spacing, ensuring better germination rates. Automated weeders detect and remove weeds without harming crops, reducing the need for chemical herbicides. Harvesting robots equipped with cameras and sensors can identify ripe fruits or vegetables and pick them gently without causing damage. This technology addresses the shortage of agricultural labour while improving efficiency. Robots work around the clock, increasing productivity and reducing production costs. Their precision contributes to sustainable farming practices.
- ☆ **Modern Technologies for Efficiency and Profitability:** The integration of modern technologies in agriculture is making environmental control more efficient, sustainable, and profitable. By combining IoT, automation, drones, AI, and robotics, farmers can achieve precision farming on a large scale. These innovations reduce wastage of resources such as water, fertilizers, and energy while improving crop yields and quality. Automation lowers dependency on manual labour, which is becoming scarce and expensive. Technologies like drones and robotics allow timely interventions, reducing losses caused by pests, diseases, or delayed harvesting. The use of data analytics ensures informed decision-making and minimizes risks. Together, these advancements support climate-smart agriculture. Ultimately, modern technologies make farming more resilient and future-ready.

4.5. Use of Artificial Lighting

Artificial lighting has become an indispensable component of modern protected cultivation systems such as greenhouses, polyhouses, and indoor vertical farms. In natural farming systems, plants depend solely on sunlight for photosynthesis and growth regulation. However, in protected cultivation, natural light availability is often limited due to structural coverings, seasonal variations, or geographical constraints. For example, during winter months in temperate regions, daylight hours are short, and solar radiation is weak, which restricts crop growth. Artificial lighting compensates for this deficit by providing plants with the required light energy to sustain physiological processes like photosynthesis, transpiration, and nutrient assimilation. In addition, light influences photomorphogenesis, flowering, and fruit setting, making it a vital factor for crop productivity. Artificial lighting allows farmers to extend the photoperiod artificially, ensuring that crops receive consistent light exposure regardless of external weather conditions. This makes year-round production of vegetables, ornamental flowers, herbs, and high-value horticultural crops possible. Furthermore, the ability to manipulate light intensity and duration gives growers greater control over crop growth stages, improving quality, yield, and uniformity. Thus, artificial lighting is not only a tool for supplementation but also a means to optimize controlled environment agriculture (CEA) for enhanced profitability.

4.5.1. Types of Artificial Lights

Several categories of artificial lights are employed in protected cultivation, each offering specific advantages depending on crop requirements and production goals. Fluorescent lamps are widely used in nurseries and small-scale operations because they emit cool, diffuse light, ideal for leafy greens and seedlings. They are cost-effective but have limited intensity and lifespan. High-Pressure Sodium (HPS) lamps are popular in commercial greenhouses due to their high efficiency in converting electricity into light. They emit a yellow-orange spectrum that promotes flowering and fruiting, making them suitable for crops like tomatoes, cucumbers, and ornamental flowers. However, they generate significant heat, which can raise greenhouse temperatures. Metal Halide (MH) lamps produce light rich in blue wavelengths, which support vegetative growth and compact plant structures, especially during early crop stages. Light-Emitting Diodes (LEDs) represent the most advanced technology, capable of producing tailored light spectra such as red, blue, far-red, or UV, depending on crop needs. LEDs consume much less energy, produce minimal heat, and have a longer operational life compared to traditional lights. Additionally, their compact design makes them ideal for vertical farming and hydroponic systems where space efficiency is crucial. Today, many commercial farms are shifting towards LEDs despite higher upfront costs, as they provide a sustainable and versatile solution for precision agriculture.

4.5.2. Role of Light Spectrum

The quality of light, defined by its spectrum, plays a critical role in influencing plant physiology and development. Blue light (400–500 nm) regulates vegetative growth by enhancing chlorophyll production, strengthening stems, and promoting compact growth. It is particularly important for seedling establishment and leafy vegetables like spinach and lettuce. Red light (600–700 nm) drives photosynthesis and plays a dominant role in reproductive development by stimulating flowering and fruiting in crops such as tomatoes, strawberries, and roses. A balanced combination of red and blue light is considered optimal for most crops, as it promotes both biomass accumulation and reproductive success. Far-red light (700–750 nm) affects photomorphogenic responses, encouraging stem elongation and influencing the shade-avoidance response, which can be manipulated to control plant architecture. Ultraviolet (UV) light has stress-inducing properties but can enhance secondary metabolite production, improving flavours, aromas, and pigmentation in crops like basil, grapes, and ornamental flowers. Modern LED systems allow precise control over these wavelengths, enabling "light recipes" tailored to each growth stage of the crop. For example, a higher ratio of blue light is used during vegetative stages, while a red-rich spectrum is applied during reproductive phases. By understanding and applying spectrum manipulation, growers can not only maximize yield but also enhance crop quality, marketability, and nutritional value.

4.5.3. Photoperiod Control

Photoperiod, or the daily duration of light exposure, is a crucial factor regulating plant growth, flowering, and dormancy. Different crops have varying photoperiodic requirements, and artificial lighting enables growers to manipulate these conditions. Long-day plants such as lettuce, spinach, and beet require extended light hours to initiate flowering, while short-day plants like chrysanthemum, poinsettia, and soybean flower under shorter day lengths. Day-neutral plants, such as tomato and cucumber, are less sensitive to day length but still respond to light intensity and quality. By controlling artificial lighting schedules, farmers can induce or delay flowering, thereby aligning crop production with market demand. For example, floriculture industries use supplemental lighting to ensure chrysanthemums bloom precisely during festive seasons. Similarly, in vegetable production, lighting is used to accelerate growth cycles and ensure continuous harvests. In research facilities, photoperiod manipulation is applied to study flowering physiology and breeding programs. Furthermore, photoperiod control also helps in avoiding premature bolting in crops such as onion and carrot, ensuring higher yields. By combining artificial lighting with blackout curtains, both long-day and short-day conditions can be simulated under the same roof, allowing diverse crop production. Thus, photoperiod control through artificial lighting is a powerful technique for enhancing flexibility, productivity, and profitability in protected cultivation.

4.5.4. Energy Efficiency and Cost Considerations

While artificial lighting significantly enhances crop productivity, its adoption involves cost and energy challenges. Traditional lighting systems such as HPS and MH lamps, though effective, consume large amounts of electricity and release excess heat, necessitating additional cooling systems in greenhouses. This increases both energy use and operational costs. In contrast, LED technology has emerged as a game-changer, offering energy savings of up to 50–70 per cent compared to conventional lamps. LEDs produce light closer to the photosynthetically active radiation (PAR) spectrum, ensuring maximum utilization by plants. Additionally, their low heat output reduces the need for ventilation and cooling, cutting down energy expenditure. Despite higher upfront costs, LEDs offer long-term economic benefits due to their durability (lasting 50,000–70,000 hours) and reduced maintenance. Farmers often adopt a hybrid strategy, using natural sunlight as the primary source while supplementing with artificial lighting during critical growth stages or low-light conditions. Government subsidies, renewable energy integration (like solar panels), and smart IoT-based lighting controls further help reduce costs. Moreover, precision lighting strategies ensure that energy is not wasted–only specific wavelengths and intensities required by crops are delivered. Thus, when implemented with efficiency considerations, artificial lighting becomes a sustainable and profitable investment for modern agriculture.

4.5.5. Applications in Protected Cultivation

Artificial lighting has widespread applications across various sectors of protected cultivation. In seedling nurseries, supplemental lighting ensures strong and uniform seedlings even in shaded conditions, enhancing transplant success rates. In vegetable production systems, particularly for crops like tomato, cucumber, and capsicum, artificial lighting accelerates growth cycles, increases yields, and improves fruit quality. Floriculture extensively relies on photoperiod manipulation through artificial lighting to regulate blooming time and improve flower colour and size, particularly in high-value ornamentals like roses, carnations, and orchids. In hydroponics and vertical farming, artificial lights act as the primary energy source for plant growth since natural sunlight is limited or absent. These controlled systems rely heavily on LEDs to provide efficient light tailored to crop needs. Furthermore, in research laboratories, scientists use artificial lighting to study plant physiology, genetics, and breeding under controlled conditions. Controlled lighting also supports production of high-value crops like medicinal plants, where quality traits such as active compound concentration are critical. Artificial lighting ensures off-season and region-independent production, thereby stabilizing supply chains. Overall, its applications extend beyond productivity to improving quality, sustainability, and scientific innovation.

4.5.6. Challenges and Future Prospects

Despite its benefits, artificial lighting adoption in protected cultivation faces several challenges. The most prominent barrier is the high initial investment, especially for LED systems, which discourages small and marginal farmers. High electricity demand also adds to operational costs, particularly in regions with expensive or unreliable power supply. In addition, lack of technical knowledge among farmers often results in inefficient use, leading to either inadequate lighting or excessive usage that causes photoinhibition (damage due to excess light). Another issue is light pollution, where continuous artificial illumination can disturb ecosystems near large-scale greenhouses. However, technological advancements are rapidly addressing these issues. The integration of smart lighting systems with IoT and Artificial Intelligence (AI) now enables real-time monitoring and automatic adjustment of light intensity, spectrum, and duration based on crop needs. The use of renewable energy sources such as solar panels and biogas generators can make lighting more sustainable. In the future, adaptive lighting solutions will become more affordable, allowing customization for small-scale farmers as well. Research is also focusing on developing dynamic light recipes to maximize plant growth efficiency while reducing energy input. With increasing demand for year-round food production and urban farming, artificial lighting is expected to play an even greater role in securing sustainable agriculture.

4.6 Automation in Greenhouses

Automation in greenhouses refers to the use of modern technologies and smart systems to manage environmental conditions, crop care, and resource utilization with minimal human intervention. Traditionally, greenhouse farming required manual monitoring of temperature, humidity, irrigation, and nutrient supply. However, with increasing demand for efficiency and precision, automation systems now integrate sensors, controllers, IoT devices, and computer-based algorithms to optimize greenhouse operations. Automated systems continuously monitor real-time conditions and make necessary adjustments instantly, ensuring crops always grow in optimal environments. This reduces human errors, saves time, lowers labour costs, and enhances productivity. By automating repetitive tasks, farmers can focus on decision-making and market strategies rather than routine maintenance. Importantly, automation makes greenhouses more climate-resilient by minimizing risks associated with sudden temperature fluctuations, water shortages, or pest outbreaks. Hence, automation is not just a luxury but a necessity for sustainable protected cultivation in the modern era (Figure 4.16).

Figure 4.16: Greenhouse Automation Systems.
(*Source*: https://climatecontrol.com)

4.6.1 Environmental Control Systems

One of the key areas of greenhouse automation is environmental regulation. Crops inside a greenhouse are highly sensitive to fluctuations in temperature, humidity, light intensity, and carbon dioxide concentration. Automated climate control systems use sensors to detect these variables and automatically trigger responses such as heating, cooling, shading, misting, or ventilation. For example, when the internal temperature rises beyond the optimal level, exhaust fans, cooling pads, or foggers are automatically activated. During colder conditions, heaters or thermal screens are switched on to conserve warmth. Humidity sensors regulate misting or dehumidifiers, ensuring disease-causing fungal conditions are minimized. Light sensors can activate artificial lighting when natural sunlight is insufficient, while CO_2 enrichment systems maintain levels required for enhanced photosynthesis. These automated adjustments create a stable microclimate, promoting uniform plant growth and maximizing yields. Unlike manual control, automation allows precise, round-the-clock regulation, which is especially important for high-value crops like tomatoes, peppers, and flowers.

4.6.2. Irrigation and Fertigation Automation

Water and nutrient management is another critical area where automation is transforming greenhouse agriculture. Traditional irrigation methods often lead to overwatering, underwatering, or nutrient imbalances, all of which reduce productivity. Automated irrigation systems employ soil moisture sensors and weather data to determine the exact water needs of plants and supply water accordingly. Similarly, fertigation systems automate the delivery of fertilizers through irrigation, ensuring accurate concentrations of nutrients reach the root zone. This precision not only saves water and fertilizer but also reduces leaching and environmental pollution. For example, drip irrigation integrated with automated timers and moisture sensors delivers water directly to the root zone only when required. Advanced fertigation controllers adjust nutrient formulations depending on crop stage–higher nitrogen during vegetative growth and more potassium during fruiting. Farmers can also monitor and control these systems remotely via mobile applications. Overall, irrigation and fertigation automation enhance efficiency, reduces costs, and improves both yield and crop quality.

4.6.3. Automation in Crop Monitoring

Greenhouse automation extends beyond environmental and irrigation control into real-time crop monitoring. Sensors, cameras, and drones are increasingly being deployed to detect plant growth patterns, nutrient deficiencies, water stress, and pest or disease infestations. Image recognition software powered by Artificial Intelligence (AI) can analyse leaf colour, shape, and texture to identify problems early, allowing timely corrective measures. For instance, yellowing leaves may indicate nitrogen deficiency, while irregular patterns could suggest pest attacks. Drones equipped with multispectral cameras can scan large greenhouse areas and provide detailed maps showing crop health variations. This level of monitoring helps farmers adopt precision agriculture practices where interventions are made only when and where necessary. By minimizing pesticide and fertilizer misuse, automation also reduces input costs and environmental impacts. Continuous crop monitoring ensures early detection of problems, leading to healthier crops, higher yields, and better product quality in the market.

4.5.4. Robotics and Automated Machinery

Robotics is playing an increasingly important role in greenhouse automation, addressing labour shortages and improving efficiency. Robots are designed to handle repetitive and labour-intensive tasks such as planting, pruning, harvesting, and weeding. For example, robotic seeders ensure uniform seed spacing and depth, enhancing germination rates. Automated transplanting machines reduce manual labour and speed up nursery operations. Harvesting robots equipped with sensors and AI can identify ripe fruits like tomatoes, cucumbers, or strawberries and pick them without causing damage. Similarly, robotic arms can prune plants or thin flowers, tasks that usually require skilled labour. Automated guided vehicles (AGVs) are used for transporting harvested produce within greenhouses, reducing manual handling and increasing efficiency. These robotic solutions not only improve precision but also reduce dependence on seasonal labour, which is often expensive and unreliable. Although initial costs are high, robotics improves long-term profitability and sustainability in commercial greenhouse operations (Figure 4.17).

Figure 4.17: Robotics in 21st Century Agriculture.
(*Source*: https://future-business.org)

4.5.5. Data Analytics and AI in Automation

Modern greenhouse automation heavily relies on data-driven technologies. The integration of data analytics and Artificial Intelligence (AI) allows farmers to make informed decisions and predict crop needs in advance. Sensors and IoT devices generate vast amounts of data on temperature, humidity, light, soil moisture, and plant health. AI algorithms analyse this data to identify trends, forecast future requirements, and recommend the best management practices. For example, predictive analytics can suggest optimal irrigation schedules based on weather forecasts or determine the best time to harvest for maximum yield and quality. Machine learning models can even predict disease outbreaks by analysing environmental conditions and historical data. AI-powered dashboards give farmers real-time updates accessible from smartphones, making greenhouse management more user-friendly. Thus, data analytics transforms raw information into actionable insights, helping optimize resource use, reduce risks, and increase profitability (Figure 4.18).

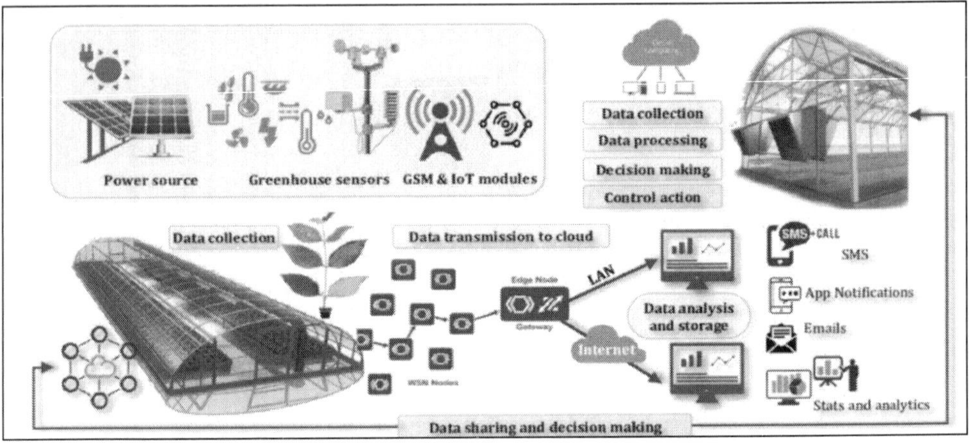

Figure 4.18: IoT Based Automated Greenhouse Monitoring System.
(*Source*: https://www.hashstudioz.com)

4.5.6. Energy Efficiency and Sustainability

Automation in greenhouses also contributes significantly to energy efficiency and sustainable farming. Automated systems ensure that resources such as water, fertilizers, and electricity are used only when needed, minimizing wastage. For instance, automated climate systems prevent excessive heating or cooling, thereby saving energy. Similarly, automated irrigation ensures that water is supplied in precise quantities, reducing losses due to evaporation or runoff. Integration of renewable energy sources such as solar panels and wind turbines further reduces dependence on non-renewable energy. Smart energy management systems can schedule operations such as lighting or pumping during low-tariff hours, lowering costs. Moreover, automation minimizes pesticide use through targeted application, reducing environmental pollution. Overall, automated greenhouses are better aligned with the principles of climate-smart and resource-efficient agriculture, ensuring sustainable food production for the future.

4.5.7. Challenges and Future Prospects

Despite its advantages, greenhouse automation faces several challenges. High initial investment costs for equipment, sensors, and robotics discourage small-scale farmers from adopting the technology. Limited technical knowledge and training also hinder effective use, leading to underutilization of advanced systems. Maintenance and repair of automated equipment require skilled technicians, which may not always be available in rural areas. Additionally, dependence on electricity and internet connectivity can be problematic in regions with poor infrastructure. However, future prospects for automation in greenhouses remain very promising. Rapid advancements in Artificial Intelligence, Internet of Things, robotics, and renewable energy will make systems more affordable, efficient, and user-friendly.

Governments and private sectors are also offering subsidies and training programs to encourage adoption. With the rising demand for high-quality food, off-season production, and climate-resilient farming, greenhouse automation is expected to expand rapidly. The future will likely see fully autonomous greenhouses where all processes–from seeding to harvesting–are managed by smart systems, revolutionizing agriculture worldwide.

Automation in Greenhouse cultivation technologies, advantages, and limitations are summarised in Table 4.3.

Table 4.3: Automation in Greenhouse Cultivation: Technologies, Advantages, and Limitations

Automation	*Focus Area*	*Key Features/ Functions*	*Benefits*	*Challenges/ Notes*
Environmental Control Systems	Regulation of temperature, humidity, CO_2, and light	Automated heating, cooling, shading, misting, ventilation, artificial lighting, CO_2 enrichment	Maintains stable microclimate, uniform growth, higher yields	Requires reliable power supply and system maintenance
Irrigation and Fertigation Automation	Water and nutrient delivery	Soil moisture sensors, automated drip systems, fertigation controllers, remote monitoring	Saves water and fertilizers, reduces pollution, ensures balanced nutrition	High setup cost, needs skilled handling
Automation in Crop Monitoring	Real-time crop health surveillance	Sensors, cameras, drones, AI image recognition, multispectral analysis	Early detection of stress, pests, and deficiencies; supports precision farming	Requires data processing tools and technical knowledge
Robotics and Automated Machinery	Replacing labor-intensive tasks	Robotic seeders, transplanting machines, harvesting robots, pruning arms, Automated Guided Vehicles (AGVs)	Reduces labor dependency, improves precision, enhances efficiency	High investment cost, limited access for small farmers
Data Analytics and AI in Automation	Decision-making and forecasting	IoT devices, AI algorithms, predictive analytics, machine learning models, smart dashboards	Data-driven farming, optimized resource use, risk reduction, higher profitability	Needs internet connectivity, data literacy among farmers

Automation	*Focus Area*	*Key Features/ Functions*	*Benefits*	*Challenges/ Notes*
Energy Efficiency and Sustainability	Resource conservation	Smart energy management, renewable integration (solar, wind), targeted input use	Reduces waste, cuts cost, supports climate-smart agriculture	Initial setup for renewables can be expensive
Challenges and Future Prospects	Barriers and opportunities	Issues: high cost, lack of training, infrastructure gaps; Prospects: smart IoT, AI, robotics, subsidies	Promises fully autonomous, efficient, sustainable greenhouses	Adoption limited among small-scale farmers presently

Soil and Substrate Management

5.0 Introduction: An Overview

Protected cultivation, which includes greenhouses, polyhouses, shade nets, and other controlled structures, has emerged as a modern approach to intensive agriculture. The success of this system largely depends on efficient soil and substrate management, as the growing medium directly influences plant growth, nutrient uptake, water retention, and overall productivity. Unlike open-field cultivation, where soil is exposed to natural weather conditions, protected environments

Soil and substrate management is the backbone of successful protected cultivation, as the growing medium directly affects crop health, yield, and quality. In soil-based systems, challenges like nutrient depletion, salinity, and soil-borne diseases are tackled through practices such as solarization, sterilization, crop rotation, and organic amendments. Substrate-based cultivation, including hydroponics and soilless media, uses materials like cocopeat, perlite, vermiculite, and rockwool to provide better aeration, drainage, and disease resistance. These substrates allow precise fertigation and efficient resource use, making them ideal for high-value crops. Nutrient and water management are closely linked with soil and substrate choice, requiring regular monitoring of pH, EC, and fertility. Sustainable practices such as biofertilizers, microbial inoculants, and recycling of substrates enhance long-term productivity. By integrating scientific methods with eco-friendly approaches, farmers can achieve year-round production with higher profitability. Thus, soil and substrate management ensure both sustainability and efficiency in protected farming systems.

demand a more precise and scientific approach to soil and substrate use. This ensures sustainability, high yield, and quality produce throughout the year. Soil management in protected cultivation involves maintaining soil health, fertility, and structure under controlled conditions. Over time, continuous cropping inside protected structures can lead to nutrient depletion, soil compaction, accumulation of salts, and increased risk of soil-borne diseases. To counter this, practices like soil solarization, crop rotation, organic amendments, and use of bio-fertilizers are adopted. Soil sterilization techniques such as steam treatment, fumigation, or use of bioagents also play a vital role in minimizing pathogens and weeds. Maintaining the right pH, electrical conductivity (EC), and organic matter levels is crucial for balancing nutrient availability. In addition to natural soils, alternative substrates have become highly popular in protected cultivation. Substrate-based cultivation systems, such as hydroponics, soilless culture, and containerized production, rely on inert or organic growing media to support plant growth. Common substrates include cocopeat, perlite, vermiculite, rockwool, peat moss, sawdust, and sand. These materials offer advantages like uniformity, better drainage, aeration, and reduced risk of soil-borne diseases. Substrates also allow precise control over nutrient and water supply through fertigation systems, making them ideal for high-value crops such as vegetables, flowers, and exotic plants. The choice of substrate depends on crop type, cost, local availability, and desired growing conditions. For instance, cocopeat is widely used due to its renewable nature, good water-holding capacity, and biodegradability. Rockwool, though more expensive, provides excellent aeration and is commonly used in commercial hydroponics. A mixture of substrates often delivers better results by combining water retention with adequate porosity. Substrate management also requires attention to physical stability, pH buffering, and cation exchange capacity to ensure consistent performance across cropping cycles.

Water and nutrient management in protected cultivation are closely linked with the growing medium. Since substrates are generally inert, all essential nutrients are supplied through fertigation in precise concentrations. This requires careful monitoring of nutrient solution composition, EC, and pH to avoid imbalances or toxicities. Similarly, in soil-based systems, integrated nutrient management involving chemical fertilizers, composts, green manures, and microbial inoculants ensures balanced nutrition. Regular testing of soil and substrate helps in timely interventions to maintain optimum fertility levels. Sustainable soil and substrate management also emphasizes minimizing chemical inputs and promoting eco-friendly practices. The use of organic substrates, biological soil amendments, and beneficial microbes helps improve plant health while reducing environmental pollution. Reuse and recycling of substrates, proper disposal of used media, and conservation of water through drip irrigation further enhance sustainability in protected cultivation. Overall, soil and substrate management form the foundation of successful protected cultivation. A well-maintained growing medium not only ensures high yields but also contributes to better quality, extended shelf life, and safe food production. With the increasing demand for pesticide-free, high-value crops, scientific soil and substrate management is becoming more critical than ever. By combining modern technologies with traditional knowledge, farmers can achieve efficient, sustainable, and profitable cultivation in controlled environments (Figure 5.1).

Figure 5.1: Soil and Substrate Management
(*Source*: http://www.agritech.tnau.ac.in)

5.1 Soil Preparation Techniques

Soil preparation is one of the most crucial steps in protected cultivation because it directly determines the health, productivity, and sustainability of crops grown inside controlled structures. Unlike open-field conditions, where natural processes can partially restore soil fertility, the intensive and continuous cropping cycles in greenhouses and polyhouses often lead to problems such as nutrient depletion, salt accumulation, compaction, and soil-borne diseases. Hence, scientific soil preparation becomes essential before each cropping season. The process begins with soil testing and analysis to determine pH, salinity, organic matter content, and nutrient availability, which helps in planning corrective measures. The soil is then levelled and shaped into raised beds or ridges to improve drainage, aeration, and root penetration. To minimize pest and disease risks, soil sterilization or solarization is practiced, wherein transparent polythene sheets or steam treatments are used to suppress harmful pathogens, nematodes, and weeds. Organic amendments such as compost, farmyard manure, vermicompost, and green manures are incorporated to improve soil fertility, structure, and microbial activity. Biofertilizers and beneficial microbes like Trichoderma and Azotobacter are introduced to enhance nutrient availability and suppress soil-borne infections. For soils affected by salinity or acidity, corrective amendments such as gypsum, lime, or dolomite are added. Basal doses of fertilizers, including primary nutrients (NPK) and essential micronutrients, are uniformly mixed into the soil for balanced nutrition. In addition, mulching with black or silver polythene films helps regulate soil temperature, conserve moisture, and reduce weed emergence. During soil preparation, irrigation systems such as drip and fertigation units are also installed to ensure efficient water and nutrient delivery. Finally, the prepared beds are sanitized and allowed to stabilize before transplanting, creating an ideal environment for young seedlings. Thus, soil preparation in protected cultivation is not a single activity but a holistic process combining physical, chemical, and biological measures to maintain long-term soil health, ensure uniform crop growth, and achieve high-quality harvests under intensive farming systems.

Soil preparation in protected cultivation is a multi-step process involving testing, sterilization, enrichment, and conditioning. A well-prepared soil ensures disease-free conditions, balanced nutrition, proper aeration, and high productivity. By integrating organic and modern scientific practices, farmers can create a sustainable soil environment that supports intensive and year-round cultivation. The following techniques are commonly adopted:

5.1.1. Soil Testing and Analysis

Soil testing and analysis is regarded as the first and most vital step in soil preparation for protected cultivation, as it lays the scientific foundation for all subsequent management practices. Protected structures like greenhouses and polyhouses involve intensive and continuous cropping cycles, which place a heavy demand on the soil. Without proper monitoring, this can lead to rapid fertility loss, nutrient imbalances, accumulation of salts, soil acidification, and declining productivity over time. Regular soil testing helps in assessing key parameters such as soil texture, structure, pH, electrical conductivity (EC), organic matter, and nutrient status, thereby enabling informed decision-making. It acts as a diagnostic tool to identify deficiencies, toxicities, or hidden imbalances that are not visible through plant growth alone. Based on the analysis, corrective measures such as liming, gypsum application, organic amendments, or precise fertilizer scheduling can be recommended. In addition, soil testing ensures that fertigation practices inside protected structures remain balanced, efficient, and crop-specific. By maintaining the soil within optimal ranges of fertility and salinity, soil testing safeguards both crop yield and quality. Moreover, it helps in sustaining soil health in the long run by preventing overuse or misuse of chemical fertilizers. Thus, soil testing and analysis not only supports high productivity but also promotes environmentally sound and sustainable protected farming.

Physical Properties

The physical properties of soil play a vital role in determining its suitability for crop production under protected cultivation. The first and most important factor is soil texture, which refers to the proportion of sand, silt, and clay present in the soil. Texture influences nutrient-holding capacity, water retention, drainage, and root penetration. A loamy soil, having a balanced proportion of sand, silt, and clay, is considered ideal for most crops inside greenhouses and polyhouses. Closely linked to texture is soil structure, which describes the arrangement of soil particles into aggregates. A well-structured soil ensures good aeration, easy root growth, and efficient water movement, while poorly structured soils tend to be compact, leading to restricted root development. Bulk density is another important property, as it indicates the degree of soil compaction. High bulk density soils restrict aeration and root expansion, whereas moderate bulk density supports healthy root systems. Porosity refers to the amount and size of pore spaces within the soil, which govern the soil's ability to hold and transmit air and water. Higher porosity is beneficial because it promotes better aeration and drainage, especially in intensive cropping systems. Similarly, water-holding capacity is critical, as it determines how much moisture is available for plant use between irrigation cycles. Soils with good water-holding capacity reduce irrigation frequency and ensure a consistent supply of water to crops. For protected cultivation, the ideal soil should be friable, loose, and well-drained, yet capable of retaining adequate moisture to support continuous crop growth. Managing these physical properties through tillage, addition of organic matter, and soil conditioners ensures a balanced growing medium. Thus, careful evaluation of physical properties is essential to prepare soils that can sustain intensive and high-value crop production under protected cultivation.

The desirable physical properties of soil for greenhouse/protected cultivation are summarised in Table 5.1.

Table 5.1: Desirable Physical Properties of Soil for Greenhouse/Protected Cultivation

Property	*Desirable Range/Condition*	*Remarks*
Soil Texture	Loam to sandy loam (balanced sand, silt, clay)	Provides good drainage, aeration, and nutrient-holding capacity.
Bulk Density	1.2 – 1.4 g/cm^3	Lower values improve root growth; >1.6 g/cm^3 indicates compaction.
Porosity	45 – 55 per cent	Adequate pore space ensures air circulation and water movement.
Water-Holding Capacity (WHC)	150 – 200 mm/m soil depth	Ensures sufficient moisture for crops between irrigation cycles.
Soil Depth	Minimum 30 – 45 cm	Provides enough rooting zone for vegetables and flowers.
Infiltration Rate	2 – 5 cm/hr	Balanced rate prevents waterlogging and allows efficient irrigation.
Soil Structure	Crumb to granular	Promotes friability, root penetration, and microbial activity.

Chemical Properties

The chemical properties of soil are crucial indicators of its fertility status and directly influence crop growth, nutrient uptake, and overall productivity in protected cultivation. The most important parameter is soil pH, which determines the acidity or alkalinity of the soil. It strongly affects nutrient solubility, microbial activity, and the chemical reactions that govern nutrient availability. For most protected crops, a pH range of 5.5 to 7.0 is considered optimal, as it ensures maximum availability of essential nutrients. Another key property is electrical conductivity (EC), which reflects the level of soluble salts present in the soil. Maintaining EC within safe limits is vital, because excess salts can lead to salinity stress, causing poor germination, stunted growth, nutrient imbalance, and reduced yields. High EC values may result from over-fertilization, poor-quality irrigation water, or insufficient leaching. In addition to pH and EC, the soil is analysed for primary nutrients like nitrogen (N), phosphorus (P), and potassium (K), which are essential for vegetative growth, root development, and flowering/fruiting. The status of secondary nutrients such as calcium (Ca), magnesium (Mg), and sulphur (S) is also important, as these support structural development, chlorophyll synthesis, and enzyme activity. Furthermore, micronutrients like zinc (Zn), iron (Fe), copper (Cu), boron (B), and manganese (Mn) are tested, since even small deficiencies can lead to significant physiological disorders and yield reduction in high-value crops. Thus, chemical property analysis provides a clear understanding of soil fertility and salinity management. It guides the application of fertilizers, soil amendments, and organic inputs in a precise manner to maintain a balanced nutrient profile. Regular monitoring of these properties ensures that the soil remains productive, nutrient-efficient, and suitable for intensive cropping systems inside greenhouses and polyhouses. The ideal chemical properties of soil for protected cultivation are summarised in Table 5.2.

Table 5.2: Ideal Chemical Properties of Soil for Protected Cultivation

Parameter	*Ideal Range/Level*	*Remarks*
Soil pH	5.5 – 7.0	Optimum range for most vegetables and flowers; ensures maximum nutrient availability.
Electrical Conductivity (EC)	< 1.5 dS/m	Higher EC indicates salinity stress; keep within safe limit.
Nitrogen (N)	280 – 450 kg/ha (available form)	Essential for vegetative growth and chlorophyll formation.
Phosphorus (P_2O_5)	25 – 50 kg/ha (available form)	Crucial for root development and flowering.
Potassium (K_2O)	250 – 400 kg/ha (available form)	Important for fruit set, size, and disease resistance.
Calcium (Ca)	1.0 – 2.5 meq/100 g soil	Needed for cell wall strength and quality produce.
Magnesium (Mg)	0.5 – 1.5 meq/100 g soil	Vital for chlorophyll and enzyme activity.
Sulfur (S)	10 – 20 ppm	Supports protein synthesis and aroma in crops like onion/garlic.

Parameter	*Ideal Range/Level*	*Remarks*
Zinc (Zn)	1.0 – 1.5 ppm	Deficiency leads to stunted growth and leaf deformities.
Iron (Fe)	4.5 – 5.0 ppm	Prevents chlorosis, especially in calcareous soils.
Copper (Cu)	0.2 – 0.5 ppm	Supports enzyme activity and lignin formation.
Boron (B)	0.5 – 1.0 ppm	Crucial for pollination, seed set, and fruit quality.
Manganese (Mn)	1.0 – 2.0 ppm	Important for photosynthesis and nitrogen metabolism.

Biological Aspects

The biological aspects of soil are increasingly recognized as equally important as its physical and chemical properties in determining soil health and crop performance under protected cultivation. Advanced soil testing today not only measures nutrient status but also assesses the level of microbial activity and the presence of beneficial organisms. A biologically active soil contains a diverse community of bacteria, fungi, actinomycetes, protozoa, and nematodes, many of which play vital roles in enhancing soil fertility. Beneficial microbes contribute to nutrient cycling by decomposing organic matter, fixing atmospheric nitrogen, solubilizing phosphates, and mobilizing micronutrients, thereby improving nutrient availability to plants. They also produce enzymes, organic acids, and growth-promoting substances that stimulate root development and overall plant vigour. Active soil biology further improves soil structure by binding soil particles into stable aggregates, which enhances aeration, water infiltration, and root penetration. At the same time, beneficial organisms act as natural protectors by suppressing soil-borne pathogens through competition, antibiosis, or parasitism. For example, fungi like Trichoderma and bacteria such as Pseudomonas fluorescens are commonly used as biocontrol agents in greenhouse soils. Bioassays, which evaluate biological activity through respiration rate, enzyme activity, or microbial biomass, provide deeper insights into soil health beyond what standard chemical tests can reveal. Thus, maintaining an active and diverse microbial population is crucial in protected cultivation, where soils are intensively cropped and more prone to depletion. By integrating organic matter, biofertilizers, compost teas, and microbial inoculants into the soil, farmers can strengthen its biological aspects, creating a self-sustaining system that ensures higher productivity and resilience against pests and diseases. The beneficial soil microorganisms and their functions is summarised in Table 5.3.

Table 5.3: Beneficial Soil Microorganisms and Their Functions

Microorganism	*Type*	*Main Function in Soil*
Rhizobium spp.	Symbiotic bacteria	Fix atmospheric nitrogen in legume root nodules.
Azotobacter spp.	Free-living bacteria	Fix nitrogen in non-leguminous crops, promote root growth.
Azospirillum spp.	Associative bacteria	Fix nitrogen, enhance root development in cereals and vegetables.
Phosphate-Solubilizing Bacteria (PSB)	Bacteria	Convert insoluble phosphates into available forms for plants.
Trichoderma spp.	Fungi	Biocontrol agent against soil-borne pathogens, improves root growth.
Pseudomonas fluorescens	Bacteria	Produces antibiotics, suppresses pathogens, promotes plant growth.
Mycorrhizal fungi (AMF)	Symbiotic fungi	Improve phosphorus uptake, increase drought resistance, enhance root surface area.
Blue-Green Algae (BGA)	Algae	Fix nitrogen in flooded soils, improve soil fertility.
Actinomycetes	Filamentous bacteria	Decompose complex organic matter, release antibiotics, improve soil structure.
Bacillus subtilis	Bacteria	Antagonistic to pathogens, induces systemic resistance in plants.

Corrective Actions Based on Results

Soil testing not only diagnoses the current condition of the soil but also provides the basis for taking appropriate corrective actions to restore balance and maintain long-term productivity in protected cultivation. If soils are found to be acidic, amendments such as lime or dolomite are applied to neutralize excess acidity and improve nutrient availability. Conversely, when soils are affected by salinity or sodality, gypsum application, leaching with good-quality water, or the use of salt-tolerant rootstocks may be adopted to reclaim the soil. To enhance fertility, organic amendments like farmyard manure, compost, vermicompost, and green manures are incorporated, which not only supply nutrients but also improve soil structure, moisture retention, and microbial activity. In cases of micronutrient deficiencies, corrective measures are taken either through foliar sprays for quick absorption or through soil application for longer-term supply. For example, zinc sulphate, ferrous sulphate, or borax is applied depending on the deficiency diagnosed. Fertility imbalances are also managed by preparing fertigation schedules tailored to soil test values and crop requirements, ensuring that plants receive nutrients in precise quantities at the right time. Integrated use of biofertilizers such as Azotobacter, Trichoderma, or phosphate-solubilizing bacteria further enhances soil biological health. Thus, corrective actions transform soil test results into practical field-level interventions, helping farmers maintain balanced nutrition, reduce input costs, and achieve sustainable high yields under intensive greenhouse and polyhouse cultivation systems. The common soil problems and corrective measures is depicted in Table 5.4.

Table 5.4: Common Soil Problems and Corrective Measures

Soil Problem	*Symptoms/Effects*	*Corrective Measures*
Acidic Soil (pH < 5.5)	Poor root growth, low nutrient availability, aluminium toxicity	Apply lime or dolomite to raise pH; use organic matter to buffer acidity.
Alkaline Soil (pH > 7.5)	Micronutrient deficiencies (Fe, Zn, Mn, B), poor nutrient uptake	Apply gypsum or elemental sulfur; add organic matter; use acid-forming fertilizers.
Saline Soil (High EC > 1.5 dS/m)	Poor germination, leaf burn, stunted growth	Leach salts with good-quality water; apply gypsum; improve drainage; adopt salt-tolerant crops.
Sodic Soil (High ESP)	Soil dispersion, poor infiltration, crusting	Apply gypsum or sulfur; leach with irrigation; add organic matter.
Low Organic Matter	Reduced microbial activity, poor soil structure	Incorporate compost, FYM, vermicompost, green manures; apply biofertilizers.
Nitrogen Deficiency	Yellowing of older leaves, stunted growth	Apply urea, ammonium sulfate, or organic N sources; use N-fixing biofertilizers.
Phosphorus Deficiency	Purpling of leaves, poor root growth	Apply single superphosphate (SSP), DAP, or phosphate-solubilizing bacteria (PSB).
Potassium Deficiency	Leaf edge burning, weak stems, poor fruit set	Apply MOP (Muriate of Potash) or SOP (Sulphate of Potash).
Micronutrient Deficiency (Zn, Fe, B, Mn, Cu)	Chlorosis, poor flowering, fruit deformities	Apply specific micronutrient salts (ZnSO, FeSO, Borax, MnSO, CuSO) through soil or foliar sprays.

Soil testing and analysis acts as a diagnostic tool that helps farmers identify nutrient status, prevent salinity build-up, and maintain soil fertility. Regular testing before each cropping season, and sometimes mid-season, ensures timely corrective measures, resulting in healthy soils, balanced nutrition, and sustainable productivity under protected cultivation.

5.1.2. Soil Levelling and Bed Formation

Soil levelling and bed preparation are among the most important preliminary operations in protected cultivation, as they directly influence crop establishment, water use efficiency, and nutrient management. Inside greenhouses or polyhouses, the soil surface is carefully levelled to eliminate depressions and uneven patches that may otherwise cause water stagnation, nutrient accumulation, or poor germination in certain spots. A uniform surface ensures that irrigation water, whether supplied through drip lines or sprinklers, spreads evenly across the crop area. After levelling, different types of beds are designed according to the crop requirement, season, and irrigation system. Raised beds are commonly preferred for vegetables, flowers, and high-value crops, as they facilitate quick drainage, prevent waterlogging, and improve aeration around the root zone. Flat beds are generally used for crops that need a larger area for spreading or where mechanised intercultural operations are planned. Ridge and furrow systems are adopted for crops like cucurbits and maize, where proper root anchorage and drainage are essential. Bed dimensions are usually standardised, with widths ranging from 1.0 to 1.2 metres and pathways of 0.4 to 0.5 metres between adjacent beds to facilitate easy movement, irrigation line placement, and intercultural practices. The height of raised beds may vary between 15 and 30 cm, depending on soil texture and drainage capacity. In sandy soils, slightly flatter beds are recommended to conserve moisture, whereas in heavy clay soils, higher raised beds are advantageous to avoid excess water retention. Bed formation also plays a role in controlling soil-borne diseases and pests by improving root health and enabling solarisation practices. In some cases, beds are mulched with black, silver, or transparent plastic films, which further help in weed suppression, moisture conservation, and maintenance of uniform soil temperature. The use of organic amendments such as compost, farmyard manure, and vermicompost is integrated during bed preparation to enrich the soil and support microbial activity. Proper levelling reduces irrigation time, prevents nutrient runoff, and ensures that fertigation inputs are uniformly distributed to all plants. It also facilitates mechanisation of operations such as sowing, transplanting, and harvesting inside the structure. Well-prepared beds ultimately create a micro-environment that favours robust seedling establishment, faster growth, and higher yields. Therefore, soil levelling (Figure 5.2) and bed formation (Figure 5.3) are not merely preparatory steps but serve as the foundation of efficient crop management in protected cultivation systems, ensuring long-term productivity, sustainability, and profitability for the farmer.

Figure 42: Soil Levelling.
(*Source*: https://jcblagri.in)

Figure 5.3: Bed Formation
(*Source*: https://www.cosmecosrl.com)

5.1.3. Soil Solarization

Soil solarization is an eco-friendly, non-chemical method of disinfecting soil, widely used in protected cultivation to reduce pest and disease pressure before planting. This technique harnesses the natural energy of the sun to sterilize the soil and create a healthier growing medium for crops. The process begins with thorough soil preparation, including ploughing and levelling, followed by irrigation to bring the soil to field capacity. Moisture plays a crucial role, as it enhances the transfer of heat through the soil profile and makes soil-borne organisms more sensitive to high temperatures. Once the soil is adequately moist, it is covered with transparent polythene sheets of 25–50 microns thickness during the peak summer months when solar radiation is at its maximum. The edges of the plastic are tightly sealed with soil to trap heat effectively and prevent loss of moisture. Under these conditions, the temperature of the upper soil layers may rise to 45–60°C, depending on ambient conditions, and is maintained continuously for 4–6 weeks. These elevated temperatures are lethal to many soil-borne pathogens such as Fusarium, Pythium, and Rhizoctonia, as well as harmful nematodes and dormant weed seeds. As a result, the soil becomes partially sterilized, leading to reduced incidence of damping-off, wilt, root-knot, and other soil-related problems in subsequent crops. Another advantage of solarization is the promotion of beneficial soil microbes such as Trichoderma and mycorrhizal fungi, which are more tolerant of higher temperatures and rebound quickly, thus establishing a natural biological balance in the soil. In addition, the decomposition of organic matter during the heating process releases nutrients in a more available form, improving soil fertility. This technique is highly sustainable, cost-effective, and does not leave behind toxic chemical residues, making it particularly suitable for organic farming and integrated pest management systems. In protected structures such as polyhouses and greenhouses, soil solarization is especially effective because the enclosed environment enhances the heating effect. Although it requires a fallow period of several weeks, the long-term benefits include healthier crops, reduced dependency on chemical pesticides, and higher yields. For small and marginal farmers, it offers a simple yet powerful tool to manage soil health and ensure sustainable crop production under protected cultivation.

Figure 5.4: Soil Solarization in Greenhouse.
(*Source*: https://www.researchgate.net)

5.1.4. Soil Sterilization and Disinfestation

Soil sterilization and disinfestation are essential practices in protected cultivation to ensure a pathogen-free and productive growing medium, particularly when soil solarization is not feasible due to climatic or operational limitations. Various physical, chemical, and biological approaches are employed depending on the type of crop, level of infestation, and availability of resources. Steam sterilization is one of the most reliable physical methods, wherein steam is passed into the soil using perforated pipes or specialized equipment. The high temperature, often exceeding 80–100°C, penetrates the soil profile and destroys a wide range of harmful organisms such as fungi, bacteria, nematodes, and weed seeds, while leaving behind a relatively clean substrate. However, steam sterilization is energy-intensive and more suitable for nurseries or high-value greenhouse crops. Where physical methods are not practical, chemical soil treatments are sometimes adopted. Formalin is commonly used as a disinfectant, applied as a soil drench and then covered with polythene sheets to allow fumigation. In the past, methyl bromide was a widely used fumigant due to its broad-spectrum efficacy, but its use has now been restricted globally because of environmental hazards and ozone-depleting effects. In its place, safer alternatives and biofumigants derived from decomposing organic matter such as mustard residues, neem cake, and brassica crops are gaining importance. Another sustainable strategy is biological disinfestation, which relies on the incorporation of beneficial microbes such as *Trichoderma harzianum, Pseudomonas fluorescens,* and mycorrhizal fungi into the soil. These bioagents act either by directly antagonizing soil-borne pathogens or by competing for nutrients and colonization sites, thus suppressing disease incidence. Biological methods are eco-friendly, improve soil health, and are compatible with organic farming systems. In practice, an integrated approach combining physical, chemical, and biological techniques is often recommended for effective soil disinfestation. For example, partial sterilization with steam or solarization may be followed by the addition of bioagents to maintain long-term disease suppression. These measures are critical in greenhouse systems, where continuous cropping without soil treatment often leads to severe pest and disease buildup. By adopting appropriate soil sterilization and disinfestation techniques, farmers can minimize crop losses, reduce pesticide dependence, and sustain high yields of vegetables, flowers, and seedlings under protected cultivation (Figure 5.5).

Figure 5.5: Soil Sterilization.
(*Source*: https://encrypted-tbn0.gstatic.com)

5.1.5. Soil Enrichment with Organic Matter

Enriching soil with organic matter is a fundamental step in protected cultivation, as it directly enhances soil fertility, structure, and biological activity, thereby creating a healthier environment for crop growth. The incorporation of farmyard manure (FYM), compost, green manure, and vermicompost supplies essential nutrients in a slow-release form while also improving the soil's physical and biological properties. Organic matter acts as a natural conditioner, increasing the soil's water-holding capacity, porosity, and aeration, which are particularly important in greenhouse conditions where crops are intensively cultivated. It also contributes to the formation of stable soil aggregates, reducing the risks of compaction and waterlogging. Decomposition of organic inputs provides a steady supply of macro- and micronutrients, while simultaneously releasing organic acids and humus that enhance nutrient availability. A major benefit of organic amendments is their role in stimulating microbial activity, as they serve as an energy source for beneficial bacteria and fungi. These microbes, in turn, support nutrient cycling, improve root health, and suppress soil-borne pathogens through competitive exclusion and the production of natural antibiotics. To further strengthen soil fertility, biofertilizers such as Rhizobium, Azotobacter, Azospirillum, and phosphate-solubilizing bacteria (PSB) are incorporated during soil preparation. These microorganisms fix atmospheric nitrogen, solubilize insoluble phosphates, and mobilize other nutrients, thus reducing dependency on chemical fertilizers. The use of organic matter also improves the cation exchange capacity (CEC) of soil, ensuring better nutrient retention and reduced leaching losses. In addition, organic inputs such as neem cake, mustard cake, and poultry manure act as natural biofumigants, reducing nematode populations and certain soil-borne diseases. In high-value protected cultivation systems, combining organic manures with microbial consortia ensures a sustainable and balanced nutrient supply throughout the crop cycle. Moreover, regular enrichment of soil with organic matter maintains long-term fertility and prevents degradation caused by continuous cropping. This practice not only enhances crop yield and quality but also supports environmental sustainability by recycling farm residues and reducing reliance on synthetic inputs. Hence, organic matter enrichment serves as the backbone of integrated soil fertility management in protected cultivation, ensuring resilient, productive, and eco-friendly farming systems (Figure 5.6).

Figure 5.6: Enriching Soil with Organic Matter in Greenhouse.
(*Source*: https://plantagreenhouses.com)

5.1.6. Soil Conditioning and Amendment

Soil conditioning and amendment are crucial practices in protected cultivation, as they help in correcting soil-related problems and improving its physical, chemical, and biological properties for better crop growth. Continuous cultivation under greenhouse or polyhouse conditions often leads to issues such as soil acidity, alkalinity, salinity, or nutrient imbalance, which must be addressed before planting high-value crops. To correct soil pH, amendments such as gypsum, lime, or dolomite are added depending on the nature of the problem. Lime and dolomite are commonly used in acidic soils to neutralize excess acidity, supply calcium and magnesium, and create a favourable environment for root growth and microbial activity. Conversely, gypsum is beneficial for reclaiming sodic soils, as it helps in replacing sodium ions with calcium, thereby improving soil structure and permeability. In saline soils, where excess soluble salts hinder water uptake and crop establishment, leaching with good-quality irrigation water is practiced to flush salts below the root zone. Apart from chemical amendments, modern soil conditioners such as zeolite, biochar, and cocopeat are increasingly used in protected cultivation. Zeolite, a natural mineral, has a high cation exchange capacity (CEC) that enhances nutrient-holding ability and reduces leaching losses. Biochar, produced by pyrolysis of organic material, improves soil porosity, water retention, and microbial activity while also sequestering carbon, making it an eco-friendly option. Cocopeat, a by-product of coconut husk processing, is lightweight, porous, and has high water-holding capacity, making it an ideal amendment for root-zone aeration and moisture regulation. These soil conditioners not only improve the physical properties such as porosity, aggregation, and bulk density, but also enhance nutrient availability and reduce the incidence of soil-borne stresses. In many cases, a combination of chemical and organic amendments is used to restore soil health and ensure long-term productivity under intensive cropping systems. Proper soil conditioning not only enhances crop yield and quality but also ensures sustainable use of resources by maintaining soil balance and resilience against degradation. Thus, soil conditioning and amendment form an integral part of soil management practices in protected cultivation, ensuring a stable and productive environment for high-value crops (Figure 5.7).

Figure 5.7: Soil Conditioners.
(*Source*: https://encrypted-tbn0.gstatic.com)

5.1.7. Basal Fertilizer Application

Basal fertilizer application is a critical step in soil nutrient management under protected cultivation, as it ensures that essential nutrients are available to the crop right from the early stages of growth. Before planting, the soil is enriched with a basal dose of major nutrients such as nitrogen (N), phosphorus (P), and potassium (K) in recommended quantities based on soil test results. Nitrogen promotes vigorous vegetative growth, phosphorus supports strong root development and early flowering, while potassium enhances plant strength, disease resistance, and fruit quality. Along with macronutrients, micronutrients such as zinc, boron, and iron are incorporated, especially when deficiencies are detected through soil testing or visible crop symptoms. These micronutrients, though required in small amounts, play vital roles in enzymatic functions, pollination, and chlorophyll synthesis, making them indispensable for balanced plant growth. The application of fertilizers in the basal stage is often done in combination with organic amendments like farmyard manure, compost, or vermicompost, which improve soil structure and act as nutrient reservoirs. This forms the basis of Integrated Nutrient Management (INM), which aims at blending chemical fertilizers with organic and biological sources to maintain soil fertility and sustainability. INM not only supplies balanced nutrition but also reduces the long-term dependence on chemical fertilizers, minimizing environmental hazards such as soil degradation, nutrient leaching, and groundwater pollution. In protected cultivation, precision in basal application is crucial because crops are grown under intensive conditions and nutrient demand is high. Fertilizers are usually incorporated into the soil during bed preparation so that nutrients are evenly distributed within the root zone. The basal application also prepares the soil for subsequent top-dressing and fertigation, ensuring a continuous nutrient supply throughout the crop cycle. In some cases, slow-release fertilizers or fortified organics are used to prolong nutrient availability. By adopting proper basal fertilizer application, farmers can ensure uniform crop establishment, improved growth, higher yields, and superior quality of produce. Thus, this practice acts as the foundation of nutrient management in protected cultivation, setting the stage for efficient crop performance and long-term soil health.

5.1.8. Soil Mulching and Surface Covering

Soil mulching and surface covering are widely practiced techniques in protected cultivation that play a vital role in enhancing soil health, conserving resources, and improving crop productivity. Mulching refers to covering the soil surface around plants with organic or inorganic materials to create a protective layer that modifies the microclimate around the root zone. Among the most effective mulching materials in protected environments are black and silver-coloured polyethylene sheets, which have multiple benefits. Black polythene mulch prevents light penetration, thereby suppressing the germination of weed seeds and drastically reducing weed competition. Silver polythene mulch, apart from controlling weeds, also reflects light, repels certain insect pests, and improves crop canopy photosynthesis by enhancing light distribution. Both types help in conserving soil moisture by reducing evaporation losses and maintaining a more stable moisture regime within the root zone. They also aid in regulating soil temperature, keeping it warmer during winter and relatively cooler during summer, which is crucial for uniform plant growth. Organic mulches such as straw, dry leaves, sugarcane trash, or crop residues may also be used and are effective in improving soil organic matter over time. However, they decompose quickly under greenhouse conditions and may harbour pests or diseases if not managed properly, making them less reliable than synthetic mulches. In contrast, polythene mulches are durable, easier to install, and provide consistent results in intensive farming systems. Another advantage of mulching is the prevention of soil erosion and nutrient leaching, as the covering protects the soil from direct impact of irrigation water. In high-value crops such as tomato, capsicum, cucumber, and cut flowers, mulching has been found to enhance yield, improve fruit quality, and reduce pesticide usage by lowering weed pressure and pest infestation. In addition, mulching supports efficient fertigation, as water and nutrients applied through drip irrigation are retained longer in the root zone. Proper installation is important, where the mulch sheet is tightly spread over raised beds and planting holes are made as per crop spacing. Overall, soil mulching and surface covering serve as a multipurpose technique in protected cultivation, offering improved resource-use efficiency, higher productivity, and better-quality produce, while also contributing to sustainable crop management (Figure 5.8).

Figure 5.8: Soil Mulching.
(*Source*: https://encrypted-tbn0.gstatic.com)

5.1.9. Irrigation System Integration

Irrigation system integration is a crucial component of soil and crop management in protected cultivation, ensuring efficient delivery of water and nutrients to plants. During the soil preparation stage, drip irrigation systems are installed to provide precise and localized water supply directly to the root zone.

Drip lines, fitted with emitters, are laid along the crop rows or raised beds, allowing uniform distribution of water while minimizing evaporation and runoff losses. This method not only conserves water but also reduces the incidence of foliar diseases that commonly occur under overhead irrigation. Along with drip systems, fertigation units are set up to integrate irrigation with nutrient application, ensuring that essential fertilizers are dissolved and delivered in soluble form directly to the root zone. Fertigation offers the dual advantage of improving nutrient-use efficiency and reducing fertilizer wastage, making it highly suitable for intensive greenhouse cultivation where crop nutrient demands are high. Proper soil preparation before system installation ensures firm placement of lateral lines, good contact between soil and emitters, and reduced risk of clogging or uneven water distribution. The integration of irrigation with soil management also allows for precise scheduling based on crop stage, soil moisture status, and prevailing climatic conditions. Automated irrigation controllers, moisture sensors, and pressure regulators are often incorporated in modern greenhouses to maintain uniformity and optimize input use. By delivering water and nutrients in a controlled manner, this system supports better root development, improved crop health, and higher yields. Additionally, irrigation system integration helps maintain a favourable microclimate inside protected structures by regulating humidity when required. The efficiency of drip and fertigation systems also makes them highly compatible with practices like mulching and soil solarization. In high-value crops such as tomato, capsicum, cucumber, and flowers, integrated irrigation ensures consistent growth, reduces input costs, and enhances both yield and quality. Thus, the integration of irrigation systems with soil preparation forms the backbone of resource-efficient and sustainable protected cultivation practices.

5.1.10. Final Soil Preparation and Bed Sanitization

Final soil preparation and bed sanitization represent the concluding stage of soil management in protected cultivation, ensuring that the growing environment is uniform, clean, and ready for planting. Once all amendments such as organic matter, conditioners, and basal fertilizers are incorporated, the soil is thoroughly tilled and homogenized to achieve a fine tilth. This operation eliminates clods, improves aeration, and ensures even distribution of nutrients and moisture within the root zone. Beds are then carefully shaped to the desired height and width, depending on the crop, with pathways left in between for intercultural operations. At this stage, sanitization of the beds becomes critical, as intensive cropping in greenhouses often leads to the accumulation of soil-borne pathogens and pests. Sanitization may be achieved through biological methods, such as the incorporation of bioagents like *Trichoderma harzianum, Pseudomonas fluorescens,* or beneficial mycorrhiza, which suppress harmful organisms while enhancing root health. In some cases, mild disinfectants or eco-friendly fumigants may be applied to reduce pathogen loads, especially in nurseries where seedlings are highly vulnerable. Care is taken to avoid harsh chemicals that can leave toxic residues and harm beneficial microbes. The beds may also be mulched or partially covered to retain the effects of sanitization and reduce weed emergence. Once sanitization is complete, planting holes or furrows are prepared at recommended crop spacings to accommodate seedlings or direct sowing, ensuring optimal plant population and uniform growth. The alignment of drip irrigation lines is also checked and adjusted at this stage so that each planting hole receives adequate water and nutrients. This final preparation guarantees that the soil is not only fertile and well-structured but also free from harmful organisms, thus creating a healthy environment for early crop establishment. A well-prepared and sanitized bed leads to vigorous seedling growth, reduced transplant shock, and better crop stand, ultimately improving productivity and quality. Therefore, final soil preparation and bed sanitization serve as the foundation for successful protected cultivation, minimizing risks and maximizing efficiency right from the start of the cropping cycle (Figure 5.9).

Figure 5.9: Nursery Seed Bed Preparation.
(*Source*: https://i0.wp.com/kitani.farm)

5.2 Soil Management Practices

Soil management is the backbone of successful protected cultivation in horticultural crops, as the entire productivity of polyhouses and greenhouses depends on the health, fertility, and structure of the soil. The first step in this process is soil testing and site selection, which ensures that the soil is suitable for high-value horticultural crops. Detailed testing of pH, electrical conductivity, texture, organic carbon, and macro- as well as micronutrients is conducted before establishing protected structures. This helps to identify potential problems such as acidity, salinity, or nutrient imbalances that may limit crop performance. A carefully chosen site with good drainage, low contamination risk, and balanced soil characteristics forms the foundation for sustainable cultivation. After selecting a site, the next stage is soil preparation and bed formation. In protected structures, deep ploughing is carried out to break hard pans and improve root penetration, while levelling ensures uniform water distribution. Raised beds are prepared to enhance drainage, prevent waterlogging, and create a favourable root environment. This also allows for easier management of irrigation and fertigation. Well-prepared soil beds inside polyhouses and greenhouses improve aeration, reduce compaction, and provide a stable base for high-value horticultural crops.

To further enrich the soil, organic matter incorporation is given special emphasis. The addition of compost, farmyard manure, vermicompost, and green manures significantly improves soil structure, fertility, and microbial activity. In protected cultivation, where cropping intensity is high, organic matter enhances water-holding capacity and provides a slow release of essential nutrients. It also improves cation exchange capacity and buffers soil against sudden pH changes, creating a healthy growing medium for horticultural crops. Since high humidity and continuous cropping in greenhouses may lead to the buildup of pathogens, soil sterilization and disinfection becomes an essential practice. Various methods such as soil solarization, steaming, and application of bioagents like Trichoderma and Pseudomonas fluorescens are used to suppress soil-borne diseases. In some cases, mild chemical disinfectants are also applied. These practices help in controlling nematodes, fungi, and harmful bacteria, thereby ensuring a disease-free soil environment. Maintaining soil hygiene in protected cultivation is crucial to avoid yield losses from recurring infections. Nutrient availability is managed carefully through nutrient management and fertigation. Integrated nutrient management that combines organic manures with inorganic fertilizers ensures both immediate and long-term nutrient supply. Fertigation, where water-soluble fertilizers are supplied through drip irrigation, increases nutrient-use efficiency and prevents wastage. Balanced nutrition helps in achieving uniform crop growth, higher yields, and better-quality produce. Monitoring soil nutrient status at regular intervals also helps in timely correction of deficiencies. Methods of soil conservation and its benefits is depicted in Figures 5.10 and 5.11 respectively.

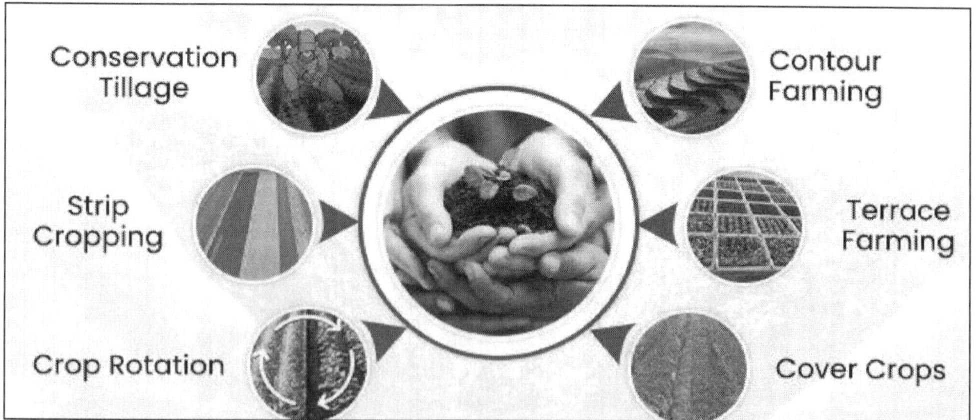

Figure 5.10: Methods of Soil Conservation
(*Source*: https://assets.tractorjunction.com)

Figure 5.11: Benefits of Soil Conservation.
(*Source*: https://assets.tractorjunction.com)

In protected cultivation systems, water is a precious resource, and therefore soil moisture conservation techniques are widely adopted. Plastic mulching prevents evaporation losses and suppresses weed growth, while drip irrigation ensures precise water delivery to the root zone. Controlled irrigation schedules are followed to maintain uniform soil moisture, preventing both waterlogging and drought stress. Proper moisture management not only improves plant growth but also reduces incidences of soil-borne diseases. Another important aspect is soil erosion and compaction control. Although protected structures reduce the direct impact of rainfall, soil erosion may still occur due to frequent irrigation or water movement. Practices like minimal tillage, use of organic mulches, and cover cropping reduce erosion risk. At the same time, continuous foot traffic and machinery movement inside greenhouses can lead to soil compaction. To address this, proper bed design and restricted pathways are maintained, ensuring healthy root aeration and better crop performance. Equally important is biological soil health improvement, which focuses on enhancing the living component of soil. Application of biofertilizers such as Rhizobium, Azotobacter, and phosphate-solubilizing bacteria improves nutrient availability. The introduction of mycorrhizal fungi enhances phosphorus uptake and strengthens plant root systems. Beneficial microbes contribute to disease suppression, organic matter decomposition, and overall soil resilience, making them an integral part of protected cultivation practices.

Despite preventive measures, soils often face constraints, and hence corrective measures are adopted. Acidic soils are treated with lime or dolomite to neutralize acidity and improve nutrient availability. Saline and sodic soils are reclaimed using gypsum or leaching with good quality water. Micronutrient deficiencies are corrected through foliar sprays or soil-applied fertilizers, depending on the specific requirement. These corrective actions ensure that crops receive a balanced supply of essential nutrients throughout the growing season. Finally, long-term success depends on adopting sustainable soil management practices. Crop rotation with legumes or other soil-enriching crops helps to break disease cycles and restore fertility. Intercropping and the use of soil-covering plants improve biodiversity and reduce pest buildup. Conservation agriculture techniques, minimal chemical usage, and increased reliance on organic inputs ensure that soil fertility is maintained for future crop cycles. In protected horticultural systems, where the intensity of cultivation is high, such sustainable practices are critical for maintaining soil productivity and ecological balance. Thus, soil management in protected cultivation is a multi-dimensional approach that integrates scientific testing, preparation, organic enrichment, sterilization, precise nutrient and water management, biological enhancement, corrective treatments, and sustainability. By following these practices, growers can ensure healthy soils, higher yields, and long-term profitability in horticultural crop production under protected environments.

5.3 Types of Substrates

5.3.1. Soil-Based Substrates

Soil-based substrates are among the oldest and most widely used growing media in protected cultivation systems (Figure 5.12). Traditionally, mineral soil forms the main component of these substrates, making them naturally abundant, cost-effective, and easy to access for farmers.

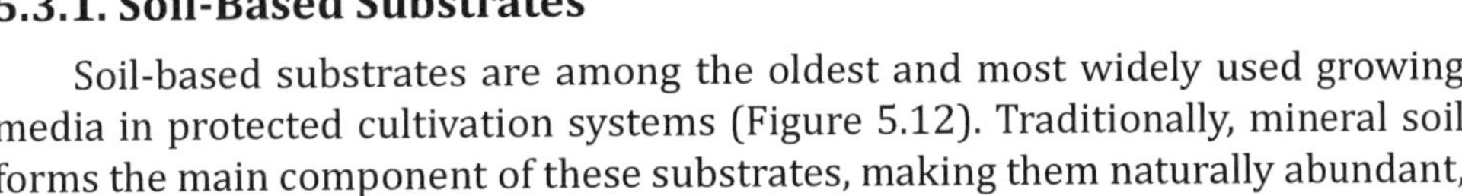

Figure 5.12: Soil-Based Substrates.
(*Source*: www.freepik.com)

Soil provides a stable medium for anchoring plant roots, offering the necessary mechanical support required for upright growth and development. It also contains a wide range of macro- and micronutrients such as nitrogen, phosphorus, potassium, calcium, magnesium, and trace elements, which are essential for plant growth and productivity. Unlike inert substrates, soil is biologically active and supports a diverse community of beneficial microorganisms including bacteria, fungi, and actinomycetes. These microbes play a vital role in nutrient cycling, organic matter decomposition, and protection against certain soil-borne pathogens. In addition to nutrient availability, soil-based substrates retain moisture effectively and buffer nutrient fluctuations, ensuring a more balanced supply of water and minerals to plants. The natural cation exchange capacity (CEC) of mineral soil helps in holding nutrients and releasing them gradually, which is beneficial for steady plant growth. The structure of soil, depending on its texture (sand, silt, and clay proportions), influences aeration, drainage, and root penetration. Loamy soils, with a balanced mixture of sand, silt, and clay, are generally preferred for use in soil-based substrates because of their good water-holding capacity and aeration. However, when used in greenhouses or polyhouses, soil-based substrates present several challenges that must be addressed to ensure high productivity and healthy crops. One of the major limitations is the potential presence of pests, pathogens, and weed seeds that can thrive in untreated soil. These organisms may cause serious diseases and yield losses if the soil is not properly sterilized before use. To minimize risks, soil solarization, steam sterilization, or chemical fumigation methods are often applied before cropping. Another limitation is the variability in soil fertility and texture across different locations, which may lead to inconsistent results in terms of plant growth and yield.

Soil compaction is also a common issue in protected cultivation, as repeated irrigation and lack of natural weathering can lead to reduced porosity and restricted root growth. This can adversely affect plant health by reducing oxygen supply to the roots. Additionally, soil-based substrates are more prone to salinity build-up under intensive fertigation practices in controlled environments, which can further hinder plant growth and nutrient uptake. Despite these challenges, farmers still prefer soil-based substrates in many regions because of their affordability, familiarity, and ability to sustain crops with minimal external inputs. To improve the performance of soil-based substrates, organic amendments such as compost, farmyard manure, vermicompost, and biofertilizers are commonly incorporated. These not only improve the nutrient status but also enhance soil structure, water retention, and microbial activity. The use of crop rotations, green manures, and organic mulches further contributes to maintaining soil fertility in protected cultivation systems. Research has also shown that integrating beneficial microbes like Trichoderma, Azotobacter, and mycorrhizal fungi into soil-based substrates enhances disease resistance and improves nutrient uptake efficiency. In modern protected cultivation, soil-based substrates are often combined with other organic or inorganic materials such as cocopeat, perlite, or vermiculite to enhance their physical and chemical properties. These mixtures help overcome the limitations of pure soil use by improving aeration, drainage, and uniformity. Overall, soil-based substrates continue to play an important role in horticultural production under protected structures, balancing tradition with modern scientific improvements. When managed properly with sterilization, amendments, and monitoring, soil remains a reliable medium that supports plant growth by providing mechanical support, nutrient supply, and a dynamic biological environment essential for sustainable crop production.

5.3.2. Cocopeat (Coir Pith)

Cocopeat, also known as coir pith or coir dust, is a natural by-product obtained during the processing of coconut husks for coir fibre extraction. It is a spongy, lightweight material that has become extremely popular in horticulture, floriculture, and protected cultivation systems due to its excellent physical and biological properties. Cocopeat is considered one of the most suitable substitutes for soil in soilless cultivation practices, particularly in polyhouses, shade nets, nurseries, and hydroponic setups. The structure of cocopeat is highly porous, which enables it to hold a significant amount of water while still allowing proper aeration. This unique balance between moisture retention and air circulation is critical for healthy root development in plants. Unlike heavy mineral soils, cocopeat does not become compacted easily; instead, it maintains a fluffy and friable texture that promotes root penetration and prevents suffocation of roots. The water-holding capacity of cocopeat is almost 8–10 times its dry weight, making it an excellent medium in areas facing irregular or limited water availability. Apart from its superior physical properties, cocopeat is also eco-friendly and renewable. Since it is derived from the husk of coconuts, which are abundantly available in tropical countries, it represents a sustainable growing medium. Instead of being discarded as waste, coir pith is processed, buffered, and used beneficially in agriculture. The use of cocopeat helps in reducing dependence on topsoil, thereby conserving natural soil resources.

Cocopeat also contributes to improved nutrient management. Though it is low in inherent nutrients, it acts as a neutral medium that can efficiently absorb and release nutrients supplied through fertilizers or fertigation systems. Its cation exchange capacity (CEC) allows it to hold essential nutrients such as calcium, magnesium, and potassium, and release them gradually to the plant roots. This controlled availability minimizes nutrient leaching and enhances fertilizer-use efficiency. In polyhouses and greenhouses, cocopeat is widely used either alone or in combination with other substrates such as perlite, vermiculite, and compost. It provides a stable root zone environment even under fluctuating external climatic conditions. The consistent moisture level maintained by cocopeat ensures uniform crop growth, higher yield, and better-quality produce. For crops like tomato, capsicum, cucumber, strawberry, and cut flowers like rose and gerbera, cocopeat has proven to be an ideal substrate. Another advantage of cocopeat is its ability to suppress soil-borne diseases and nematode attacks to some extent, since it is usually free from pathogens when properly processed and sterilized. Its lightweight nature makes handling, transportation, and filling of pots or grow bags much easier compared to traditional soil. Furthermore, cocopeat improves root aeration, leading to enhanced uptake of water and nutrients, which is particularly beneficial in high-density planting systems. The pH of cocopeat generally ranges between 5.5 and 6.8, which is considered suitable for most horticultural crops. However, raw cocopeat may contain high levels of soluble salts, especially sodium and potassium, which can adversely affect plant growth if not treated properly. Therefore, commercial cocopeat products are usually washed, buffered, and sieved to make them fit for agricultural use. This preprocessing ensures that the electrical conductivity (EC) remains within the safe range for sensitive crops.

Cocopeat is also valued for its reusability. With proper sterilization and replenishment of nutrients, the same cocopeat medium can be used for multiple crop cycles, thereby reducing production costs for growers. Its biodegradability further ensures that once discarded, it does not contribute to environmental pollution, but instead enriches the soil as an organic amendment. In the context of sustainable agriculture, cocopeat plays a vital role as a renewable, eco-friendly, and effective substrate. It not only reduces the dependency on soil but also supports water conservation, efficient fertilizer use, and healthier plant growth. Its widespread adoption in polyhouse cultivation is a testimony to its reliability and adaptability across diverse crops and climates. Thus, cocopeat is rightly considered a modern, versatile, and sustainable substrate that bridges the gap between traditional soil-based cultivation and advanced soilless farming systems. Its contribution to improving yield, quality, and resource efficiency makes it indispensable in present-day protected cultivation practices (Figure 5.13).

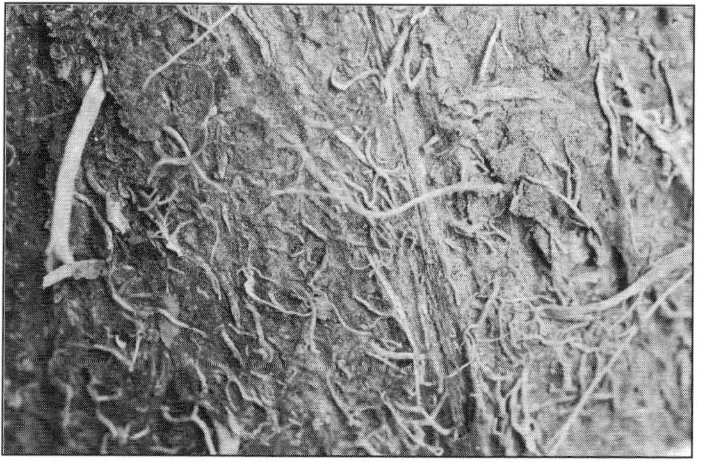

Figure 5.13: Coir Pith or Coir Dust
(*Source*: www.freepik.com)

5.3.3. Vermiculite

Vermiculite is a naturally occurring mineral that belongs to the mica family and is widely recognized for its unique property of expanding when heated at high temperatures. This expansion process produces lightweight, golden-brown flakes that possess a spongy texture, making vermiculite highly suitable as a substrate in horticultural and agricultural practices. It is sterile, non-toxic, and odourless, making it a safe medium for plant cultivation, particularly in controlled environments such as greenhouses, polyhouses, and nurseries. One of the most important features of vermiculite is its high water-holding capacity. It can absorb and retain moisture several times its weight, ensuring a constant supply of water to plant roots. This property prevents seedlings from drying out quickly, which is especially beneficial during early germination stages when consistent moisture levels are critical. Along with water, vermiculite also helps in holding dissolved nutrients, allowing plants to access them more efficiently and reducing the chances of nutrient leaching. Another vital advantage of vermiculite is that it provides excellent aeration to the root zone. Its porous structure creates air spaces in the growing medium, preventing soil compaction and ensuring that roots receive sufficient oxygen for respiration and growth. This balance between water retention and aeration is one of the main reasons vermiculite is widely used in propagation trays, seedling blocks, and potting mixes (Figure 5.14).

Figure 5.14: Vermiculite.
(*Source*: https://upload.wikimedia.org)

Vermiculite is also valued for its neutral to slightly alkaline pH, which makes it compatible with most horticultural crops. It does not degrade or decompose over time, allowing it to maintain its structure and efficiency for long periods. Moreover, being lightweight, it reduces the overall weight of potting mixtures, making handling, transportation, and application easier for growers. In protected cultivation systems, vermiculite is often blended with other substrates such as perlite, cocopeat, or compost to create a balanced growing medium. While vermiculite alone tends to hold excessive water, mixing it with perlite enhances drainage, and combining it with cocopeat improves the organic content and microbial activity of the substrate. These mixtures provide a more stable environment for crop roots, ensuring balanced moisture, aeration, and nutrient availability. Beyond seed germination and nursery raising, vermiculite is also used in hydroponics and soilless culture. It can serve as a medium for direct rooting of cuttings, vegetable seedlings, and ornamental plants. Its sterile nature minimizes the risk of soil-borne pathogens, pests, and weed seeds, which is particularly advantageous in polyhouse production where plant health and hygiene are critical. From a practical perspective, vermiculite is easy to handle and apply. It can be used as a top dressing to retain surface moisture, mixed into soils to improve porosity, or incorporated in seedling trays for uniform root development. In addition to horticulture, vermiculite is also applied in mushroom cultivation, floriculture, and specialized nurseries where moisture control is essential. In summary, vermiculite is a versatile and eco-friendly substrate that plays a significant role in protected cultivation. Its ability to retain water, provide aeration, and support root health makes it highly valuable for seedlings, hydroponics, and container gardening. When used in combination with other media, it creates an ideal growing environment for a wide range of crops, ensuring vigorous growth, better transplant survival, and improved productivity in protected systems.

5.3.4. Perlite

Perlite is a naturally occurring volcanic mineral that undergoes a special heat expansion process, transforming it into a lightweight, highly porous medium widely used in horticulture and protected cultivation. When volcanic glass is heated rapidly at high temperatures, it expands like popcorn, creating tiny air pockets within its structure. This expanded form of perlite is sterile, inert, and exceptionally light in weight, making it easy to handle and mix with other substrates. Its unique structure enhances aeration by preventing compaction of the growing medium, while simultaneously improving drainage and reducing the risk of waterlogging, which is particularly important in polyhouse and hydroponic systems. Unlike soil, perlite contains no organic matter and does not support the growth of pathogens, weeds, or pests, ensuring a clean and disease-free root environment (Figure 5.15). In cultivation, perlite plays a critical role in balancing the water-to-air ratio around plant roots. Its porous surface absorbs a small quantity of water while allowing excess water to drain away, maintaining an optimum root zone environment that encourages vigorous growth. Due to its sterile and inert nature, perlite does not alter the pH of the growing medium and provides a stable substrate for nutrient uptake. Farmers and nursery growers often blend perlite with cocopeat, vermiculite, or compost to create a customized substrate that combines moisture retention with superior aeration. For example, when mixed with cocopeat, perlite prevents excessive compaction and improves root respiration, while in combination with vermiculite, it balances water-holding capacity with free drainage, making it highly suitable for seed germination, cuttings, and hydroponic culture.

Figure 5.15: Perlite
(*Source*: www.freepik.com)

Perlite's lightweight nature reduces transportation and handling costs, and it can be reused multiple times after proper sterilization, adding to its economic and environmental advantages. In hydroponics, it is frequently used as a standalone growing medium in containers or as a component in nutrient film techniques and drip irrigation systems, where its excellent drainage properties prevent salt accumulation in the root zone. Additionally, perlite is chemically inert and free from toxic substances, ensuring safe use in food crop production. Beyond horticulture, it is also applied in landscaping, green roofs, and soil conditioning, where it loosens heavy clay soils and enhances water infiltration.

Overall, perlite is a versatile, eco-friendly substrate that ensures strong, healthy root development, uniform crop growth, and efficient water and nutrient use under protected cultivation systems. Its ability to provide both aeration and drainage, while being sterile and reusable, makes it an indispensable component in modern horticultural practices, especially in polyhouses and hydroponic farming setups.

5.3.5. Rockwool (Mineral Wool) as a Hydroponic Substrate

Rockwool is a highly versatile growing medium manufactured from natural basaltic rock, which is first melted at extremely high temperatures, often exceeding 1500°C, and then spun into fine fibres resembling cotton-like wool. These fibres are later compressed into slabs, blocks, or cubes of different sizes suitable for plant propagation and cultivation. Its unique origin from volcanic rock makes it inert, sterile, and free from pathogens, pests, or weed seeds, which is a highly desirable property in protected and hydroponic farming systems. One of the most notable characteristics of Rockwool is its exceptional water-holding capacity combined with superior aeration. The fibrous structure of the material allows it to absorb and retain a large volume of water while simultaneously maintaining sufficient air-filled pore space to ensure proper oxygen supply to the plant roots. This dual property prevents both drought stress and waterlogging, creating a balanced root-zone environment for optimal growth.

In hydroponics, where soil is absent, Rockwool provides strong physical support to the plants. The roots can easily penetrate and anchor themselves within the fibres, stabilizing the plant structure, while the substrate itself does not compact over time. This makes it especially suitable for crops that require long growing periods, such as tomatoes, cucumbers, peppers, roses, and gerbera. Another key advantage is that Rockwool is chemically inert and neutral, meaning it does not react with nutrient solutions or alter their composition. However, fresh Rockwool has a naturally high pH (around 7.0–8.0), which must be conditioned before use by soaking in a slightly acidic solution, usually adjusted to pH 5.5–6.0. Once stabilized, it provides a consistent medium where nutrients supplied through fertigation can be precisely managed and efficiently utilized by the plant roots. Rockwool is manufactured in various forms–such as propagation cubes for seedlings and cuttings, grow blocks for transplanting, and larger slabs for mature plants–offering flexibility across different stages of crop production. Seedlings started in small cubes can be easily transferred to bigger blocks or slabs without disturbing the root system, ensuring uniform and healthy plant development (Figure 5.16).

Figure 5.16: Rockwool (Mineral Wool) as a Hydroponic Substrate.
(*Source*: www.freepik.com)

In commercial hydroponic vegetable and flower cultivation, Rockwool has become one of the most widely used substrates due to its reliability, high yields, and ease of handling. Large-scale greenhouse industries across Europe, North America, and Asia extensively use Rockwool in drip-irrigated systems, where excess nutrient solution drains freely while the substrate retains sufficient moisture for continuous plant uptake. This promotes vigorous root growth, faster vegetative development, and higher yields with consistent quality. Apart from its agronomic benefits, Rockwool is also valued for its durability and reusability. With proper management, it can sustain multiple crop cycles, although sterilization between crops is essential to avoid disease buildup. Disposal, however, can be a concern since Rockwool is non-biodegradable. To address this, recycling methods are being developed, such as crushing used Rockwool for reuse in construction materials, insulation, or soil conditioning. Overall, Rockwool is regarded as a standard hydroponic medium, offering a perfect balance of water retention, aeration, structural support, and adaptability for a wide range of crops. It allows precise control over the root-zone environment, enabling growers to achieve higher productivity, superior quality produce, and efficient resource utilization. Despite some challenges in disposal and initial pH adjustment, its advantages have made it indispensable in modern protected cultivation and hydroponic farming systems worldwide.

5.3.6. Sand

Sand is one of the oldest and most widely used inert growing media in horticulture and protected cultivation systems. It is a naturally occurring granular material composed of finely divided rock and mineral particles, primarily silica in the form of quartz. Depending on its origin, sand can be river sand, pit sand, or desert sand, each type varying in texture, particle size, and cleanliness. The physical properties of sand make it unique and suitable for specific horticultural practices, though it is often used in combination with other media to balance its limitations. One of the most important advantages of sand is its role in improving soil drainage. When mixed with heavy clay soils, sand helps loosen compacted layers, allowing excess water to move through more freely. This prevents waterlogging, root suffocation, and the proliferation of fungal diseases caused by excess moisture. Its coarse texture creates macropores that enhance aeration around the root zone, ensuring that oxygen reaches the developing root system. Because of this property, sand is frequently included in potting mixtures, raised beds, and nursery substrates. Sand also plays a crucial role in nursery bed preparation. It provides a clean, loose, and well-aerated environment for the germination of seeds and the establishment of young seedlings. Its granular nature reduces the risk of soil-borne pathogens and ensures that delicate seedlings can be uprooted and transplanted easily without severe root damage. Similarly, sand is highly valued in the rooting of cuttings. Softwood and hardwood cuttings often require a medium that prevents water stagnation while maintaining moderate moisture levels. Sand provides just that, offering an ideal balance of aeration and moisture for the initial stages of root initiation. Despite its usefulness, sand has several limitations when used as a sole growing medium. It has very limited nutrient-holding capacity because it lacks cation exchange sites. This means that essential plant nutrients like nitrogen, potassium, and magnesium are easily leached away with irrigation water. Additionally, its water-holding capacity is low compared to other substrates like cocopeat or vermiculite, which makes frequent irrigation necessary in sand-based cultivation. To overcome this drawback, sand is often mixed with organic matter such as compost, cocopeat, or farmyard manure, which improve both water retention and nutrient availability (Figure 5.17).

Figure 5.17: Sand Used Inert Growing Media in Horticulture and Protected Cultivation Systems.
(*Source*: www.freepik.com)

Cleanliness and particle size are also important considerations when using sand. Coarse, washed river sand is preferred in horticulture because it is relatively free from clay, silt, salts, and organic impurities that may affect plant growth. Fine sand, by contrast, may compact and reduce aeration, defeating the purpose of its use. In hydroponic systems, sterilized sand has been used as a growing medium where nutrient solutions are supplied externally, compensating for its poor nutrient-holding ability. Sand is also important in specialized horticultural practices. For example, in turf grass management, a sand-based root zone is preferred because it ensures rapid drainage and firm surfaces, especially in golf courses and sports fields. In floriculture, sand beds are used for bulbous plants, gladiolus, and ornamental cuttings to encourage healthy root systems. In fruit crop propagation, especially for mango, guava, and citrus, sand-filled polybags or beds are commonly used to raise rootstocks. In terms of sustainability, sand is a naturally abundant resource, but its use must be balanced with environmental concerns. Excessive mining of river sand can lead to ecological degradation, loss of aquatic habitats, and soil erosion. Therefore, alternatives like manufactured sand or combinations with other renewable media are encouraged in sustainable horticulture. Overall, sand is a simple, inexpensive, and highly effective medium for specific horticultural applications. It is best suited for nursery beds, seed germination, rooting of cuttings, and as a soil amendment to improve drainage and aeration. However, due to its inherent limitations of low nutrient and water-holding capacity, it is rarely used alone. Instead, it serves as a complementary component in substrate mixtures, ensuring physical balance in growing media. When used judiciously, sand continues to remain a reliable and essential medium in both traditional and modern cultivation practices.

5.3.7. Gravel and Expanded Clay Pebbles

Gravel and expanded clay pebbles are among the most widely recognized inert substrates used in hydroponic and protected cultivation systems due to their unique physical characteristics and long-term reusability. Gravel, which is naturally occurring crushed stone or coarse aggregate, provides excellent drainage, firm support, and effective root anchorage for plants. On the other hand, expanded clay pebbles are manufactured by heating natural clay at very high temperatures until it expands, forming lightweight, porous, and rounded pellets with a hard outer shell and spongy interior. These properties allow them to maintain both aeration and water retention in balance, making them highly suitable for recirculating hydroponic systems like Nutrient Film Technique (NFT), Deep Water Culture (DWC), and aquaponics. The inert nature of gravel and clay pebbles ensures that they do not chemically react with nutrient solutions or release any toxic compounds into the root zone. Their structural stability prevents decomposition, compaction, or breakdown over time, which makes them reusable for multiple crop cycles after proper cleaning and sterilization. This long lifespan makes them cost-effective despite their relatively high initial investment. Moreover, the uniform particle size and rounded shape of expanded clay pebbles facilitate easy transplanting, root inspection, and nutrient circulation around the plant base. Gravel, due to its heavier weight, provides more robust anchorage, making it particularly useful for tall or fruit-bearing plants that require stable support, such as tomato, cucumber, and capsicum. However, its high bulk density increases system weight and handling effort, which limits its application in portable or rooftop hydroponics.

Figure 5.18: Gravel and Expanded Clay Pebbles are Widely Used in Hydroponic and Protected Cultivation Systems.

(*Source*: www.freepik.com)

In contrast, expanded clay pebbles are lightweight, making them easier to handle, transport, and integrate into modular hydroponic units. Their porous structure improves capillary action, ensuring that moisture and nutrients are evenly distributed to plant roots while excess water drains away quickly to prevent waterlogging (Figure 5.18).

These substrates are commonly used in ebb-and-flow (flood-and-drain) hydroponic systems, where the nutrient solution periodically floods the root zone and then drains back into the reservoir. The ability of expanded clay pebbles to absorb a small amount of nutrient solution while still maintaining aeration helps in creating a balanced root environment. They are also favoured in aquaponics because their porous surface area provides an ideal habitat for beneficial nitrifying bacteria that convert fish waste into plant-available nutrients, thus maintaining water quality and system efficiency. Another key advantage of both gravel and clay pebbles is their ease of cleaning and sterilization. After each crop cycle, they can be rinsed to remove root residues and biofilm, then disinfected using mild chemical solutions, steam, or solarization before reuse. This reduces the cost of replenishing substrates and contributes to sustainable farming practices. Despite these advantages, there are some limitations. Gravel, if not properly washed before use, may contain fine dust or impurities that can clog irrigation lines or interfere with nutrient delivery. Similarly, clay pebbles, though lightweight and efficient, are more expensive than locally available alternatives such as sand or cocopeat. Their water-holding capacity is moderate, so they often need frequent irrigation cycles compared to substrates like cocopeat or vermiculite. Nevertheless, when integrated into well-designed hydroponic systems, they remain one of the most reliable and efficient growing mediums for high-value crops. In summary, gravel and expanded clay pebbles provide a stable, inert, and reusable substrate option with excellent drainage and aeration characteristics. Gravel ensures strong anchorage and structural support, while expanded clay pebbles combine lightweight handling with high porosity and bacterial colonization potential. Together, they play a crucial role in modern recirculating hydroponic and aquaponic systems, making them indispensable for growers aiming for sustainable, high-yield, and resource-efficient cultivation practices.

5.3.8. Sawdust and Wood Chips

Sawdust and wood chips are organic substrates derived as by-products of timber processing, wood cutting, or carpentry industries, and are widely used in different forms of horticulture, mushroom cultivation, and hydroponic or semi-hydroponic systems. These materials are highly valued due to their porous structure, which provides excellent aeration and creates a favourable environment for root respiration. The loose and fibrous nature of sawdust allows roots to penetrate easily, while wood chips offer additional structural support to the growing medium. Both substrates contribute to better drainage, thereby preventing waterlogging and root diseases, which are common problems in compact or poorly aerated soils. One of the key features of sawdust and wood chips is their ability to maintain a light and airy texture when mixed with other media such as soil, compost, cocopeat, or sand. This blending not only enhances the physical properties of the substrate but also balances water-holding capacity with aeration. In horticultural nurseries, sawdust is often used as a rooting medium for cuttings of ornamental plants, vegetables, and tree species because of its fine texture and ability to hold sufficient moisture around the delicate root zone. Wood chips, being coarser, are frequently applied as mulching material on the surface of growing beds, helping in temperature regulation, weed suppression, and gradual release of organic matter during decomposition. However, one of the limitations of sawdust and wood chips is their tendency to decompose over time. Microbial activity breaks down the cellulose and lignin present in wood, leading to the gradual loss of volume and changes in the substrate structure. During decomposition, microorganisms compete with plants for available nitrogen, which can cause nutrient deficiency in crops if not properly managed. Therefore, when sawdust or wood chips are used as substrates, it is often recommended to supplement the medium with nitrogen-rich fertilizers or organic manures to balance nutrient availability and support healthy plant growth. The replacement of these substrates at regular intervals becomes necessary to maintain a stable growing environment, especially in commercial systems where consistency in crop performance is crucial (Figure 5.19).

Figure 5.19: Sawdust and Wood Chips are Widely Used in different Forms of Horticulture, Mushroom Cultivation, and Hydroponic or Semi-hydroponic Systems. (*Source*: www.freepik.com)

Sawdust and wood chips are eco-friendly materials since they utilize waste products from the wood industry that might otherwise contribute to environmental burden. Their biodegradability makes them suitable for sustainable cultivation practices, particularly when integrated into organic farming systems. Additionally, these substrates improve soil organic matter content when incorporated into field soil after use, thereby enhancing soil fertility and structure. In protected cultivation systems such as polyhouses or greenhouses, sawdust is often mixed with inert materials like perlite or vermiculite to create a balanced substrate that supports uniform root growth. In mushroom farming, sawdust, especially from hardwood species, is a standard substrate used for the cultivation of edible mushrooms like oyster and shiitake. Its fibrous nature and ability to retain moisture while providing adequate aeration make it an ideal medium for fungal colonization. Wood chips are also used in landscaping and gardening as a cost-effective medium for ground cover, moisture conservation, and improvement of soil microflora. Overall, sawdust and wood chips are versatile organic substrates that contribute to sustainable horticultural practices. While their tendency to decompose requires careful management and periodic replacement, their benefits in terms of aeration, drainage, organic matter enrichment, and eco-friendliness make them valuable components of both traditional and modern cultivation systems.

5.3.9. Peat Moss

Peat moss is one of the most widely used soilless substrates in horticulture and hydroponics, derived from partially decomposed sphagnum moss accumulated over centuries in waterlogged peat bogs. It forms naturally in wetland ecosystems where low oxygen levels slow the decomposition of organic matter, resulting in a fibrous, lightweight, and spongy material. Its structure makes it highly effective for retaining water while still maintaining good porosity for air circulation around plant roots. One of the most notable features of peat moss is its high water-holding capacity, often absorbing up to 10–20 times its dry weight in water, which ensures a constant moisture supply for plants and minimizes irrigation frequency. Despite being lightweight, it provides excellent support for root systems, making it suitable for seed germination, nursery raising, container gardening, and greenhouse cultivation. Chemically, peat moss is acidic in nature, with a pH range typically between 3.0 and 4.5, which makes it particularly beneficial for acid-loving crops such as blueberries, strawberries, azaleas, rhododendrons, and camellias.

Figure 5.20: Peat Moss is Widely Used Soilless Substrates in Horticulture and Hydroponics.
(*Source*: www.freepik.com)

To balance pH for general crops, it is often amended with lime or mixed with neutral substrates like perlite or vermiculite. When blended with perlite, it enhances aeration and prevents compaction; when combined with vermiculite, it improves both moisture retention and nutrient-holding capacity. These mixtures create a well-structured growing medium suitable for a wide range of horticultural applications (Figure 5.20).

Nutritionally, raw peat moss is low in essential plant nutrients, but it functions as an excellent medium for holding fertilizers and nutrients supplied through irrigation or fertigation. Its cation exchange capacity (CEC) is relatively high, enabling it to absorb and slowly release nutrients to plants, which reduces leaching losses in container or hydroponic systems. This property is especially valuable in modern greenhouse management, where precise nutrient delivery is crucial for uniform crop growth. In terms of usage, peat moss is a key component in commercial potting mixes, seedling trays, plug propagation, mushroom cultivation, and hydroponic systems. It provides uniform texture, cleanliness, and a sterile environment, reducing the risk of soil-borne diseases, pests, and weed contamination. Because it is naturally free from harmful pathogens, it is often preferred over traditional soil in high-value crop production under protected cultivation. However, there are environmental concerns associated with peat moss harvesting. Since peat bogs take thousands of years to form, excessive extraction can deplete natural ecosystems, release stored carbon, and contribute to environmental degradation. For this reason, sustainable peat management and alternative renewable substrates such as cocopeat are being encouraged. Still, peat moss remains a dominant component of horticultural substrates in many parts of the world due to its unique balance of water retention, aeration, and acidity. Overall, peat moss is an indispensable growing medium for horticulturists, nurserymen, and hydroponic growers, providing a reliable, versatile, and effective substrate when used alone or in combination with other inert materials. Its properties support vigorous root growth, consistent plant establishment, and high-quality yields, making it a cornerstone in modern protected cultivation practices.

5.3.10. Organic Composts

Organic composts represent one of the oldest and most sustainable soil amendments used in agriculture and horticulture, offering a natural means of improving soil fertility and plant growth. They include materials such as farmyard manure (FYM), vermicompost, and bio-compost, each with distinct properties and benefits. Farmyard manure is derived from the decomposed mixture of animal dung, urine, bedding materials, and crop residues, and is valued for its ability to enrich soils with both macro and micronutrients while simultaneously improving physical structure. Vermicompost, produced through the activity of earthworms on organic residues, is a finely granulated, nutrient-rich, and biologically active product that enhances soil aeration, moisture-holding capacity, and microbial diversity. Bio-composts, often generated using microbial inoculants and crop residues, provide a concentrated source of nutrients along with beneficial microorganisms that support plant health. The primary advantage of organic composts lies in their ability to release nutrients slowly and steadily over time, ensuring that plants receive a continuous supply of essential elements without the risk of nutrient leaching that often occurs with chemical fertilizers. Their use contributes significantly to the buildup of soil organic matter, which in turn improves cation exchange capacity, water infiltration, and aggregation of soil particles. These properties make composts invaluable in both open-field cultivation and protected cultivation systems such as greenhouses, polyhouses, and nurseries, where maintaining soil health and structure is critical for achieving high productivity.

Figure 5.20: Organic Composts.
(*Source*: www.freepik.com)

Organic composts also play a vital role in promoting beneficial microbial activity within the root zone. The decomposed organic matter serves as food for a wide range of microorganisms, including bacteria, fungi, and actinomycetes, which in turn help decompose complex organic compounds into plant-available nutrients. This microbial activity not only boosts nutrient cycling but also suppresses harmful pathogens through competitive exclusion and the production of natural antibiotics. Additionally, composts act as a reservoir of humic and fulvic acids, which stimulate root growth, enhance nutrient uptake, and improve overall plant vigour. Another advantage is their environmental sustainability. Unlike synthetic fertilizers, which require high energy inputs during production and often cause pollution when over-applied, organic composts recycle farm and household wastes, thereby reducing environmental load. They also help in sequestering carbon in soils, mitigating greenhouse gas emissions, and contributing to climate resilience in agriculture. When blended with inert media such as perlite, vermiculite, cocopeat, or sand, composts provide a balanced growing substrate that combines nutrient richness with good drainage and aeration. This makes them particularly suitable for potting mixes, nursery seedbeds, hydroponic blends, and containerized cultivation of vegetables, ornamentals, and fruit crops. Composts, however, need to be properly matured before use. Immature or poorly decomposed composts can release phytotoxic compounds, harbor weed seeds, or even cause nutrient imbalances. Therefore, adequate composting time, aeration, and moisture management are necessary to ensure that the final product is stable, odourless, and safe for plants. Depending on the feedstock used and composting method adopted, the nutrient profile of composts can vary, but in general, they are rich in nitrogen, phosphorus, potassium, calcium, and trace elements, making them a comprehensive soil amendment.

In protected cultivation, where soils are often intensively used and prone to nutrient depletion and salinity, composts help restore balance by buffering soil pH, reducing salt accumulation, and enhancing resilience against abiotic stresses. They also improve the tilth and workability of soils, facilitating easier root penetration and better root-to-soil contact. For organic farming systems, composts are indispensable, as they are compliant with certification standards and provide a sustainable alternative to synthetic fertilizers. Overall, organic composts serve as an integral part of modern substrate management strategies. They not only improve fertility and microbial activity but also restore degraded soils, reduce dependency on chemical inputs, and support long-term agricultural sustainability. By combining the strengths of farmyard manure, vermicompost, and bio-compost, and by blending them judiciously with inert media, growers can create ideal substrate mixtures that support healthy plant growth, higher yields, and improved quality of produce in diverse cultivation systems.

5.4 Substrate Preparation and Management

Protected cultivation systems such as polyhouses, greenhouses, and shade nets are designed to create favourable conditions for crop production, where substrate management plays a central role. Substrates act as the foundation for plant growth by replacing or supplementing natural soil, offering a cleaner and more controlled medium. Unlike field soils, artificial substrates are free from many soil-borne diseases, pests, and weeds, thereby ensuring healthier crop establishment. They provide stable root anchorage, which is essential for supporting vigorous plant growth and preventing lodging. Properly prepared substrates also regulate moisture by retaining sufficient water for plant use while simultaneously allowing excess drainage to avoid waterlogging. Their porous structure improves aeration, enabling roots to breathe and facilitating beneficial microbial activity. Additionally, substrates help in balancing nutrient supply, ensuring that essential elements are readily available for uptake. In high-value horticultural crops such as vegetables, flowers, and fruit seedlings, this precise control enhances productivity and quality. Different materials like cocopeat, perlite, vermiculite, peat moss, sand, and compost are blended to achieve the desired balance of properties. Each substrate offers unique benefits–some excel in moisture retention, others in aeration, and some in nutrient-holding capacity. The choice of substrate depends on crop type, stage of growth, and management practices adopted in the protected structure. Through careful preparation and management, substrates significantly reduce production risks and maximize yield potential. They also support year-round cultivation, making horticulture more profitable and sustainable. Overall, substrate management is not merely a technical step but a key strategy for successful protected cultivation systems.

5.4.1. Selection of Substrate

The selection of a suitable substrate is one of the most important steps in protected cultivation, as it directly influences plant establishment, growth, and productivity. A good substrate must provide an ideal physical, chemical, and biological environment for the root system. Substrates should always be free from pathogens, insect pests, nematodes, and weed seeds, since these can severely affect plant health in the confined space of protected structures. The physical properties of the substrate, such as porosity, bulk density, and texture, determine how well the roots can anchor and access air and water. Sufficient porosity ensures that roots receive adequate oxygen for respiration, while proper water-holding capacity helps maintain continuous moisture availability without causing waterlogging. Chemically, the substrate should have a balanced pH, usually between 5.5 and 7.0, which allows optimum nutrient uptake by most horticultural crops. Electrical conductivity (EC) should also be within safe limits to avoid salt stress. Biologically, the substrate should support beneficial microorganisms that enhance nutrient cycling and suppress harmful pathogens. Common substrates used in protected cultivation include soil, cocopeat, perlite, vermiculite, rockwool, peat moss, sand, sawdust, compost, and various combinations of these materials. Soil, though traditional, requires sterilization before use to eliminate diseases and weeds. Cocopeat is lightweight, renewable, and excellent in water retention. Perlite and vermiculite are inert minerals that improve aeration and moisture balance. Rockwool provides uniform structure and high water-holding capacity, making it popular in hydroponics. Peat moss is valued for its lightweight and acidic nature, while sand is often added to improve drainage. Sawdust and compost serve as organic components that add nutrients and enhance microbial activity. Often, mixtures of two or more substrates are preferred, as they combine the strengths of different materials to create an ideal growing medium. Thus, careful selection of substrate, based on crop requirements, environmental conditions, and management practices, is essential to achieve high yield and quality in protected cultivation.

5.4.2. Pre-treatment and Sterilization

Pre-treatment and sterilization of substrates are essential steps in protected cultivation to ensure that the growing medium remains free from harmful organisms and chemical imbalances that could affect crop performance. Soil-based substrates, though naturally rich in minerals and microbial life, often harbor pathogens, nematodes, weed seeds, and insect pests. Therefore, sterilization is carried out before use. One common method is solarization, where moist soil is covered with transparent polyethylene sheets for 4–6 weeks under direct sunlight, raising the temperature to lethal levels for many pathogens and weeds. Steam sterilization is another effective practice, where soil or substrate is exposed to high-temperature steam, killing harmful fungi, bacteria, and nematodes without leaving toxic residues. In some cases, chemical sterilants like formaldehyde are used, though their application must be followed by aeration to avoid phytotoxic effects on plants. For soilless substrates, different treatments are required. Cocopeat, widely used in nurseries and hydroponics, contains naturally high levels of salts, particularly sodium and potassium, which can interfere with nutrient uptake. To correct this, cocopeat is thoroughly washed and buffered with calcium or magnesium solutions to bring the electrical conductivity (EC) to safe levels. Rockwool and perlite, being inert and sterile by manufacture, usually do not require elaborate sterilization. However, they should be rinsed before use to remove fine dust particles and to stabilize their pH within the desirable range of 5.5–6.5. This step also ensures uniform wetting and reduces initial hydrophobicity in rockwool slabs. Regular pre-treatment not only ensures a disease-free start for seedlings and crops but also enhances uniform germination, strong root growth, and balanced nutrient uptake. In commercial cultivation, sterilization practices are strictly followed to minimize risks, reduce chemical pesticide use, and maintain sustainability. Therefore, careful pre-treatment and sterilization of substrates form the backbone of healthy and productive horticultural systems under protected conditions.

5.4.3. Substrate Mixing

Substrate mixing is a crucial step in preparing a balanced growing medium for protected cultivation, as no single substrate can provide all the desired properties required for optimum plant growth. Individual substrates may excel in certain aspects such as water retention, aeration, or nutrient supply, but blending them together helps in overcoming their limitations and creating a more balanced medium. For instance, cocopeat is known for its high water-holding capacity but may sometimes restrict aeration; when combined with perlite, which is highly porous, the mix provides both adequate moisture and air space for roots. A widely used mixture in vegetable and flower cultivation is Cocopeat + Perlite in a 3:1 ratio, which supports vigorous root growth and uniform plant development. Similarly, in nursery beds, a traditional mixture of Soil + Sand + Farmyard Manure (2:1:1) is preferred, as it provides structural support from soil, drainage improvement from sand, and organic nutrients from FYM. For delicate seedlings grown in trays, a lighter medium is essential to ensure easy root penetration and balanced water-air dynamics. A common choice is Vermiculite + Perlite + Cocopeat in equal proportions (1:1:1), as this blend ensures excellent germination, soft texture, and optimum nutrient availability during the early growth stages. The process of mixing must be done carefully to achieve proper homogenization, which ensures that every part of the substrate has a uniform distribution of aeration, moisture, and nutrients. Uneven mixing can lead to irregular growth, poor seedling establishment, or localized nutrient deficiencies. In commercial practice, mixing is often done on clean surfaces or with mechanical mixers to save labour and achieve consistency. Balanced substrate mixtures not only support healthy root development but also reduce plant stress, improve transplant success, and enhance overall yield quality. By tailoring specific blends to crop type, growth stage, and environmental conditions, farmers and nursery managers can optimize resource use and productivity. Thus, substrate mixing serves as a scientific and practical approach to ensure a reliable growing medium in protected cultivation systems.

- Substrates are often blended to improve structure and nutrient balance.
- Typical mixes:
- Cocopeat + Perlite (3:1) for vegetables and flowers.
- Soil + Sand + FYM (2:1:1) for nursery beds.
- Vermiculite + Perlite + Cocopeat (1:1:1) for seedling trays.
- Proper homogenization ensures uniform aeration, moisture, and nutrient distribution.

5.4.4. Bed and Container Preparation

Bed and container preparation is a vital step in protected cultivation, as it creates the immediate environment where roots develop and crops establish. Depending on the crop type and production system, substrates may be placed in raised beds, grow bags, nursery trays, or pots. Raised beds are commonly used in soil-based systems inside polyhouses, as they improve drainage, prevent water stagnation, and make intercultural operations easier. The beds are usually 30–45 cm high and 60–120 cm wide, designed to facilitate aeration and root penetration. For soilless cultivation, grow bags filled with cocopeat, perlite, or substrate mixtures are widely adopted, especially for crops like tomato, cucumber, and capsicum. Similarly, seedling trays made of plastic are filled with fine-textured substrate mixtures such as cocopeat, vermiculite, and perlite, which provide the ideal conditions for uniform germination and easy transplanting. Pots or containers are used for flowers, ornamentals, and small-scale vegetable cultivation, offering flexibility in handling and arrangement. Once the substrate is filled, the surface is levelled and lightly compacted to ensure uniform distribution and to prevent the formation of air pockets. Over-compaction is avoided, as it reduces aeration and hampers root growth, whereas loose filling can lead to uneven water distribution and poor anchorage. After preparation, irrigation and nutrient delivery systems are integrated. Drip irrigation lines are placed along the beds or inserted into grow bags and pots to deliver water directly to the root zone, minimizing wastage. In advanced systems, fertigation units are attached to supply dissolved fertilizers along with irrigation water, ensuring precise nutrient availability. This integration is essential for substrates that have little or no inherent nutrient content, such as perlite, rockwool, or cocopeat. Properly prepared beds and containers ensure that plants start with a uniform and healthy growing medium, receive balanced water and nutrients, and remain free from waterlogging or nutrient stress. Thus, this step forms the structural and functional base of successful protected cultivation practices.

5.4.5. Nutrient Management in Substrates

Nutrient management in substrates is one of the most critical aspects of protected cultivation, as many substrates used in modern systems are inert and contain little to no inherent fertility. Materials such as cocopeat, perlite, vermiculite, sand, or rockwool primarily serve as support structures for the root system and provide aeration and water retention, but they cannot independently meet the nutritional needs of crops. To overcome this limitation, nutrients are supplied externally, mostly in the form of water-soluble fertilizers delivered through fertigation systems. This allows precise and timely application of essential macro- and micronutrients directly to the root zone, ensuring efficient uptake by plants. For successful fertigation, the pH and Electrical Conductivity (EC) of the substrate solution must be carefully monitored. The optimum pH generally ranges between 5.5 and 6.5 for most horticultural crops, allowing maximum nutrient availability. Deviations from this range may cause deficiencies or toxicities; for example, iron deficiency at high pH or calcium deficiency at low pH. Similarly, EC reflects the concentration of soluble salts in the root zone, and maintaining it within safe limits (usually 1.5–2.5 dS/m for vegetables) is essential to prevent salt stress. Excessive EC can result in stunted growth, leaf burn, or poor fruit quality, while very low EC indicates nutrient deficiency. Apart from chemical fertilizers, the integration of organic amendments into substrate management helps improve sustainability. Materials such as vermicompost, compost, biofertilizers, and compost teas not only provide supplementary nutrients but also enhance microbial activity in the root zone. These beneficial microorganisms help in nutrient mineralization, disease suppression, and improved root health. In some systems, slow-release fertilizers and biofertilizer inoculants are added to reduce dependency on synthetic inputs. Efficient nutrient management ensures uniform plant growth, higher productivity, and better quality of produce in protected cultivation. By balancing fertigation with organic inputs and maintaining ideal pH and EC levels, farmers can optimize crop performance while maintaining soil and environmental health.

5.4.6. Moisture and Aeration Management

Moisture and aeration management are vital components of substrate-based protected cultivation, as they directly influence root health, nutrient uptake, and overall plant performance. Since different substrates vary in their water-holding capacity and porosity, irrigation must be carefully scheduled based on crop stage, substrate type, and prevailing climatic conditions. Young seedlings generally require frequent but light irrigation to maintain consistent moisture around delicate roots, while mature plants need deeper and more regulated watering to support higher transpiration and nutrient demand. Automated systems like drip irrigation and micro-sprinklers are widely used to achieve precision and minimize water wastage. One of the most common mistakes in substrate cultivation is over-irrigation, which can lead to nutrient leaching, loss of fertilizers, and development of anaerobic conditions around the root zone. This deprives roots of oxygen, reduces microbial activity, and predisposes plants to root diseases such as Pythium and Phytophthora. On the other hand, under-irrigation causes water stress, wilting, reduced photosynthesis, and poor fruit set, ultimately lowering yield and quality. Therefore, maintaining an optimum moisture balance is essential. Aeration is equally important because roots require oxygen for respiration. Substrates such as cocopeat, vermiculite, and peat moss retain good amounts of water but may restrict aeration if not mixed with porous materials. In contrast, substrates like sand, gravel, and perlite provide excellent aeration but drain rapidly, making it essential to ensure frequent irrigation and proper nutrient supply. Good drainage systems are necessary to prevent water stagnation and salt buildup, especially in inert media. Periodic flushing with clean water is often practiced in hydroponic systems to remove excess salts and restore balance in the root zone. Maintaining the right balance of moisture and aeration ensures that crops receive adequate water and oxygen simultaneously, promoting vigorous root growth, efficient nutrient absorption, and high productivity. Thus, precise irrigation management combined with careful substrate selection forms the backbone of sustainable and profitable protected cultivation

5.4.7. Monitoring and Maintenance

Monitoring and maintenance of substrates in protected cultivation are continuous processes that ensure long-term productivity, sustainability, and crop health. Since substrates provide the foundation for root development, any imbalance in their physical, chemical, or biological properties can directly affect plant performance. Therefore, regular monitoring of microbial health, nutrient availability, and salt accumulation is essential. Microbial health assessment helps in maintaining a balance between beneficial and harmful organisms; beneficial microbes improve nutrient cycling and disease suppression, while pathogenic microbes may cause root rots or wilt diseases if left unchecked. In soilless systems, excessive use of water-soluble fertilizers often leads to salt accumulation in the root zone. High salinity can interfere with water uptake, cause nutrient imbalances, and reduce plant Vigor. To prevent this, growers practice periodic flushing with clean water, which washes away accumulated salts and restores the electrical conductivity (EC) of the substrate to safe levels. This is particularly important in substrates such as cocopeat, perlite, and rockwool, which are prone to salt buildup due to their inert nature. Another important aspect of maintenance is the reuse and recycling of substrates. Materials like rockwool and perlite can be reused for multiple crop cycles, but only after thorough sterilization to eliminate residual pathogens and contaminants. However, caution must be exercised, as improper sterilization may carry forward diseases to the next crop cycle. Cocopeat and compost-based mixes can also be partially recycled, provided they are refreshed with nutrients and sterilized if necessary. Proper disposal of worn-out substrates is equally important to avoid environmental pollution. Continuous monitoring of parameters such as pH, EC, drainage, and root zone temperature further ensures stable growing conditions. By combining regular observation with corrective maintenance practices, farmers can prevent crop failures, reduce input losses, and improve the efficiency of protected cultivation systems. Hence, monitoring and maintenance act as the backbone of sustainable substrate management in horticultural crop production. Proper substrate preparation and management are vital for successful protected cultivation. A well-prepared substrate ensures optimum root health, balanced nutrition, efficient water use, and higher productivity of horticultural crops under controlled environments.

Benches and Containers in Protected Cultivation

6.0 Introduction: An Overview

Protected cultivation has emerged as one of the most advanced and sustainable approaches in modern agriculture, enabling growers to produce high-quality crops under controlled environmental conditions. Within this system, benches and containers play a vital role in ensuring optimum plant growth, efficient space utilization, and effective resource management. Benches serve as elevated platforms that provide better aeration, drainage, and hygiene for plants, while containers act as flexible growing units that can be

For successful protected cultivation, the careful selection of suitable structures, benches, and containers is of utmost importance. These components must be chosen based on the type of crops to be grown, their growth habit, duration, and management requirements. Durable and easily accessible benches should be prioritized to ensure efficient use of space and smooth cultural operations. Containers must be selected according to the vigour, root system, and cropping duration of the plants to support healthy growth. Choosing appropriate systems not only reduces production costs but also improves productivity and crop quality. When combined with the cultivation of high-value crops, these decisions can significantly enhance profitability. Thus, proper planning in the selection of infrastructure ensures maximum returns for farmers under protected cultivation.

designed in various sizes and materials to suit different crops and cultivation systems. Together, they form the structural backbone of protected cultivation practices, particularly in polyhouses, greenhouses, shade-net houses, and other controlled environments. The use of benches and containers allows for systematic arrangement of plants, leading to uniform growth, easy monitoring, and reduced risks of soil-borne diseases. Elevated benches help in improving accessibility for operations like irrigation, pruning, spraying, and harvesting, thus minimizing labour fatigue and maximizing efficiency. Containers, on the other hand, provide growers with the flexibility to cultivate crops in soilless media such as cocopeat, perlite, or vermiculite, enabling precise control over nutrient and moisture supply. This makes them especially useful in hydroponics and other advanced cultivation methods. Another advantage of benches and containers is their contribution to maintaining cleanliness and sanitation within protected structures. Since plants are grown above ground level, the chances of weed infestation, waterlogging, and pest attack are significantly reduced. The modularity of containers also supports crop rotation, movement, and replacement without disturbing the overall setup. Moreover, different types of benches–such as fixed benches, rolling benches, and tiered benches–cater to diverse production requirements, from ornamental plants to vegetables and medicinal crops. In commercial horticulture, the choice of suitable benches and containers is guided by factors such as crop type, available space, investment capacity, and cultivation objectives. For instance, cut flowers like gerbera and carnation often require elevated benches with precise fertigation systems, whereas leafy vegetables can be successfully grown in containers arranged in vertical stacks. With the rising emphasis on resource efficiency and sustainable farming, the role of benches and containers has become more critical than ever before. Thus, benches and containers not only enhance productivity and quality but also support eco-friendly and scientific cultivation practices. They bridge the gap between traditional soil-based farming and modern protected agriculture, making them indispensable components in the journey toward higher yields, better crop management, and year-round production.

6.1 Protected Structures

In protected cultivation, different types of protected structures are designed and constructed to create a controlled or semi-controlled environment for crop production. These structures are developed according to location-specific factors such as temperature, rainfall, humidity, wind velocity, and solar radiation. The choice of structure also depends on the climatic conditions of the region, ranging from arid zones to high rainfall or cold regions. Protected structures provide a favourable microclimate for crops, enabling off-season and year-round cultivation of high-value horticultural produce. They safeguard plants from adverse weather conditions like heavy rain, frost, hail, and scorching heat. By modifying light, temperature, and humidity, these structures enhance crop productivity and quality. Depending on the requirement, structures may range from low-cost, naturally ventilated polyhouses to highly sophisticated, climate-controlled greenhouses. They are also customized according to the crop to be grown, whether vegetables, flowers, fruits, or nursery plants. Thus, protected structures serve as the backbone of modern horticultural practices, ensuring higher income, efficient resource use, and sustainable production. In protected cultivation of horticultural crops (polyhouses, greenhouses, shade net houses, *etc.*), benches are widely used for raising nursery plants, cut flowers, potted ornamentals, vegetables, and tissue culture-raised seedlings. Benches help in better space utilization, facilitate management, and improve microclimate control. The major types of benches are:

6.1.1 Walk-in-Tunnels

Walk-in tunnels are simple and economical protected structures, widely used in modern horticultural production systems (Figure 6.1). They are generally constructed in an arc or semicircular shape with a height of about 2.0 to 2.5 meters at the centre and an average width of around 4 meters. Due to their convenient size and design, these tunnels can easily accommodate two to three vegetable beds, making them highly suitable for small and medium-scale growers. Walk-in tunnels are particularly ideal for cultivating low-canopy crops such as capsicum, lettuce, spinach, coriander, and bush-type beans, which perform well under moderate protection.

Figure 6.1: Walk-in Tunnels.
(*Source*: www.freepik.com)

They are also extensively used for raising healthy seedlings in nurseries, ensuring protection from excessive rainfall, wind, and pests during the early growth stages. Additionally, these tunnels serve as effective units for the hardening and acclimatization of tissue culture plants and other delicate planting materials. Their controlled environment creates favourable microclimatic conditions, helping in better germination, uniform growth, and higher survival rates of seedlings. Another significant advantage is their use in the off-season cultivation of cucurbits like melons, cucumber, and bottle gourd, allowing farmers to fetch premium market prices. The structures are generally low-cost, easy to install, and require minimal maintenance, which makes them accessible even to resource-poor farmers. They offer partial protection from frost, hail, and scorching sunlight, thereby extending the growing season. Walk-in tunnels also help in reducing irrigation requirements and allow efficient utilization of fertilizers and plant protection chemicals. Being naturally ventilated, they provide good air circulation, minimizing the chances of disease incidence. Their flexibility and portability make them highly adaptable for different cropping systems. Overall, walk-in tunnels are a cost-effective and practical solution for enhancing productivity, improving crop quality, and achieving sustainable horticultural production under protected cultivation.

6.1.2 Plastic Low Tunnels/Row Covers

Plastic low tunnels, also known as row covers, are an important component of protected cultivation practices in horticultural crops. These structures consist of a transparent polyethylene sheet, usually 30–50 micron thick, stretched over a simple steel frame of about 50 cm height and 1 m width. They act as miniature greenhouses, covering the crop rows directly in open field conditions. The primary objective of these tunnels is to modify the microclimate around plants by trapping solar radiation during the day, thereby raising the temperature inside. This warmth not only supports better germination and early seedling establishment but also provides protection against frost injury during cold nights. Plastic low tunnels are particularly useful in temperate and subtropical regions where low temperatures and frost otherwise restrict crop growth. By creating a favourable micro-environment, these tunnels enable early cultivation and extend the growing season of several vegetables. Crops such as cucurbits, melons, beans, and leafy vegetables respond positively to this technique. Studies have shown that low tunnels hasten crop growth, resulting in early flowering, fruit set, and harvesting. This early maturity allows farmers to capture premium market prices, particularly when off-season vegetables are in high demand.

In snap beans, cultivation under low tunnels has demonstrated significant improvement in plant height, number of leaves, and overall yield compared to open-field conditions. The technique allows successful spring production of beans without the need for expensive heating systems. Low-cost construction materials and easy installation make these tunnels highly economical and farmer-friendly. In addition to plastic cover, integration with plastic mulch on the soil surface further conserves soil moisture, suppresses weeds, and enhances temperature regulation. Moreover, scientific results suggest that foliar spraying of protective chemicals like calcium chloride ($CaCl_2$) or potassium chloride (KCl) enhances the stress-tolerance capacity of plants grown under tunnels. These sprays help in mitigating chilling injury and contribute to the synergistic effect of tunnels and mulching in safeguarding crops. Thus, plastic low tunnels serve as a low-cost, effective, and sustainable technique to improve productivity, ensure early harvest, and reduce climatic risks in vegetable production systems (Figure 6.2).

Figure 6.2: Plastic Low Tunnels/Row Covers.
(*Source*: www.freepik.com)

6.1.3 Net Houses

Net houses are specialized protective structures widely adopted in horticultural crop production to create a favourable microclimate and safeguard crops from adverse biotic and abiotic factors. Broadly, they are of two types–shade nets and insect-proof nets, each serving distinct purposes in crop management. Shade nets are primarily used to reduce the intensity of direct sunlight falling on plants, thereby preventing leaf scorching, excessive transpiration, and wilting during hot weather. These nets are available in different colours such as black, green, and white, with shading intensities ranging from 25 per cent to 75 per cent, depending on the crop requirement and season. Insect-proof nets, on the other hand, are designed with specific mesh sizes ranging between 25 and 60 mesh to restrict the entry of insect pests. A mesh size of 40 or higher is generally sufficient to control most common flying pests, including aphids, whiteflies, and thrips. These nets not only reduce pest incidence but also minimize the need for repeated chemical sprays, thereby promoting eco-friendly cultivation practices. In addition, net houses are extensively used for controlled pollination, which is highly beneficial in hybrid seed production and maintaining genetic purity of planting material.

Net houses have proven particularly successful in the cultivation of vegetables such as tomato, brinjal, and capsicum. For instance, under Punjab conditions, off-season tomato cultivation in net houses extends the availability of fresh tomatoes from the last week of January to the first week of June. Trials have shown a significant reduction in fruit damage caused by *Spodoptera litura*, with complete absence of *Helicoverpa armigera* and *Aphis gossypii*, pests that are otherwise highly destructive in open-field cultivation. This not only ensures higher yields but also improves the quality and shelf life of harvested produce. In brinjal cultivation, net houses have been found to enhance marketable yield while providing effective control over fruit and shoot borer infestation. Similarly, net house technology is highly advantageous in virus-free potato seed production, where insect-proof nets prevent the entry of virus vectors, ensuring healthy planting material for future generations. By modifying temperature, reducing pest load, and preventing sun injury, net houses offer a low-cost, efficient, and sustainable technology for improving productivity and quality in horticultural crops (Figure 6.3).

Figure 6.3: Net Houses
(*Source*: www.freepik.com)

6.1.4 Plastic Mulch

Plastic mulch is an important component of modern protected cultivation practices, where the soil surface around the base of plants is covered with a thin layer of plastic film. This practice, known as mulching, serves multiple agronomic purposes, making it highly beneficial in vegetable and fruit production systems. The primary advantage of plastic mulch is moisture conservation, as it reduces evaporation losses from the soil surface and ensures a more stable supply of water to the root zone. This results in improved water-use efficiency, especially in arid and semi-arid regions. Another significant role of plastic mulch is in weed suppression. By blocking sunlight, black polythene sheets prevent weed seed germination and growth, thereby reducing competition for nutrients, water, and space. In addition, mulching improves soil temperature regulation, allowing faster seed germination and better root development during early stages of crop growth. Enhanced microbial activity and organic matter decomposition beneath the mulch also lead to gradual improvement of soil structure and fertility.

Plastic mulch also facilitates better gaseous exchange in the rhizosphere, particularly the release and diffusion of CO_2, which supports photosynthesis and overall plant vigour. In fruit-bearing crops, mulching prevents direct contact of fruits with soil, thus reducing rotting, soiling, and microbial contamination, leading to cleaner and market-preferred produce. This is particularly valuable in crops like strawberry, watermelon, and tomato. Different coloured mulches are used for specific purposes. Silver and yellow-coloured films are effective in repelling insect pests such as aphids, whiteflies, and thrips, which act as vectors of viral diseases. Black polythene mulch remains the most popular in vegetable cultivation due to its opacity, which ensures effective weed control and soil warming. Transparent mulch, on the other hand, is sometimes used for soil solarization to kill harmful soil-borne pathogens before planting. Thus, plastic mulching not only enhances crop growth and yield but also reduces the reliance on herbicides and pesticides, making it a cost-effective and eco-friendly technique. Its adoption has become an integral part of precision farming, particularly in high-value horticultural crops where quality, cleanliness, and early maturity are critical for profitability (Figure 6.4).

Figure 6.4: Plastic Mulch.
(*Source*: www.freepik.com)

6.1.5 Floating Plastic Covers

Floating plastic covers are an effective, low-cost method of protecting vegetable crops from frost, snow, and chilling injury under open-field conditions. These covers are made of lightweight, transparent polyethylene sheets that are directly spread over the crop canopy without the support of frames. The covers "float" on the plants, allowing flexibility and easy handling while creating a protective barrier between the crop and the harsh external environment. The primary function of floating covers is to trap solar radiation during the day and retain the accumulated heat during night hours. This process significantly increases the air and soil temperature around the plants, thereby minimizing frost damage and promoting better growth during cold spells. Unlike conventional low tunnels or greenhouses, floating plastic covers are simple to install, easy to remove, and reusable, making them highly suitable for small and marginal farmers.

These covers are especially valuable in the early establishment stages of sensitive crops like tomato, cucumber, beans, lettuce, and leafy greens, which are highly prone to frost injury. By creating a warmer microclimate, floating covers not only protect plants but also promote faster germination, quicker seedling growth, and early maturity of crops. The use of such covers helps in advancing the harvesting season, enabling farmers to secure premium prices in the market. In addition to temperature regulation, floating plastic covers reduce direct damage from snow accumulation and prevent desiccation of plants caused by cold winds. The covers also reduce soil moisture evaporation, thus improving water-use efficiency. Since they are lightweight, the covers do not hinder normal plant growth and can be lifted temporarily during cultural practices like irrigation or weeding. Overall, floating plastic covers represent a simple yet highly effective protective technology that extends the growing season, improves crop quality, and reduces production risks in cold-prone areas. Their adoption has become increasingly popular in regions facing frequent frost events, providing farmers with a practical solution for sustainable vegetable production in open fields (Figure 6.5).

Figure 6.5: Fresh Produce in a Plastic Bag Tomatoes and Greens.
(*Source*: www.freepik.com)

6.1.6 Soil Trenches

Soil trenches, also known as underground solar greenhouses, represent one of the cheapest and most efficient structures for growing vegetables during extreme winter conditions. The technique involves digging a rectangular trench into the soil, which acts as a naturally insulated chamber for plant growth. These structures are especially popular in cold desert regions like Ladakh, where sub-zero temperatures make conventional open-field cultivation almost impossible during winter. A typical soil trench is about 5–6 m in width and 2–3 m in depth, with its orientation kept along the north–south direction to maximize exposure to sunlight throughout the day. The design usually follows a dimension of around 30 feet (length) × 10 feet (width) × 3 feet (depth). The four walls are constructed with locally available stone masonry, which provides stability and helps in retaining the absorbed solar heat.

The basic principle of soil trenches relies on the combined effect of solar radiation during the daytime and soil heat storage at night. The thick stone walls and the earth surrounding the trench act as thermal buffers, trapping heat energy and maintaining a comparatively warmer microclimate inside. This ensures optimum temperatures for vegetable growth even when outside temperatures remain freezing. Inside the trench, a furrow of about 1.5 feet width is made along the east wall to facilitate irrigation. The growing space is further divided into 6–8 plots, which makes crop management and water distribution more efficient. Crops like spinach, coriander, fenugreek, and leafy vegetables thrive particularly well under this system, while hardy vegetables such as carrot, radish, and turnip can also be grown successfully. Soil trenches offer multiple benefits: they are low-cost, require minimal construction materials, conserve heat naturally, and utilize local resources effectively. They extend the growing season, provide fresh vegetables to cold desert populations during harsh winters, and reduce dependency on outside supplies. Due to their simplicity, affordability, and adaptability, underground solar greenhouses have emerged as a sustainable solution for ensuring year-round vegetable availability in high-altitude cold desert ecosystems (Figure 6.6).

Figure 6.6: Soil Trenches
(*Source*: www.freepik.com)

6.1.7 Hot Beds

Hot beds are one of the oldest and simplest protected cultivation techniques, widely used to raise vegetable seedlings even under sub-zero temperature conditions. The basic principle behind hot beds is the utilization of biological heat generated during the decomposition of organic matter, primarily cattle dung, mixed with straw or other crop residues. This natural heating mechanism provides a warm and favourable microclimate for seed germination and seedling growth during harsh winters. To prepare a hot bed, alternating layers of half-rotten farmyard manure or dung and dry straw are spread on the ground, usually within a wooden or brick frame to maintain shape and insulation. As the organic matter begins to decompose, microbial activity generates significant heat, which raises the temperature of the bed. A thin layer of fertile soil is then spread over the top of the heap to serve as the seedbed, where vegetable seeds are directly sown.

Hot beds are particularly suitable for off-season nursery raising of vegetables such as tomato, brinjal, chili, and early cucurbits, where strong and healthy seedlings are required before transplanting into the main field or protected structures. This technique allows farmers in cold regions to advance their sowing schedule by several weeks, ensuring early crop availability and better market prices. In addition to temperature regulation, hot beds improve moisture retention and nutrient availability due to the gradual decomposition of organic matter. They also provide partial protection against frost and snow when covered with transparent plastic sheets or glass frames, further enhancing the efficiency of the system. Hot beds are low-cost, farmer-friendly, and can be prepared with locally available materials, making them highly popular among small and marginal farmers. By combining traditional knowledge with simple scientific principles, hot beds serve as an eco-friendly and sustainable method of nursery raising, ensuring a reliable supply of seedlings during unfavourable seasons (Figure 6.7).

Figure 6.7: Hot Beds
(*Source*: www.freepik.com)

6.1.8 Greenhouses

A greenhouse is a framed or inflated structure, covered with a transparent or translucent material such as polyethylene, polycarbonate, or glass, designed to create a favourable environment for plant growth. The primary purpose of a greenhouse is to modify and control the microclimate around crops, thereby enabling higher productivity and quality compared to open-field conditions. The covering material allows the penetration of solar radiation while reducing heat loss, which helps in maintaining optimum temperature, humidity, and light conditions inside. Greenhouses are large enough to allow a person to easily enter, move around, and carry out various horticultural operations such as sowing, irrigation, training, pruning, spraying, and harvesting. This makes them highly practical for commercial-scale production. Depending on design, they may be naturally ventilated or equipped with climate-control systems that regulate temperature, humidity, carbon dioxide concentration, and light intensity. Such modifications ensure that crops receive the most suitable growth conditions throughout their life cycle. The flexibility of greenhouses lies in their ability to alter the microclimate as per crop requirements. For instance, heating systems can be used in cold climates, while evaporative cooling systems may be employed in hot regions. Supplementary lighting, CO_2 enrichment, and fertigation can further enhance crop growth and yield.

Greenhouse technology is particularly valuable for achieving off-season cultivation of high-value crops such as tomato, capsicum, cucumber, lettuce, and flowers like gerbera, rose, and carnation. By regulating environmental factors, greenhouses allow round-the-year production and help farmers target premium markets. Additionally, greenhouse cultivation ensures better resource-use efficiency, reduces crop losses from adverse weather, and protects plants from pests and diseases when combined with insect-proof netting. It also facilitates the adoption of modern production practices such as hydroponics, vertical farming, and precision irrigation. Thus, greenhouse technology is considered the most effective method of protected cultivation, ensuring higher yields, better quality, and year-round availability of vegetables, fruits, and flowers. It has become an essential tool for modern horticulture, particularly in regions where climatic extremes limit open-field production.

Principle of Greenhouse

The working of a greenhouse is based on the fundamental principle of the greenhouse effect, a natural process that helps maintain a warmer environment inside the structure compared to the outside atmosphere. When solar radiation falls on the greenhouse, a large proportion of the shortwave radiation passes through the transparent or translucent cladding material, such as polyethylene film, polycarbonate sheets, or glass. This transmitted radiation is absorbed by the soil, plants, and other internal surfaces inside the greenhouse. These internal surfaces, after absorbing solar energy, re-radiate heat in the form of longwave infrared radiation. However, the cladding material has low transparency to infrared rays, which means that most of this heat cannot escape back into the atmosphere. As a result, heat becomes trapped inside the structure, leading to a gradual rise in the internal air temperature compared to the outside. This phenomenon creates a warmer and more stable microclimate, suitable for plant growth even under harsh external conditions. In regions with severe winters, cold arid deserts, or high hills, where outdoor cultivation is extremely challenging, this temperature rise of about 10–12 °C above ambient conditions becomes highly beneficial. The elevated temperature not only prevents frost injury but also supports the germination, growth, and flowering of temperature-sensitive crops.

This principle of controlled heat retention enables greenhouses to provide an extended growing season, facilitate off-season production, and ensure consistent yield and quality. The degree of heating inside depends on factors such as cladding material transmittance, greenhouse design, ventilation, and insulation efficiency. Modern greenhouses often integrate additional heating, cooling, and shading systems to optimize this natural effect and maintain desired crop-specific microclimates. Thus, the principle of the greenhouse effect forms the scientific basis of greenhouse technology, allowing it to serve as one of the most reliable methods of protected cultivation for vegetables, flowers, and fruits in diverse agro-climatic conditions.

Attributes of Greenhouse

There are four important components in a green house. These are:

- ☆ **Framed or Inflated Structure:** A greenhouse is essentially a framed or inflated structure that provides mechanical support to the covering material. The frame may be constructed from materials like galvanized iron (GI) pipes, aluminium, or bamboo, depending on cost and durability requirements. In some cases, double-layer inflated plastic films are used to improve insulation. The strength and stability of the structure are critical for withstanding wind, snow, and other environmental stresses while providing a secure environment for crop growth.
- ☆ **Transparent or Translucent Cladding Material:** The covering material of a greenhouse plays a key role in regulating the entry of solar radiation. Transparent or translucent materials such as polyethylene, polycarbonate, or glass are used to ensure optimum light penetration for photosynthesis. These materials allow shortwave solar radiation to enter while restricting longwave infrared radiation from escaping, thus maintaining warmth inside. The choice of cladding material depends on local climate, durability, cost, and crop requirements.
- ☆ **Partially or Completely Controlled Micro-Climate:** One of the most important attributes of a greenhouse is its ability to create and regulate a micro-climate around the crops. Factors such as temperature, humidity, ventilation, carbon dioxide concentration, and light intensity can be modified to suit the physiological needs of plants. In simple greenhouses, natural ventilation and shading are used, while in advanced structures, automated systems for heating, cooling, irrigation, and fertigation are installed. This controlled environment ensures higher productivity and quality irrespective of external weather conditions.

- ☆ **Adequate Size for Human Operations:** A greenhouse must be large enough to allow a person to enter and carry out routine intercultural operations like irrigation, pruning, training, weeding, fertilization, and harvesting. Proper height and width are important not only for ease of operations but also for efficient air circulation and uniform distribution of heat and light. Walkways, bed arrangements, and working space are designed to facilitate labour efficiency and crop management without causing damage to plants.

Advantages of Greenhouse Technology

- ☆ **Year-round Cultivation of Crops**: Greenhouses provide a controlled micro-climate, enabling the production of vegetables, flowers, and fruits throughout the year, irrespective of external climatic conditions. This ensures a continuous supply of produce and stabilizes market availability.
- ☆ **Higher Productivity Compared to Open-field Cultivation**: Due to regulated temperature, humidity, and CO_2 levels, crops grown in greenhouses exhibit faster growth, higher yields, and better resource-use efficiency compared to conventional open-field systems.
- ☆ **Higher Income for Farmers**: Off-season cultivation and the ability to produce high-value crops lead to premium market prices, thereby enhancing farmers' profitability and economic sustainability.
- ☆ **Good Quality Produce**: The controlled conditions reduce exposure to dust, heavy rains, and soil splashes, resulting in cleaner, more uniform, and superior-quality produce with longer shelf life.
- ☆ **High-quality Virus-free Seed Production**: Greenhouses with insect-proof netting help in maintaining pest- and vector-free environments, making them highly suitable for virus-free seed production of vegetables and other crops.
- ☆ **Effective Control of Diseases and Pests**: By restricting insect entry and enabling better sanitation, the greenhouse environment significantly reduces the incidence of pests and diseases, minimizing the need for chemical pesticides.
- ☆ **Higher Percentage of Seed Germination**: The favourable temperature and humidity levels inside a greenhouse create ideal conditions for seed germination, leading to higher and more uniform seedling establishment.
- ☆ **Hardening and Acclimatization of Tissue-cultured Plantlets**: Greenhouses serve as an intermediate stage for gradually exposing tissue-cultured plantlets to external conditions, ensuring better survival rates and adaptability after field transplantation.
- ☆ **Suitable for Small Landholding Farmers**: With the potential to produce high yields on limited space, greenhouses are highly suitable for farmers

with small landholdings, enabling them to maximize productivity per unit area.

- **Employment Generation for Rural Youth**: As greenhouse cultivation requires regular care, management, and labour-intensive operations, it provides significant employment opportunities for rural youth, thereby supporting livelihoods.

Disadvantages of Greenhouse Technology

- **High Initial Investment**: One of the major limitations of greenhouse cultivation is the high establishment cost. Construction of frames, cladding material, irrigation, and environmental control systems requires substantial investment, which is often unaffordable for small and marginal farmers without external financial support.
- **Limited Affordability for Poor Farmers**: In India, where the majority of farmers belong to smallholder or resource-poor categories, the high capital requirement restricts the adoption of greenhouse technology, unless subsidies or institutional support are available.
- **Requirement of Skilled Labour**: Greenhouse operations involve technical practices such as fertigation, irrigation scheduling, climate regulation, and pest monitoring, which demand trained and skilled manpower. A lack of skilled labourers often hampers efficient operation.
- **Dependence on Technical Knowledge**: Successful greenhouse management requires proper awareness of plant physiology, protected cultivation techniques, and equipment handling. Farmers lacking adequate training may fail to obtain expected returns from their investment.
- **Lack of Awareness and Education**: In many rural areas, illiteracy and limited exposure to modern farming technologies pose a barrier to greenhouse adoption. Lack of awareness leads to underutilization or mismanagement of greenhouse resources.
- **High Maintenance and Operational Costs**: Greenhouses require continuous maintenance of cladding material, irrigation systems, and ventilation equipment. The recurring costs of replacement, electricity, and skilled supervision add to the financial burden.
- **Vulnerability to System Failures**: In highly automated or climate-controlled greenhouses, failure of cooling or heating systems can result in sudden crop loss, especially during extreme weather conditions.
- **Limited Suitability in Harsh Climates without Add-ons**: In very hot or cold regions, additional heating or cooling arrangements may be necessary, further increasing operational expenses.
- **Market Risks**: While greenhouse cultivation enables off-season production, market saturation or price fluctuations can reduce profitability, particularly for perishable crops like vegetables and flowers.

- ☆ **Dependence on External Inputs**: The system often requires good quality hybrid seeds, fertilizers, and protective chemicals, making farmers dependent on external resources, which may not always be easily accessible.

6.2. Benches

Benches are one of the most important components of a greenhouse, as they provide a stable, elevated, and well-drained platform for growing plants in an organized manner. They help in the systematic arrangement of crops, ensuring maximum utilization of greenhouse space. Raised benches improve aeration around the root zone and facilitate efficient drainage, thereby minimizing the chances of waterlogging and soil-borne diseases. They also make cultural operations like watering, pruning, spraying, and harvesting more convenient and less labour-intensive. Different designs and arrangements of benches, such as longitudinal, peninsular, or movable types, can be adopted depending on the size of the greenhouse and the type of crops grown. Benches also allow growers to segregate plant species easily, which is particularly beneficial in nurseries or small-scale cultivation. Modern benching systems, including rolling or ebb-and-flow benches, enhance resource efficiency by saving water, fertilizers, and labour costs. Their use results in better hygiene inside the greenhouse as plants are kept off the ground, reducing disease incidence. In addition, benches support uniform crop growth and improved crop quality by providing equal exposure to light and air circulation. The adoption of well-designed benching systems increases productivity, ensures comfort for workers, and contributes to sustainable greenhouse management. Thus, benches are indispensable structural elements for efficient and scientific cultivation under protected conditions.

Figure 6.8: Benches of Greenhouse.
(*Source*: www.freepik.com)

Bench Space Efficiency

Bench space efficiency refers to how effectively the floor area of a greenhouse is utilized for plant growth through benches. It is calculated as the ratio of the square feet of bench surface area to the total square feet of the greenhouse floor area, expressed in percentage. Higher benching efficiency indicates better utilization of available space for crop production. Efficiency depends on bench design, arrangement (longitudinal, peninsular, or movable), and aisle spacing. Thus, optimizing bench space efficiency ensures maximum crop yield with minimum structural cost.

Bench Design

Ebb and flow benches, also known as flood or sub-irrigation benches, are modern alternatives to conventional wooden or metal benching systems widely used in protected cultivation. These benches are designed with watertight, moulded plastic trays that facilitate an efficient and uniform irrigation method. At regular intervals, nutrient-enriched water (fertigation solution) is allowed to flood the trays, enabling plants to absorb moisture and nutrients through the capillary action of the growing medium. Once the absorption process is complete, the excess water drains back into a reservoir, making it possible to re-circulate and reuse the solution. This system not only conserves resources but can save up to 50 per cent of water and fertilizers compared to traditional practices. Since all benches are irrigated simultaneously, labour requirements are greatly reduced, improving operational efficiency. Another major advantage is that water is applied directly to the root zone, keeping foliage dry, which minimizes the risk of foliar diseases and enhances overall plant health. The system also ensures uniform distribution of nutrients and water, leading to consistent growth and higher production levels. Additionally, ebb and flow benches reduce leaching losses and prevent contamination of groundwater by recycling the runoff.

Figure 6.9: Dendrobium Orchid Cultivation on Metal Bench.
(*Source*: https://www.researchgate.net)

Their hygienic design also decreases the chances of soil-borne diseases as plants are not in direct contact with the ground. Moreover, these benches improve space utilization, allow cleaner working environments, and provide ergonomic advantages for growers. Thus, the adoption of ebb and flow bench designs not only increases productivity and crop quality but also promotes sustainable resource management, making them highly suitable for commercial greenhouse operations (Figures 6.9 and 6.10).

Figure 6.10: Vertical Cultivation of Leafy Vegetables on Shelves
(*Source*: https://www.researchgate.net)

Common Materials Used for Benches

- **Wooden Benches:** Wooden benches are widely used in greenhouses due to their easy availability, low initial cost, and flexibility in construction. For durability, highly resistant woods such as locust, cedar, redwood, and cypress are preferred, as these species naturally resist decay and insect attack. To further enhance resistance against rotting, benches are coated with preservatives like copper naphthenate, which provide long-term protection from moisture and fungal damage. Wooden benches are lightweight, easy to move, and can be customized in size and shape depending on crop requirements. They are particularly suitable for small-scale growers who frequently change plant arrangements. Despite being cost-effective, wooden benches require periodic maintenance, including repainting or treatment, to maintain their strength and durability. Their biggest advantage lies in their natural insulation property, which helps regulate temperature around plant roots. However, they may not withstand prolonged exposure to excessive humidity without treatment. Overall, wooden benches are economical, flexible, and farmer-friendly, especially for short-term and small-scale greenhouse cultivation.

- **Concrete Benches:** Concrete benches are strong, durable, and permanent structures commonly adopted in large-scale or commercial greenhouses. Unlike wood, concrete is naturally resistant to decay, rot, or insect attack, eliminating the need for additional preservative treatments. These benches provide a stable platform with high load-bearing capacity, making them suitable for heavy pots and long-term use. For added strength and longevity, steel rods are often embedded in the concrete. To ensure proper drainage and prevent water stagnation, holes are drilled at regular intervals across the surface. Although these benches lack flexibility in terms of rearrangement, their robustness makes them ideal for fixed cultivation systems. They also offer a hygienic surface, reducing the chances of disease spread compared to wooden benches. Their maintenance requirement is minimal, and they withstand both environmental stress and chemical exposure during cleaning or fertilization. The only drawback is their rigidity, as they cannot be relocated once installed. Overall, concrete benches are highly durable, maintenance-free, and suitable for permanent greenhouse setups.
- **Metal Benches:** Metal benches, especially those made of steel or galvanized iron, are a modern choice for greenhouse benching systems. They are highly durable, resistant to rot, pests, and decay, making them a reliable alternative to wooden structures. Galvanized metals, in particular, provide long-lasting performance as they resist rusting even in humid greenhouse conditions. Metal benches can be used alone or combined with other materials like plastic trays for better functionality. Although their initial cost of installation is higher compared to wood or concrete, they are considered a one-time investment due to their minimal maintenance needs. Their sleek design and light weight also make them easy to clean and disinfect, promoting better hygiene. Metal benches provide excellent strength and can support heavy pots, making them suitable for both small and large-scale cultivation. They also allow for modern designs like movable or rolling bench systems, maximizing space efficiency. Overall, metal benches combine longevity, strength, and low maintenance, making them an efficient and professional choice for greenhouse cultivation.

Temporary or Portable Benches

- **Plastic:** Plastic benches are lightweight, easy to install, and widely used as temporary benching options in small-scale or short-duration greenhouse setups. They are popular because of their low cost, ease of handling, and resistance to moisture, rot, and corrosion. Plastic benches can be cleaned and disinfected easily, making them suitable for maintaining hygiene in plant nurseries. However, they are less durable compared to wood, metal, or concrete benches and are prone to damage under heavy loads. Their limited strength makes them unsuitable for large pots or commercial-scale cultivation where weight-bearing capacity is essential. Despite these drawbacks, they are useful in experimental setups, small nurseries, or seasonal cultivation practices. Plastic benches are also portable and can be shifted without much effort, providing flexibility to growers. Overall, they are ideal for temporary arrangements but not recommended for long-term or heavy-duty use.

- ☆ **Pallets:** Pallet benches are one of the most inexpensive and practical types of temporary benches, often used by small-scale growers or hobbyists. They are created by placing wooden or plastic pallets over cement blocks to provide a raised and stable platform for plant growth. This simple design requires minimal investment and can be assembled or dismantled quickly, making it highly flexible for short-term use. Pallet benches allow growers to raise plants above ground level, improving drainage and reducing direct soil contact, which helps prevent disease. Their modular nature makes them easy to expand or rearrange based on crop requirements. However, they may not offer the durability or strength needed for heavy loads and can wear out faster with continuous exposure to moisture. Still, they are a cost-effective solution for small nurseries, seasonal crops, or research purposes where temporary arrangements are sufficient. Thus, pallet benches combine affordability, portability, and simplicity, making them suitable for growers with limited resources or short-term needs.

Use of Benches

The use of benches in greenhouses plays a crucial role in simplifying and improving crop management practices. Benches provide elevated platforms that make operations such as irrigation, weeding, hoeing, pruning, spraying, and harvesting much easier and more convenient for workers. They reduce the physical strain on labourers by minimizing the need for constant bending and also improve the efficiency of daily cultural practices. By keeping plants raised above the ground, benches reduce the requirement for costly land preparation, levelling, and soil treatments inside the greenhouse. They also allow for systematic arrangement of plants, leading to efficient utilization of available space and better crop organization. Benches improve drainage and aeration, helping to maintain healthier root zones and reducing the incidence of soil-borne diseases. Moreover, they provide flexibility for segregating different plant species, which is particularly useful in nurseries or multi-crop cultivation. Their use ensures more hygienic conditions, reduces weed growth, and promotes uniform crop development. Before installing benches, aspects like type of material, durability, crop load, irrigation system, and greenhouse design must be carefully considered. Overall, benches not only enhance productivity but also support sustainable, efficient, and scientific management of greenhouse crops.

6.2.1 Types of Benches

Floor as Benches

Floor as benches represent one of the most commonly used and simplest methods for crop cultivation in protected structures. In this system, the floor itself acts as the cultivation bed, eliminating the need for elevated or specially designed benches. The floor is usually cemented and levelled carefully to ensure uniform water distribution, particularly when flood irrigation is practiced. A properly levelled floor prevents water stagnation and ensures even moisture supply across the bed. For improved functionality, porous concrete floors may also be used. These allow partial drainage and aeration, thereby reducing the risk of waterlogging and root diseases. The width of the beds on such floors can be extended depending on the type of crop being cultivated and the available space inside the structure. This makes floor-based benches highly adaptable to a variety of horticultural crops. The main advantage of using floors as benches is their low construction cost compared to raised benches, making them suitable for farmers with limited resources. They are easy to prepare, require minimal materials, and can be used for both vegetable and ornamental crops. For some species with robust root systems and low maintenance requirements, floor benches provide satisfactory growth conditions. However, there are also disadvantages associated with this system. Since crops are grown at ground level, workers are required to bend repeatedly while performing intercultural operations such as sowing, weeding, spraying, and harvesting. This leads to increased physical strain and reduced labour efficiency. Additionally, sanitation management is more difficult at floor level, as soil-borne pathogens and weeds are harder to control. Despite these limitations, floor as benches remain a practical and economical choice, especially in small-scale nurseries, low-cost greenhouses, and in situations where crop species do not demand highly specialized benching systems. With proper design, floor levelling, and water management, this method continues to be widely used in protected cultivation systems (Figure 6.11).

Figure 6.11: Floor as Bench.
(*Source*: www.freepik.com)

Raised Benches

Raised benches in protected cultivation of horticultural crops play an important role in ensuring better management, efficient space utilization, and ease of cultural operations. These benches are arranged in such a way that maximum use of facilities and growing areas can be achieved. Peninsular arrangements of benches allow more growing space as compared to longitudinal ones, although longitudinal benches are more convenient for routine tasks such as watering, harvesting, and inspection. Raised benches also help in easy segregation of plant species, which is highly beneficial for small-scale growers dealing with diverse crops. Recently, non-stationary or movable benches have become more popular among progressive growers because they further increase floor space efficiency up to 90 per cent by allowing rolling movement. These movable benches are mounted on long steel poles and can be shifted easily with the help of a crank, thereby creating temporary aisles for working comfort. Such arrangements provide flexibility in operations and are particularly advantageous where optimum use of land and infrastructure is required. Raised benches also improve aeration around plants, prevent soil-borne diseases by keeping crops away from direct soil contact, and enhance drainage management. However, movable benches may not be suitable in cases where plants are frequently shifted or transferred, hence conventional stationary benches are more appropriate for small-scale nurseries. Overall, raised benches ensure systematic arrangement, better crop supervision, ergonomic advantages for workers, and improved crop quality under protected cultivation (Figure 6.12).

Figure 6.12: Raised Benches.
(Source: www.freepik.com)

6.3 Containers and their Uses

Under greenhouse cultivation, the selection of suitable containers is a critical factor that directly influences plant growth, crop quality, and overall productivity.

The choice of containers depends largely on the type of crop to be grown, its growth stage, duration of cropping cycle, plant vigour, root system, and growing habit. Long-duration and deep-rooted crops such as tomatoes, cucumbers, and capsicum require larger containers with adequate depth and volume to support extensive root growth and nutrient uptake. On the other hand, short-duration and shallow-rooted crops like lettuce, spinach, and herbs can be successfully grown in smaller containers, reducing space and input costs. Containers not only provide physical support for plants but also serve as reservoirs for water and nutrients, ensuring better management of fertigation practices. They play an important role in controlling root zone environment, improving aeration, and preventing waterlogging when equipped with proper drainage holes. Containers also allow flexibility in greenhouse operations, as they can be moved, rearranged, or replaced based on crop rotation or seasonal requirements. Different materials such as clay, plastic, metal, and biodegradable pots are used, each with its own advantages and limitations. Plastic containers are lightweight, durable, and easy to handle, while clay pots provide better aeration but are heavy and fragile. Biodegradable containers offer an eco-friendly option but are less durable. For commercial greenhouse production, stackable trays, grow bags, and hydroponic channels are also popular choices, enhancing space efficiency and enabling high-density planting. Proper container selection helps minimize root diseases, optimize resource use, and improve crop uniformity. Additionally, containers make greenhouse cultivation more organized by segregating different crops and facilitating easy monitoring and management. Thus, the careful choice and use of containers not only support healthy plant growth but also contributes to sustainable and profitable greenhouse farming.

6.3.1 Advantages

- ✰ Increases plant productivity as well as production time: Use of containers and benches in greenhouse cultivation helps extend the growing season, allowing multiple crop cycles per year, which ultimately increases overall productivity and ensures continuous supply of produce.
- ✰ Production of good quality produce: Controlled conditions in containers provide balanced nutrition and proper root zone management, leading to uniform, healthy, and market-grade quality produce.
- ✰ Uniform and vigorous plant growth: Since each plant has its own controlled growing space, competition for nutrients and water is reduced, resulting in vigorous, balanced, and uniform plant growth across the crop stand.
- ✰ Easy to transport and handling of plants also no transplanting shocks: Container-grown plants can be easily shifted from one place to another without disturbing root systems, thus minimizing transplanting shocks and ensuring better establishment.

- Easy to maintain and sanitation of greenhouse: Containers make it simpler to maintain hygiene inside the greenhouse by reducing soil-borne infections, facilitating cleaning, and supporting better sanitation practices.
- Easy to harvest, grading and shifting of produce: Harvesting from container-grown plants is more organized, and grading as well as shifting of harvested produce becomes convenient due to systematic plant arrangements.
- Good drainage and aeration facilities: Well-designed containers provide adequate drainage holes, preventing waterlogging and ensuring sufficient aeration to roots, which promotes healthy root development.
- By help of micro-irrigation like drip it's easy to analyse plant nutrient efficacy: Containers integrate well with micro-irrigation systems such as drip or fertigation units, making it possible to precisely analyse nutrient uptake efficiency and adjust inputs accordingly for optimum crop performance.

6.3.2 Different Types of Containers and their Uses

In greenhouse cultivation, containers serve as essential tools for supporting plant growth and ensuring efficient management of crops. They act as controlled units for root development, water retention, and nutrient supply, making them highly versatile in diverse cropping systems. Different types of containers such as clay pots, plastic pots, metal containers, grow bags, seedling trays, and biodegradable pots are commonly used depending on crop requirements. Each type has its own unique properties in terms of durability, aeration, cost, and suitability for particular crops. For instance, large and deep containers are preferred for long-duration, deep-rooted crops, while smaller, shallow containers are ideal for short-duration leafy vegetables. Modern options like grow bags and hydroponic channels maximize space use and enable high-density planting. The choice of container also affects drainage, aeration, and ease of handling in greenhouse operations. Besides supporting plant growth, containers reduce transplanting shock, allow crop rotation flexibility, and improve sanitation within the greenhouse. Thus, selecting the right type of container is vital for achieving uniform growth, higher productivity, and sustainable greenhouse management.

Plastic

Plastic containers are the most commonly used and economical type of planter in greenhouse cultivation due to their wide availability, light weight, and ease of handling. They are manufactured in different sizes, shapes, colours, and designs, allowing growers to select the most suitable option based on crop type and production system. Depending on quality and thickness, the price of plastic containers can range from very cheap disposable pots to more durable and expensive models suitable for repeated use. These containers are non-porous and water-tight, which means they do not absorb moisture, thus preventing water loss and ensuring efficient irrigation management. They are available with or without drainage holes, giving flexibility for different cropping needs; containers with holes provide excellent drainage, while those without are useful for hydroponics or special growing systems. Their smooth or textured surfaces also make them aesthetically appealing and easy to clean.

In terms of uses, plastic containers are ideal for raising vegetable seedlings, ornamental plants, and short-duration crops that require frequent handling and transportation. They are especially useful in nurseries for germination and initial plant growth, as they allow easy shifting and transplanting without causing root damage. Large-sized plastic planters are also employed for growing long-duration crops such as tomato, capsicum, or cucumber in greenhouse conditions. Additionally, they can be reused multiple times if handled carefully, making them cost-effective in the long run. Their lightweight nature allows growers to rearrange crops easily for better spacing, light interception, and management practices. Furthermore, plastic containers are compatible with micro-irrigation systems like drip fertigation, ensuring precise delivery of water and nutrients. Although less durable compared to metal or clay, their convenience, versatility, and affordability make them the most popular choice among greenhouse growers (Figure 6.13).

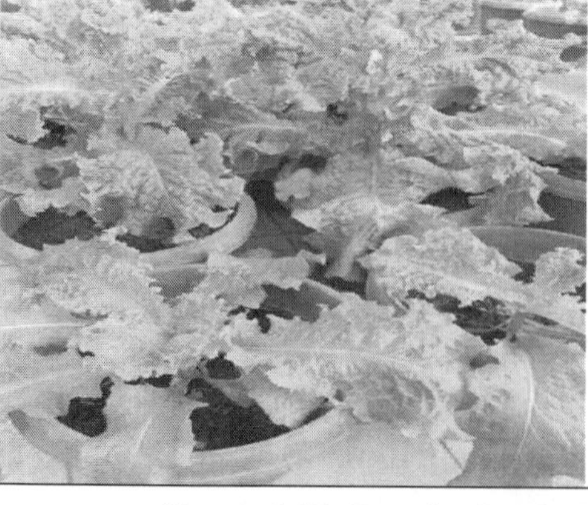

Figure 6.13: Growing Leafy Vegetables in Plastic Containers.
(*Source*: https://www.researchgate.net)

Ceramic

Ceramic containers are elegant and durable options for greenhouse and nursery cultivation, though they are relatively more expensive compared to plastic containers. These containers are available in a wide variety of shapes, sizes, and artistic designs, making them attractive for both functional and decorative purposes. They are often sold with or without drainage holes, allowing growers to choose them for different types of crops and growing systems. The surface of ceramic containers may be textured, painted, or given a glossy glaze that enhances their appearance and makes them resistant to moisture absorption. Some ceramic pots are hand-made and intentionally designed with artistic imperfections, which add aesthetic value and uniqueness.

In terms of functionality, ceramic containers are heavier than plastic, which provides greater stability and prevents plants from tipping over in windy or high-traffic greenhouse environments. They also provide good insulation to the root zone, protecting plants from sudden temperature fluctuations. Since they are semi-porous (if unglazed), they can help in moderate aeration and evaporation, thereby improving root health. Glazed ceramic pots, on the other hand, are non-porous and retain moisture longer, making them suitable for crops that require consistent soil moisture. These containers are commonly used for ornamentals, flowering plants, and high-value crops where appearance and presentation are equally important as growth performance. Their aesthetic appeal makes them especially suitable for greenhouse displays, research collections, and retail nurseries. However, they are fragile and can break if mishandled, and their high cost limits their use in large-scale commercial production. Despite these drawbacks, ceramic containers are valued for their durability, insulation capacity, and decorative appeal, making them an excellent choice for specialty crops and ornamental plant cultivation (Figure 6.14).

Figure 6.14: Ceramic Containers
(*Source*: www.freepik.com)

Wood

Wood containers are not commonly used in large-scale greenhouse cultivation but are often chosen for their naturalistic appearance and decorative value. They are typically crafted from different types of wood and finished with various designs or coatings to enhance aesthetics. Wooden containers provide a rustic and traditional look, making them attractive for ornamental and display purposes. However, since wood is naturally porous and prone to rotting when exposed to constant moisture, these containers are usually lined with non-porous materials such as plastic sheets or metal linings to improve their durability and prevent water damage. This lining also helps retain moisture in the growing medium while protecting the wooden surface from decay.

In terms of uses, wooden containers are mostly preferred for ornamental plants, shrubs, and decorative displays inside greenhouses, retail nurseries, or landscape settings rather than for intensive crop production. They are suitable for medium to large-sized plants that require good root space and stability. Wood provides a natural insulation effect, protecting roots from rapid changes in temperature, which is an added advantage in greenhouse environments. Due to their heavy structure, wooden containers are stable and less likely to tip over, making them suitable for tall or bushy plants. While they are more expensive and require maintenance such as painting or treating with preservatives, their visual appeal makes them a popular choice for specialty crops, bonsai cultivation, and exhibition plants. However, their limited durability and weight make them less practical for commercial-scale greenhouse farming. Overall, wooden containers are valued more for their decorative charm and stability than for efficiency, making them ideal for aesthetic greenhouse setups and ornamental horticulture (Figure 6.15).

Figure 6.15: Wood Containers.
(*Source*: www.freepik.com)

Fibre Glass

Fiberglass containers are among the commonly used options in greenhouse cultivation because of their strength, durability, and attractive appearance. These containers are manufactured using reinforced fiberglass materials, which make them lightweight yet strong enough to withstand greenhouse conditions. They are available in a wide variety of shapes, sizes, and designs, allowing growers to select styles that suit both functional and decorative needs. To enhance their aesthetic value, fiberglass containers are often painted in different colours or finished with textures that mimic stone, ceramic, or other materials. Their water-tight construction prevents leakage, while ridges in their design provide additional strength and resistance to cracking. Most fiberglass containers are manufactured with drainage holes, ensuring proper water outflow and preventing root zone waterlogging.

In terms of uses, fiberglass containers are versatile and suitable for both ornamental and crop cultivation inside greenhouses. They are particularly beneficial for long-duration crops and perennials because of their stability and resistance to environmental wear and tear. Their lightweight structure makes them easier to transport and rearrange compared to concrete or ceramic containers, while still offering a premium, decorative appearance. Fiberglass pots also insulate plant roots against temperature fluctuations, providing a favourable microenvironment for growth. Since they are non-porous, they retain moisture better than clay containers, reducing irrigation frequency. These features make them ideal for flowering plants, foliage plants, and high-value crops grown under controlled conditions. However, their higher cost compared to plastic containers may limit their use in large-scale commercial farming. Overall, fiberglass containers combine durability, aesthetic appeal, and functionality, making them a preferred choice for greenhouse growers who seek both practicality and style (Figure 6.16).

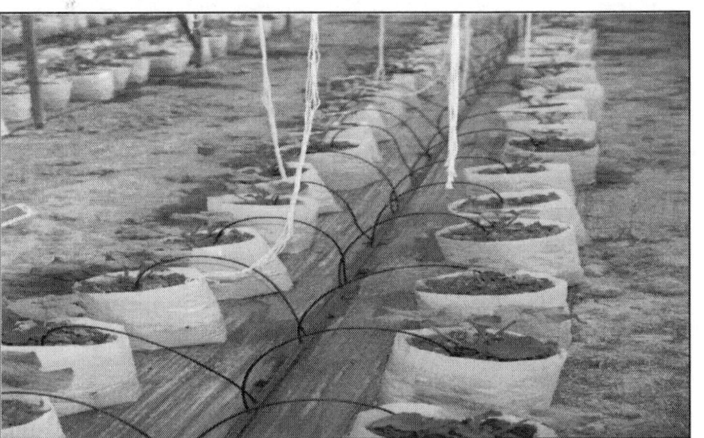

Figure 6.16: Fiberglass Containers.
(*Source*: www.freepik.com)

Metal

Metal containers are attractive and sturdy planters that provide a distinctive accent for plants in greenhouse and ornamental cultivation (Figure 6.17). The most common metals used in their construction include brass, copper, stainless steel, and aluminium, while wrought iron is often employed as a decorative material. These containers are available in a variety of surface finishes, ranging from antique and rustic looks to highly polished or brushed textures, making them visually appealing for both traditional and modern greenhouse settings. Metal planters are valued for their durability and strength; however, they require more maintenance compared to plastic, clay, or fiberglass containers, as they may tarnish, rust, or discolour over time when exposed to moisture and fertilizers.

Figure 6.17: Metal Containers.
(*Source*: www.freepik.com)

In terms of functionality, many metal containers are manufactured without drainage holes, which makes them unsuitable for direct planting. Instead, they are often used as outside covers or decorative sleeves for holding plastic or clay pots inside. This dual arrangement not only enhances aesthetics but also provides protection for inner containers. Metal containers, especially aluminium and stainless steel, are lightweight and easier to handle, while copper and brass options are heavier and mostly used for decorative, high-value plants. Wrought iron containers are particularly popular for ornamental displays, climbing plants, or hanging baskets due to their strength and artistic designs. Although they are not widely used for large-scale greenhouse crop production, they serve an important role in nurseries, retail displays, exhibitions, and ornamental horticulture. Their biggest advantages are durability, elegance, and ability to complement greenhouse interiors, while their limitations include higher cost, susceptibility to rust (in some types), and the need for protective coatings or liners. Overall, metal containers are best suited for decorative purposes, specialty crops, and display plants where both appearance and stability are important.

Clay/Terra Cota

Clay or terra cotta containers are among the oldest and most traditional types of planters used in greenhouse cultivation and ornamental displays (Figure 6.18). These containers are made from natural clay, making them porous in nature, which allows air and water to move through the walls of the pot. This porosity enhances root aeration and helps prevent waterlogging, thereby creating a healthier root environment. However, the same property also causes quick drying of soil, requiring frequent irrigation compared to non-porous materials like plastic or glazed ceramic. Since excess water can drain and sometimes bleed through the walls, it is essential to place clay pots on waterproof trays or saucers to avoid staining floors and surfaces. Clay containers are typically heavier than plastic, providing good stability for tall or bushy plants.

Figure 6.18: Metal Containers.
(*Source*: www.freepik.com)

In terms of uses, clay or terra cotta pots are ideal for indoor ornamental plants, succulents, cacti, and crops that prefer dry root zones because of their excellent aeration capacity. They are also widely used in nurseries and greenhouse displays for their aesthetic appeal and natural earthy look. The weight and sturdiness of clay pots make them less likely to tip over, which is beneficial in windy greenhouse environments or outdoor extensions. Since they are available in different sizes and shapes, they can be adapted for seedlings, medium-sized crops, or decorative plant arrangements. Their eco-friendly nature and biodegradability add sustainability value, although they are fragile and can break easily if mishandled. Despite the need for frequent watering and careful handling, clay containers remain a popular choice for both greenhouse growers and plant enthusiasts due to their natural appearance, breathability, and suitability for specific crops like ornamentals and succulents.

Irrigation and Fertigation Management

7.0 Introduction

Efficient irrigation and fertigation management are vital components of protected cultivation systems, as they directly influence crop productivity, resource use efficiency, and profitability. Under greenhouse conditions, water and nutrients can be precisely supplied to plants according to their requirements, ensuring optimal growth and minimizing losses. Unlike open-field cultivation, protected structures restrict natural rainfall, making artificial irrigation essential for sustaining crops. At the same time, fertigation–the application of fertilizers through irrigation water–offers a scientific and economical way of delivering nutrients directly to the root zone. A well-designed irrigation and fertigation strategy ensures uniform plant growth, improved quality of produce, and significant savings in water, fertilizers, and labour.

7.0.1 Importance of Irrigation in Protected Cultivation

- ✰ Irrigation in greenhouses is absolutely indispensable because crops are cultivated under controlled structures where natural rainfall does not contribute, making artificial water supply the only source for plant survival and growth.
- ✰ A well-managed irrigation system ensures a continuous and adequate supply of water in precise quantities, preventing both water stress and excessive watering, which can otherwise reduce yield and quality.

- Scientific irrigation enhances physiological processes such as photosynthesis, nutrient absorption, and metabolic activities in plants, thereby ensuring vigorous growth, higher productivity, and superior quality produce.
- Modern irrigation methods like drip, sprinkler, foggers, and mist systems ensure uniform distribution of water, conserve resources, and maintain the soil or substrate at optimum moisture levels required for protected crops.
- Efficient irrigation also minimizes the risk of waterlogging and root-zone diseases, reduces wastage of water and nutrients, and improves water-use efficiency, making it highly suitable for high-value crops in protected cultivation.

7.0.2 Importance of Fertigation in Protected Cultivation

- Fertigation is an advanced technique that combines irrigation and nutrient management, supplying water-soluble fertilizers directly to the crop root zone through irrigation systems. This ensures that plants receive essential nutrients in readily available form exactly where they are needed. By targeting the root zone, nutrient availability is maximized, minimizing wastage. It is highly effective for high-value crops grown under controlled environments. Thus, fertigation forms the backbone of precision farming in protected cultivation.
- One of the major advantages of fertigation is the enhancement of nutrient-use efficiency by reducing nutrient losses due to leaching, runoff, or volatilization. Since fertilizers are supplied in small and frequent doses, crops can utilize them effectively. This prevents soil nutrient imbalances and reduces environmental pollution. It also ensures that expensive fertilizers are used judiciously, thereby cutting down input costs. Consequently, fertigation ensures economic and sustainable nutrient management.
- A balanced and timely nutrient supply through fertigation promotes uniform crop growth and development across the greenhouse. It helps maintain steady vegetative and reproductive growth, ensuring better flowering, fruit set, and overall plant health. Controlled nutrient delivery also shortens the crop cycle and leads to early maturity. This results in higher yields and better-quality produce. In commercial protected cultivation, such uniformity is crucial for meeting market standards.
- Fertigation also significantly reduces labour costs, as irrigation and fertilizer application are carried out simultaneously through the same system. This automation minimizes manual work like broadcasting or foliar spraying of fertilizers. It also reduces errors in nutrient application and ensures timely supply without dependence on labour availability.

In large-scale greenhouses, this is especially beneficial in saving time and cost. Hence, fertigation makes crop management more efficient and reliable.

- Another important advantage of fertigation is the flexibility it provides in adjusting nutrient concentration according to crop stage, growth habit, and vigour. For example, nitrogen can be increased during vegetative stages and potassium during flowering and fruiting. This stage-specific nutrient regulation maximizes productivity and quality. Moreover, deficiencies can be corrected quickly by altering fertigation schedules. Such precise control is not possible in conventional fertilizer application methods.

7.0.3 Objectives of Irrigation and Fertigation Management

- **To optimize water and nutrient use efficiency:** Ensure precise delivery of water and fertilizers to the crop root zone, minimizing wastage and maximizing resource utilization.
- **To reduce production costs and labour dependency:** Adopt automation and advanced fertigation systems that save time, energy, and manpower while lowering input costs.
- **To enhance yield, quality, and profitability of high-value crops:** Provide balanced nutrition and adequate moisture for achieving uniform growth, higher productivity, better quality, and premium market returns.
- **To maintain soil health and fertility:** Prevent salinity buildup, leaching, and nutrient imbalances through controlled irrigation and fertigation practices that sustain long-term soil productivity.
- **To support sustainable greenhouse and protected farming:** Minimize water pollution, nutrient runoff, and greenhouse gas emissions, contributing to eco-friendly and climate-resilient agriculture.
- **To achieve precision and consistency in crop management:** Facilitate uniform application of water and nutrients, reducing variability in crop performance and ensuring predictable harvests.
- **To improve resource sustainability and food security:** Conserve scarce resources such as freshwater and fertilizers, ensuring their availability for future generations while meeting rising food demand.

7.0.4 Challenges in Irrigation and Fertigation Management

- The selection of a suitable irrigation system for specific crops and greenhouse designs is a major challenge. Each crop has different water and nutrient requirements, and greenhouse layouts also influence system efficiency. A wrong choice can lead to poor growth, water wastage, or high operational costs. Hence, matching irrigation systems with crop needs and structure design is crucial for successful protected cultivation.

- Maintaining correct water quality, pH, and electrical conductivity (EC) for fertigation solutions is another critical challenge. Poor-quality water with high salts, bicarbonates, or unsuitable pH affects nutrient solubility and crop uptake. Imbalanced EC can result in nutrient deficiencies or toxicities. Regular testing and monitoring of water quality are therefore essential for ensuring healthy crop growth.
- Avoiding clogging of drip lines and ensuring uniform distribution of water and nutrients across all plants is a common issue in fertigation. Clogging may occur due to sediment, algae, or precipitation of fertilizers inside pipes and emitters. This leads to uneven supply, causing variability in plant growth and yield. Regular maintenance, filtration, and system cleaning are necessary to overcome this challenge.
- The requirement of technical knowledge for precise scheduling, monitoring, and operation of irrigation and fertigation systems is another limitation. Growers need to understand crop water needs, nutrient dynamics, and system calibration to achieve precision. Lack of training or errors in management can lead to under- or over-irrigation and nutrient imbalances. Thus, capacity-building and technical support are essential for farmers adopting modern systems.

7.1 Irrigation Systems in Protected Cultivation

7.1.1 Hand Watering

Hand watering is the most traditional method of irrigation and is still practiced in specific situations. Growers generally rely on it where crops are grown at very high densities, such as in seedbeds, or when only a few selected pots or areas dry out more quickly than others. Although simple and inexpensive to set up, hand watering is both labour-intensive and costly. In fact, the labour saved by installing an automated irrigation system typically offsets its initial investment within a year, making hand watering an uneconomical choice in modern protected cultivation. When hand watering is used, it is essential to attach a water breaker at the end of the hose. This device softens the water stream, allowing a higher flow rate without washing away or compacting the root substrate in pots or benches. It also helps preserve the integrity of the substrate surface, thereby protecting delicate root systems. Despite these precautions, hand watering remains time-consuming, monotonous, and inefficient compared to automated irrigation systems. Automated methods not only save labour but also ensure greater uniformity, precision, and resource efficiency, which are crucial for achieving higher productivity and sustainability in horticultural crop production (Figure 7.1).

Figure 7.1: Hand Watering.
(*Source*: https://www.greenhousegrower.com)

Materials Used

- **Hose pipe –** standard garden or nursery hose for delivering water.
- **Water breaker** – a device attached to the hose end to reduce the force of water and prevent substrate washout.
- **Buckets or watering cans** – for small-scale operations or where hoses are unavailable.
- **Optional tools** – funnels or small nozzles for more controlled water application.

Suitability

- **Seedbeds and nurseries** – ideal for high-density crops where precise watering of individual seedlings is needed.
- **Small-scale or home gardens –** convenient for limited areas with few pots or plants.
- **Spot irrigation** – suitable for watering only selected pots or areas that dry out faster.
- **Backup irrigation –** useful when automated irrigation systems are not installed or temporarily fail.

Advantages

- **Low initial cost** – requires no special equipment or infrastructure.
- **Simple and easy to practice –** does not require technical skills.
- **Targeted application –** allows watering of specific pots, seedbeds, or dry areas.
- **Flexible use –** can be done at any time and adapted to different crop stages.

- **Useful for small-scale cultivation –** suitable for nurseries, home gardens, and limited areas.
- **Emergency backup method –** effective when automated systems fail or are not yet installed.

Limitations

- **Labour-intensive** – requires significant manual effort, especially for large areas.
- **Time-consuming –** can take a long time to water all crops uniformly.
- **Uneconomical for large-scale production –** high labour costs make it costly over time.
- **Inconsistent water distribution –** may result in uneven crop growth.
- **Risk of substrate or root damage** – improper watering can wash away or compact the substrate.
- **Monotonous and physically demanding –** can lead to worker fatigue and reduced efficiency.
- **Limited precision –** does not support efficient fertigation or automated nutrient delivery.

7.1.2 Perimeter Watering System

The perimeter watering system is a traditional and widely adopted irrigation method for crop production in benches and beds under protected cultivation. In this system, a plastic pipe is installed around the perimeter of the bench, fitted with multiple nozzles that spray water directly onto the substrate surface beneath the crop foliage. This ensures surface-level moisture and supports early plant growth, particularly in nursery beds and shallow-rooted crops (Figure 7.2).

Materials Used

- **Polythene pipes:** Flexible, lightweight, and easy to install, but prone to rolling or shifting if not properly anchored. This misalignment often causes nozzles to rise or fall, leading to uneven orientation and irregular water distribution.
- **PVC pipes:** Rigid, stable, and more durable than polythene, providing a stationary setup that maintains correct nozzle positioning and uniform application.

Advantages

- Simple design and relatively low installation cost.
- Easy to operate and maintain with minimal technical skills.
- Suitable for watering shallow-rooted crops or seedling trays where precise placement is less critical.
- Can be combined with automation systems to reduce labour requirements.

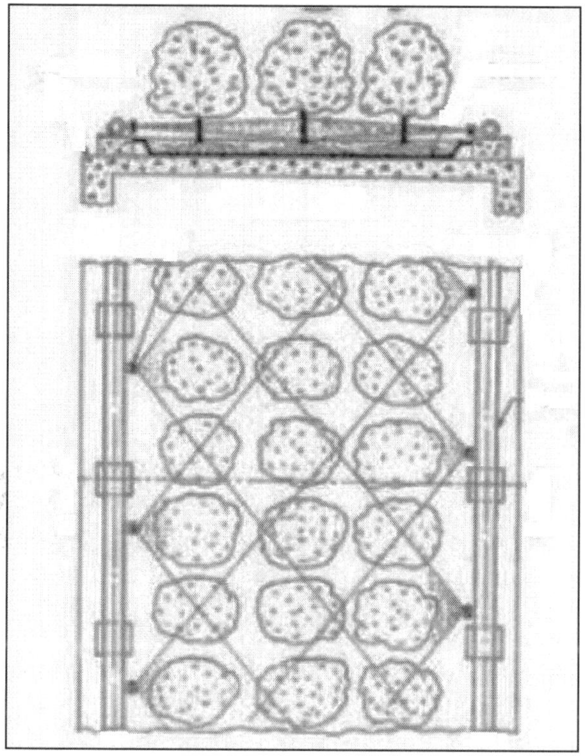

Figure 7.2: Perimeter Watering.

Limitations

- ☆ Water distribution may be uneven if pipes are not fixed properly, particularly with polythene types.
- ☆ Spraying over the foliage can increase the risk of foliar diseases, especially under humid greenhouse conditions.
- ☆ Inefficient in terms of water use compared to drip or subsurface irrigation, as evaporation and runoff losses are common.
- ☆ Not ideal for high-value crops requiring precise water and nutrient delivery.

Suitability

The perimeter watering system is best suited for nurseries, low-value crops, and situations where low-cost irrigation is prioritized over precision. For high-value horticultural crops in protected cultivation, more advanced systems such as drip or fertigation units are preferred for efficiency, precision, and sustainability.

7.1.3 Overhead Sprinklers

Overhead sprinklers are a widely used irrigation method in horticultural practices, offering uniform water distribution across the crop area. In this system, a main pipe is installed along the centre of each bed, ensuring easy access and support for the laterals. From this main pipe, riser pipes are positioned at regular intervals and extended to a height that exceeds the anticipated final height of the crop. For bedding plant flats, a riser height of approximately 0.6 meters is sufficient, whereas taller crops such as fresh flowers require risers up to 1.8 meters to prevent foliage interference and ensure effective coverage. Each riser is fitted with a nozzle at the top, which is the key component controlling the water pattern and distribution. Nozzles vary in design, ranging from those providing a fixed 360° spray to rotating types that cover a full circle, thus catering to different plant spacing and canopy requirements. The adjustable nature of these nozzles allows growers to modify the flow rate and coverage, optimizing water use efficiency. In some setups, trays are strategically placed beneath pots to capture any water that would otherwise fall between plants and be wasted, reducing runoff and conserving resources. The system not only ensures uniform irrigation but also reduces labour requirements compared to manual watering methods. Overhead sprinklers can be adapted to deliver fertilizers through fertigation, integrating nutrition with irrigation for enhanced plant growth. They are particularly useful in greenhouse environments, where precise water application is crucial to maintaining optimal humidity and preventing soil erosion. The height and spacing of risers can be tailored based on crop type, growth stage, and density, ensuring that all plants receive adequate moisture. Additionally, the system minimizes leaf wetting for crops sensitive to fungal diseases by controlling droplet size and spray intensity. Careful planning of riser positions and nozzle types can reduce energy and water consumption while promoting healthy crop development. Overall, overhead sprinklers provide a flexible, efficient, and labour-saving solution for irrigating a wide range of horticultural crops, from delicate bedding plants to high-value cut flowers, ensuring uniform growth, improved yield, and sustainable resource management (Figure 7.3).

Materials Used

- ☆ **Main Pipe:** Typically made of durable PVC or galvanized iron (GI), the main pipe runs along the centre of the bed, forming the backbone of the sprinkler system.
- ☆ **Riser Pipes:** Constructed from PVC, GI, or HDPE, these pipes are installed at regular intervals and extended above the crop canopy to ensure proper water distribution.
- ☆ **Nozzles:** Precision-engineered nozzles come in fixed 360° patterns or rotating types, allowing uniform spray coverage and flexibility for different crop requirements.

Figure 7.3: Overhead Sprinklers in Greenhouse Irrigation.
(*Source*: https://greenhouseemporium.com)

- ☆ **Trays (Optional):** Used beneath pots or containers to catch excess water, preventing wastage and runoff.
- ☆ **Clamps and Fittings:** Strong connectors secure the main pipe, risers, and nozzles, ensuring a leak-free system.
- ☆ **Support Structures:** Wires, stakes, or frames hold risers upright and stable, even under high water pressure.

Advantages

- ☆ **Uniform Irrigation:** Ensures water reaches all plants evenly, preventing dry spots or waterlogging.
- ☆ **Labor-Saving:** Reduces the need for frequent manual watering, especially in large nurseries or greenhouse setups.
- ☆ **Fertigation Compatible:** Allows nutrients to be delivered along with water, enhancing plant growth and health.
- ☆ **Water Efficiency:** Reduces water wastage when trays or catchment systems are placed beneath pots.
- ☆ **Customizable Coverage:** Adjustable nozzle types and riser heights allow tailored irrigation for different crops and plant densities.
- ☆ **Greenhouse Humidity Management:** Helps maintain ideal humidity levels, promoting healthy growth and reducing plant stress.
- ☆ **Soil Protection:** Minimizes soil erosion and compaction by controlling droplet size and spray intensity.
- ☆ **Adaptable for Plant Height:** Suitable for various plant heights, from small bedding flats to tall cut flowers, ensuring effective coverage.

Limitations

- **Leaf Wetting:** Overhead sprinklers wet the foliage, which can increase the risk of fungal and bacterial diseases in sensitive crops.
- **Water Wastage:** Some water may fall between pots or beds if trays or catchments are not used, leading to inefficiency.
- **Evaporation Losses:** Fine droplets may evaporate before reaching the plant canopy, especially in hot or dry conditions.
- **Uneven Coverage:** Improper riser spacing, nozzle selection, or pressure variations can result in dry spots or over-irrigation.
- **High Initial Cost:** Installation requires investment in pipes, risers, nozzles, and fittings, which can be expensive for large greenhouses.
- **Maintenance Requirements:** Nozzles can clog, pipes may leak, and risers can get misaligned, requiring regular inspection and maintenance.
- **Energy Dependence:** The system often relies on pumps and electricity to maintain adequate pressure, increasing operational costs.
- **Limited Suitability for Certain Crops:** Crops sensitive to wet foliage, such as tomatoes or certain ornamentals, may not perform well with overhead irrigation.
- **Humidity Management Issues:** Continuous spraying can raise humidity excessively in enclosed greenhouses, encouraging disease and slowing growth.
- **Potential Soil Erosion:** If water is applied too heavily or at high pressure, it may disturb the soil or substrate surface.

Suitability

- **Bedding Plants and Flats:** Ideal for small plants; risers of ~0.6 m provide optimal coverage without wetting foliage excessively.
- **Fresh Flowers and Tall Crops:** Riser height of ~1.8 m ensures water reaches taller stems and dense canopies.
- **High-Density Planting Beds:** Ensures uniform moisture distribution in closely spaced plants or pots.
- **Greenhouse Cultivation:** Supports controlled irrigation and humidity, critical for sensitive crops.
- **Fertigation-Ready Systems:** Perfect for integrating nutrient delivery with irrigation.
- **Flexible Crop Management:** Height and nozzle adjustments make the system suitable for multiple crop cycles and varying growth stages.

7.1.4 Boom Watering

Boom watering is an efficient irrigation method widely used in protected cultivation, particularly for the production of seedlings grown in plug trays. This system can function either as an open or a closed setup, depending on the requirements of the crop and greenhouse design. A typical boom watering system consists of a long water pipe, or "boom," which spans the width of a greenhouse bay. The pipe is equipped with multiple nozzles that can spray either water alone or a nutrient-rich fertilizer solution directly onto the crop, allowing simultaneous irrigation and fertigation. The boom is mounted at its central point onto a carriage that moves along rails, often suspended above the central walkway of the greenhouse. This design enables the boom to traverse smoothly from one end of the bay to the other, ensuring uniform coverage across all seedlings. By automating the movement and distribution of water, boom watering reduces labour requirements, enhances precision in water application, and improves overall crop quality. The system is especially advantageous for high-density seedling production, as it provides consistent moisture without disturbing delicate plants. Additionally, it allows growers to adjust nozzle type, spray pattern, and flow rate according to crop size, growth stage, and spacing, making it a highly versatile and efficient irrigation solution for modern protected horticulture (Figures 7.4 and 7.5).

Figure 7.4: Boom Watering Irrigation.
(*Source*: https://www.openpr.com)

Materials Used

- ☆ **Boom Pipe:** Typically made of PVC or galvanized iron, spanning the greenhouse bay.
- ☆ **Nozzles:** Adjustable types for spraying water or fertilizer solutions uniformly.
- ☆ **Carriage and Rails:** Allow smooth movement of the boom along the bay; usually made of steel or aluminium.

Figure 7.5: Boom Watering Irrigation System.
(*Source*: www.freepik.com)

- ☆ **Pump and Piping System:** Provides required water pressure for consistent spraying.
- ☆ **Fittings and Clamps:** Secure all connections and maintain leak-free operation.

Advantages

- ☆ **Uniform Irrigation:** Ensures even water distribution across all seedlings in plug trays.
- ☆ **Fertigation Friendly:** Delivers nutrients along with water, promoting faster and healthier growth.
- ☆ **Labor-Saving:** Reduces manual watering, particularly in high-density seedling production.
- ☆ **Adjustable Spray:** Nozzle type, pattern, and flow rate can be modified according to crop requirements.
- ☆ **Minimizes Plant Disturbance:** Delicate seedlings remain undisturbed during irrigation.
- ☆ **Time-Efficient:** Covers large greenhouse bays quickly compared to hand watering.
- ☆ **High Precision:** Ensures controlled water application, reducing wastage and runoff.

Suitability

- ☆ **Seedling Production:** Ideal for plug trays and nursery beds.
- ☆ **High-Density Greenhouses:** Suitable for densely planted crops where manual irrigation is labor-intensive.

- **Fertigation Systems:** Perfect for crops requiring simultaneous irrigation and nutrient delivery.
- **Delicate or Small Plants:** Ensures plants are not damaged by direct water impact.
- **Commercial Greenhouses:** Large-scale protected cultivation operations benefit most.

Limitations

- **High Initial Cost:** Installation of pipes, carriage, rails, and nozzles requires significant investment.
- **Maintenance Needs:** Nozzles can clog, rails may need alignment, and pumps require regular inspection.
- **Energy Dependence:** Requires electricity for pump operation and boom movement.
- **Limited Crop Types:** Not suitable for crops sensitive to leaf wetting if overhead spray is used.
- **Space Requirement:** Rails and carriage need sufficient central walkway space in the greenhouse.
- **Complexity**: Requires skilled personnel for operation and calibration of the system.

7.1.5 Drip Irrigation

Drip irrigation is a highly efficient and precise method of delivering water directly to the root zone of plants, making it ideally suited for protected cultivation in greenhouses and polyhouses. This system uses a network of pipes, tubes, and emitters to supply water in small, controlled quantities, minimizing wastage and maximizing water use efficiency. It is particularly beneficial in areas with limited water availability or high evaporation rates, as water is applied directly to the soil or growing medium. Drip irrigation also allows fertigation, enabling nutrients to be delivered simultaneously with water, which promotes uniform growth and improves crop yield. In greenhouses, the system can be customized to suit the spacing, size, and growth stage of different horticultural crops. By maintaining consistent soil moisture, drip irrigation reduces plant stress, prevents waterlogging, and promotes healthier root development. The system is highly adaptable for various crops, including vegetables, ornamentals, flowers, and seedlings grown in trays or pots. It reduces labour requirements significantly, as manual watering becomes unnecessary once the system is installed and programmed. Drip irrigation also helps maintain optimal greenhouse humidity, reducing the risk of fungal diseases caused by excessive leaf wetting. The emitters can be spaced according to plant density, and flow rates can be adjusted to meet the specific water requirements of each crop. In addition, it minimizes nutrient leaching, ensuring efficient use of fertilizers while protecting the environment. The system can be integrated with timers, sensors, and automation devices, allowing precise control over irrigation schedules. Drip irrigation also conserves energy by reducing pumping requirements compared to overhead or boom systems. It is suitable for both soil-based beds and containerized cultivation, offering flexibility in protected environments. Overall, drip irrigation provides a sustainable, resource-efficient, and high-precision solution for modern horticultural practices, supporting enhanced productivity, improved quality, and reduced operational costs in protected cultivation systems.

Figure 7.6: Drip Irrigation System.
(*Source*: www.freepik.com)

Materials Used

- ✰ **Main Pipes:** PVC or HDPE pipes for water distribution.
- ✰ **Lateral Pipes/Tubing:** Flexible polyethylene tubes delivering water to emitters.
- ✰ **Emitters/Drippers:** Devices controlling water flow directly to the plant root zone.
- ✰ **Filters**: Sand or screen filters to prevent clogging of emitters.
- ✰ **Fertilizer Tank/Injector:** For fertigation, supplying nutrients with irrigation water.
- ✰ **Fittings and Connectors:** Tees, elbows, clamps, and connectors for network assembly.
- ✰ Pressure Regulators: Maintain uniform water flow across the system.
- ✰ Support Stakes: Hold lateral tubing in place over beds or pots.

Advantages

- ✰ **Water-Efficient:** Delivers water directly to roots, minimizing evaporation and runoff.
- ✰ **Labor-Saving:** Reduces manual watering, especially in high-density planting.

- **Fertigation Compatible:** Nutrients can be applied directly with water for uniform growth.
- **Consistent Soil Moisture:** Prevents water stress and promotes healthy root development.
- **Customizable:** Emitter spacing, flow rate, and tubing layout can be adapted for different crops.
- **Disease Prevention:** Reduces leaf wetting, minimizing fungal and bacterial infections.
- **Energy-Efficient:** Requires lower pumping pressure compared to overhead systems.
- **Environmentally Friendly:** Minimizes nutrient leaching and conserves water.

Suitability

- **Seedling Production:** Ideal for plug trays, potted plants, and nursery beds.
- **High-Value Crops:** Vegetables, flowers, ornamentals, and medicinal plants.
- **Protected Cultivation:** Greenhouses, polyhouses, and shade-net houses.
- **Containerized or Bed Cultivation:** Suitable for soil beds, grow bags, or pots.
- **Fertigation Systems:** Compatible with integrated nutrient management practices.
- **Varied Plant Densities:** Adjustable for close-spacing or high-density planting.

Limitations

- **High Initial Cost:** Installation of pipes, emitters, filters, and regulators requires investment.
- **Clogging Issues:** Emitters can clog due to suspended particles or chemical fertilizers.
- **Maintenance Intensive:** Requires regular cleaning of filters, emitters, and pipes.
- **Energy Dependence:** Pumps and automated systems require electricity.
- **Skill Requirement:** Requires technical knowledge for design, installation, and management.
- **Limited Surface Coverage:** Only irrigates root zones, unsuitable for crops needing leaf wetting.
- **Damage Risk:** Tubing may be damaged by farm operations or pests.

Comparison of overhead, boom, and drip irrigation systems in protected cultivation of horticultural crops is summarised in Table 7.1.

Table 7.1: Comparison of Overhead, Boom, and Drip Irrigation Systems in Protected Cultivation of Horticultural Crops

Irrigation System	*Materials Used*	*Advantages*	*Suitability*	*Limitations*
Overhead Irrigation	Main pipe, risers, nozzles, trays (optional), clamps, support stakes	Uniform irrigation; Labor-saving; Fertigation compatible; Reduces water wastage; Adjustable coverage; Maintains humidity; Minimizes soil erosion; Suitable for various plant heights	Bedding plants/ flats (~0.6 m); Fresh flowers (~1.8 m); High-density beds; Greenhouses; Fertigation systems	Leaf wetting may increase disease; Evaporation losses; Uneven coverage; High initial cost; Maintenance needed; Energy dependent; Not ideal for sensitive crops
Boom Irrigation	Boom pipe, adjustable nozzles, carriage and rails, pump, fittings	Uniform irrigation; Fertigation friendly; Labor-saving; Adjustable spray; Minimal plant disturbance; Time-efficient; High precision	Seedling production; High-density greenhouses; Fertigation systems; Delicate/ small plants; Commercial greenhouses	High initial cost; Maintenance required; Energy dependent; Limited for leaf-sensitive crops; Needs central walkway; Operational complexity
Drip Irrigation	Main/lateral pipes, emitters, filters, fertilizer injector, fittings, pressure regulators, stakes	Water-efficient; Labor-saving; Fertigation compatible; Consistent soil moisture; Customizable; Reduces leaf diseases; Energy-efficient; Environmentally friendly	Seedlings; High-value crops; Greenhouses/ polyhouses; Container/ bed cultivation; Fertigation systems; Varied plant densities	High initial cost; Clogging issues; Regular maintenance; Energy dependent; Requires skill; Limited leaf coverage; Risk of tubing damage

7.1.6 Misters and Foggers

Foggers and misters are widely used in greenhouses as an effective method for cooling and maintaining optimal humidity levels. Foggers are usually connected to lateral pipes through micro-tubes and suspended from iron wires at a height of about 3 meters, while misters are installed on stakes directly within the crop beds. Both systems produce very fine water droplets, approximately 70 microns in size, which evaporate before reaching the plant canopy, ensuring uniform cooling without wetting the plants. Modern foggers and misters are equipped with anti-leak devices that prevent water from dripping after the system is switched off. They operate at a pressure of 3 to 5 kg/cm^2 and have a relatively low discharge rate, covering a small area at a time. To maintain ideal greenhouse conditions, these systems are typically operated for 30 to 60 seconds, three to four times every hour at regular intervals. By providing consistent cooling and humidity control, misters and foggers play a crucial role in promoting healthy plant growth and maximizing crop productivity (Figures 7.7–7.9).

Figure 7.7: Greenhouse Irrigation Misting System.
(*Source*: https://www.exportersindia.com)

Figure 7.8: Fogging System.
(*Source*: https://5.imimg.com)

Figure 7.9: High Pressure Fog Systems in Horticulture
(*Source*: https://media.licdn.com)

Advantages of Misters and Foggers

- ☆ Provide uniform cooling and increase relative humidity in the greenhouse.
- ☆ Water droplets evaporate before reaching the crop, reducing the risk of waterlogging.
- ☆ Equipped with anti-leak devices to prevent dripping after operation.
- ☆ Operate efficiently with minimal water consumption.
- ☆ Easy to install and adjust according to crop height and greenhouse layout.

Materials Used and Suitability

- ☆ **Foggers:** Connected via micro-tubes to lateral pipes, suspended from iron wires; suitable for overhead cooling and larger greenhouse areas.
- ☆ **Misters:** Mounted on stakes within beds; ideal for localized cooling near plant canopy and high-density crops.
- ☆ Both systems are suitable for vegetable nurseries, ornamentals, and microgreens where precise humidity and temperature control are required.

7.2 Fertigation: Concepts and Practices

7.2.1 Introduction to Fertigation

Fertigation is the technique of applying fertilizers along with irrigation water directly to the crop root zone. It integrates two critical agricultural practices–irrigation and fertilization–into a single, efficient system. In protected cultivation, fertigation ensures precise delivery of water-soluble nutrients in the right concentration, at the right time, and at the right place, leading to improved resource use efficiency, reduced wastage, and higher crop yields.

7.2.2 Objectives of Fertigation

- To supply nutrients in synchrony with crop growth stages.
- To reduce nutrient losses caused by leaching, volatilization, and surface runoff.
- To minimize labour and operational costs by combining irrigation and fertilization.
- To achieve uniform distribution of nutrients across the crop stand.
- To enhance fertilizer, use efficiency and improve crop quality.

7.2.3 Components of a Fertigation System

- **Water Source and Pump:** A reliable water source is the foundation of any fertigation system. Pumps are installed to provide the required pressure and ensure a steady, uniform flow of water throughout the irrigation network. In protected cultivation, pumps must be carefully selected to maintain consistent discharge and avoid pressure fluctuations, which could otherwise result in uneven nutrient distribution.
- **Fertilizer Tank or Injector:** The fertilizer tank or injector is the core of the fertigation unit. It dissolves water-soluble fertilizers and injects them into the irrigation system in precise proportions. Depending on the system design, injectors may operate manually, hydraulically, or electronically. Advanced computerized injectors can automatically adjust fertilizer concentrations based on crop requirements, ensuring high accuracy and efficiency.
- **Mainline and Laterals:** The mainline acts as the primary channel, carrying nutrient-rich water from the pump to different sections of the greenhouse. Laterals are connected to the mainline and deliver water directly to individual crop rows or beds. Pipes are usually made of PVC or HDPE, which are durable, resistant to corrosion, and suitable for maintaining steady flow. Proper design of mainline and lateral layout is essential to minimize pressure losses and guarantee uniform distribution.
- **Emitters/Drippers/Nozzles:** Emitters, drippers, or micro-nozzles are the delivery points that release water and nutrients directly to the plant root zone or over the canopy, depending on the system used. They are engineered to release water at a controlled rate, preventing wastage and ensuring precise application. Different crops and stages of growth may require specific emitter discharge rates, which can be customized for efficient fertigation.
- **Filters (Sand, Screen, Disc):** Filters are critical components in fertigation systems, especially in drip irrigation, where clogging is a common

problem. Sand filters remove suspended particles, screen filters trap larger debris, and disc filters catch fine impurities. Clean water not only prevents emitter blockages but also extends the life of the system and ensures uninterrupted fertigation. Regular maintenance of filters is essential for smooth operation.

- **Control Units:** Control units regulate water pressure, flow rate, and fertigation schedules to meet crop requirements accurately. Modern control systems may include pressure regulators, flow meters, timers, and even computer-based controllers. In advanced protected cultivation, automated sensors linked with climate control systems adjust fertigation dynamically according to soil moisture, crop growth stage, and weather conditions inside the greenhouse.

7.2.4 Fertilizers Used in Fertigation

- **Nitrogen Sources:** Nitrogen is the most vital nutrient for vegetative growth and is widely supplied through fertigation. Common nitrogen sources include Urea, Calcium Ammonium Nitrate (CAN), Ammonium Nitrate, and UAN (Urea Ammonium Nitrate) solution. These fertilizers are highly soluble and can be easily injected into irrigation systems. Urea provides quick-release nitrogen, while CAN and ammonium nitrate offer more balanced, stable nitrogen availability, reducing the risk of leaching in sandy soils.
- **Phosphorus Sources:** Phosphorus plays a key role in root development, energy transfer, and early plant establishment. In fertigation, Mono-Ammonium Phosphate (MAP) and Phosphoric Acid are the preferred sources due to their high solubility and compatibility with drip and micro-irrigation systems. MAP also supplies nitrogen along with phosphorus, while phosphoric acid not only provides phosphorus but also helps maintain acidic pH in irrigation water, reducing clogging risks.
- **Potassium Sources:** Potassium enhances fruit quality, size, colour, and disease resistance. Fertigation-friendly potassium sources include Potassium Nitrate (KNO_3), Potassium Sulphate (K_2SO), and Potassium Chloride (KCl). Potassium nitrate supplies both potassium and nitrogen, making it highly efficient. Potassium sulphate is suitable for chloride-sensitive crops like strawberry, grape, and tomato. Potassium chloride, though cost-effective, is used sparingly in sensitive crops due to the risk of chloride toxicity.
- **Micronutrients:** Micronutrients, though required in small quantities, are essential for enzyme activation, photosynthesis, and metabolic processes. In fertigation, micronutrients are applied in chelated forms of Iron (Fe), Zinc (Zn), Manganese (Mn), Copper (Cu), Boron (B), and Molybdenum (Mo) to ensure maximum solubility and uptake efficiency. Chelates

prevent micronutrients from precipitating or binding with soil particles, ensuring their availability to plants throughout the crop cycle.

- ✰ **Calcium and Magnesium Sources:** Calcium and magnesium are secondary nutrients that play crucial roles in strengthening cell walls, improving fruit firmness, and aiding chlorophyll synthesis. Calcium Nitrate is the most common source of calcium in fertigation, supplying both calcium and nitrate-nitrogen. Magnesium Sulphate (Epsom Salt) is widely used as a magnesium source, ensuring proper photosynthesis and preventing leaf chlorosis. Both nutrients are easily soluble and safe for use in drip systems. Comparison of overhead, boom, and drip irrigation systems in protected cultivation of horticultural crops is summarised in Table 7.2.

Table 7.2: Fertilizer Sources Used in Fertigation for Protected Cultivation

Nutrient Group	*Fertilizer Sources*	*Role in Crop Growth*	*Advantages in Fertigation*	*Crop Suitability*
Nitrogen	Urea, Calcium Ammonium Nitrate (CAN), Ammonium Nitrate, UAN solution	Promotes vegetative growth, leaf development, protein synthesis	Highly soluble; Rapid plant uptake; Flexible in dosing	Vegetables (tomato, cucumber, capsicum), leafy greens, nursery crops
Phosphorus	Mono-Ammonium Phosphate (MAP), Phosphoric Acid	Root development, energy transfer (ATP), early crop establishment	High solubility; MAP supplies both N and P; Phosphoric acid lowers pH, prevents clogging	Seedlings, root crops, early growth stages of vegetables and flowers
Potassium	Potassium Nitrate (KNO_3), Potassium Sulphate (K_2SO), Potassium Chloride (KCl)*	Improves fruit size, color, taste, and disease resistance	KNO_3 provides both N and K; K_2SO chloride-free; Flexible dosing	Fruit vegetables (tomato, capsicum, cucumber), strawberries, grapes, cut flowers (*KCl only in tolerant crops)
Micro-nutrients	Chelated forms of Fe, Zn, Mn, Cu, B, M	Enzyme activation, chlorophyll synthesis, flowering, and fruit set	Chelates prevent precipitation; Ensure availability; Effective even in alkaline soils	All high-value horticultural crops in greenhouses, especially sensitive crops like strawberry, capsicum, rose

Nutrient Group	*Fertilizer Sources*	*Role in Crop Growth*	*Advantages in Fertigation*	*Crop Suitability*
Calcium and Magnesium	Calcium Nitrate, Magnesium Sulphate	Calcium strengthens cell walls, fruit firmness; Magnesium is essential for chlorophyll and photosynthesis	Fully soluble; Provide secondary nutrients along with N and S; Correct deficiencies effectively	Tomatoes, peppers, cucumbers, strawberries, leafy greens, floriculture crops

7.2.5 Methods of Fertigation

- **Tank Mixing Method:** In this basic method, fertilizers are dissolved manually in a large tank and then allowed to flow into the irrigation line along with water. It is the simplest and most economical approach, requiring minimal technical expertise and infrastructure. However, uniform nutrient concentration may not always be achieved without continuous mixing, and precision is relatively low. It is best suited for small-scale greenhouses, nurseries, and polyhouses where cost-effectiveness is prioritized over high accuracy.
- **Venturi Injector Method:** The Venturi injector utilizes the principle of pressure differential. As water passes through a narrow constriction in the pipe, suction is created that draws fertilizer solution into the irrigation line. This method is compact, inexpensive, and has no moving parts, making it easy to maintain. It allows fertilizers to be injected continuously during irrigation, ensuring better nutrient distribution than tank mixing. However, it requires a minimum water pressure to function properly, limiting its use in systems with low-pressure pumps.
- **Bypass or Parallel Tank Method:** In this system, a portion of irrigation water is diverted through a fertilizer tank, where it dissolves the fertilizers before rejoining the mainline. Compared to tank mixing, it offers better uniformity and more control over nutrient application. It is relatively simple to install and operate, making it a popular choice among medium-scale greenhouse growers. However, the nutrient concentration may still fluctuate depending on water flow, and periodic monitoring is required to avoid under- or over-fertilization.
- **Fertilizer Pump/Injector System:** In this method, specialized fertilizer pumps or injectors are installed to push fertilizer solution directly into the irrigation line under controlled pressure. These pumps ensure a precise and consistent supply of nutrients, regardless of fluctuations in water flow or pressure. The system is more accurate than tank or bypass methods and is often used in medium to large-scale operations. Although

the cost is higher and requires skilled handling, it offers greater reliability and efficiency for commercial protected cultivation.

- ☆ **Advanced Automated Systems:** Automated fertigation systems use computerized controllers integrated with sensors to regulate fertilizer delivery based on crop stage, soil moisture, and environmental conditions. These systems can adjust dosage, timing, and nutrient composition dynamically, ensuring optimal crop growth and maximum input efficiency. They are highly suitable for hydroponics, large commercial greenhouses, and high-value horticultural crops such as tomatoes, cucumbers, capsicum, strawberries, and cut flowers. While the initial investment and technical requirements are high, the long-term benefits in terms of yield, quality, and resource savings make them the most advanced option available.

7.2.6 Scheduling and Management

- ☆ **Fertilizer Doses and Crop Growth Stages:** Fertigation must be synchronized with the physiological needs of the crop. During the vegetative stage, higher nitrogen levels are required to promote leaf and stem growth. In the flowering phase, phosphorus becomes critical for strong root development and better flower initiation. During fruiting and maturation, potassium plays a dominant role in improving fruit size, colour, and quality. A stage-wise nutrient plan ensures maximum crop performance and avoids nutrient wastage.
- ☆ **Frequency of Application:** Applying fertilizers in small, frequent doses is far more effective than delivering large, single applications. Frequent fertigation ensures a consistent nutrient supply to the root zone, improves absorption, reduces leaching losses, and prevents sudden surges of salts that may damage roots. This approach also provides flexibility to adjust nutrient levels according to crop response.
- ☆ **Timing of Fertigation:** The timing of nutrient delivery significantly influences uptake efficiency. Fertigation should be carried out during the cooler hours of the day, such as early morning or late afternoon, when plant transpiration is steady and evaporation losses are minimal. Midday fertigation should be avoided, as high temperatures can cause rapid evaporation, salt accumulation, and reduced nutrient absorption.
- ☆ **Monitoring pH of Solution:** Maintaining the correct pH of irrigation water and fertilizer solution is essential for nutrient availability. Most crops thrive when the fertigation solution has a pH between 5.5 and 6.5. Outside this range, certain nutrients may precipitate or become unavailable to plants. Acidic fertilizers such as phosphoric acid or nitric acid are often used to adjust and stabilize the pH of the fertigation solution.

- **Monitoring Electrical Conductivity (EC):** The electrical conductivity (EC) of the fertigation solution indicates the concentration of dissolved salts. Monitoring EC helps avoid both under-fertilization and excessive salt buildup, which can stress plants or cause root burn. Crops have specific EC requirements, and exceeding these levels can reduce yield and quality. Regular EC monitoring ensures precise control over nutrient concentration and optimal crop health.

Do's in Fertigation Scheduling

- Apply nutrients according to crop growth stages (N for vegetative, P for flowering, K for fruiting).
- Use frequent, small doses instead of large, infrequent applications.
- Schedule fertigation during early morning or late afternoon to reduce evaporation losses.
- Regularly monitor pH (5.5–6.5) of irrigation water and fertigation solution.
- Check and adjust EC to match the crop's requirement for optimal nutrient uptake.
- Flush the system with clean water after fertigation to prevent salt buildup in pipes and emitters.
- Keep detailed records of fertigation schedules to make stage-wise adjustments in future crops.

Don'ts in Fertigation Scheduling

- Do not apply large fertilizer doses at once, as it may cause root burn or nutrient imbalance.
- Avoid fertigation during the hottest hours of the day, when uptake is minimal and losses are higher.
- Do not mix incompatible fertilizers (*e.g.*, calcium nitrate with sulphates or phosphates) in the same tank.
- Never ignore filter cleaning and maintenance, as clogging disrupts uniform nutrient delivery.
- Avoid using low-quality or insoluble fertilizers, which may cause precipitation and block emitters.
- Do not rely on guesswork–always base fertigation schedules on soil/water analysis and crop stage needs.

7.2.7 Advantages of Fertigation

- **Enhances Nutrient and Water Use Efficiency:** Fertigation ensures that both water and nutrients are supplied directly to the root zone, minimizing wastage and losses through leaching or runoff, thereby maximizing resource utilization.

- ✰ **Provides Flexibility in Application:** Growers can adjust the timing, concentration, and composition of fertilizer solutions according to crop growth stages and environmental conditions, ensuring precise nutrient management.
- ✰ **Improves Crop Yield and Quality:** By delivering nutrients in readily available forms at critical stages, fertigation promotes vigorous growth, uniform development, and higher yields with superior quality produce.
- ✰ **Reduces Soil Compaction and Labor Costs:** Since irrigation and fertilization are integrated into one system, the need for repeated field operations is reduced, preventing soil compaction and lowering overall labour requirements.
- ✰ **Facilitates Uniform Nutrient Distribution:** Fertigation systems distribute nutrients evenly across the crop area, ensuring that all plants receive balanced nourishment regardless of their position in the greenhouse or field.
- ✰ **Suitable for High-Value Crops in Protected Cultivation:** Fertigation is particularly beneficial for sensitive, high-value crops like tomato, cucumber, capsicum, and cut flowers, where precise nutrient management is essential for maximizing profitability.

7.2.8 Limitations of Fertigation

- ✰ **High Cost of Water-Soluble Fertilizers:** Fertigation requires the use of fully water-soluble fertilizers, which are generally more expensive than conventional granular forms, increasing production costs for growers.
- ✰ **Risk of Emitter Clogging:** If water is not properly filtered or if fertilizers contain insoluble impurities, emitters and drippers may clog, leading to uneven nutrient distribution and poor crop performance.
- ✰ **Need for Skilled Management:** Fertigation demands technical knowledge to prepare accurate nutrient solutions and schedules; improper management may result in over-fertilization, salt accumulation, or nutrient imbalances.
- ✰ **High Initial Installation Cost:** The cost of installing fertigation units, pumps, filters, and automation systems is relatively high, making it less affordable for small-scale farmers without external support.
- ✰ **Requirement for Continuous Monitoring:** Regular monitoring of water quality, pH, and electrical conductivity (EC) is essential for effective fertigation; neglect can cause crop stress, poor nutrient uptake, or system damage.

7.2.9 Best Practices for Fertigation in Protected Cultivation

- **Use of High-Quality, Fully Water-Soluble Fertilizers:** Always select fertilizers that are 100 per cent water-soluble and free from impurities to ensure smooth flow through emitters and avoid precipitation or clogging in the system.
- **Regular Flushing of the System:** Flush irrigation lines with clean water at regular intervals to prevent salt deposits, scaling, and chemical buildup that can damage equipment and affect nutrient delivery.
- **Calibration of Fertigation Units:** Ensure that injectors and dosing equipment are calibrated frequently for precise nutrient application, preventing under- or over-fertilization.
- **Maintain Irrigation Water Quality:** Monitor and maintain irrigation water at an optimal pH (5.5–6.5) and electrical conductivity (EC) suited for the crop, as deviations can hinder nutrient uptake and crop growth.
- **Apply Nutrients in Split Doses:** Divide the total fertilizer requirement into smaller, frequent applications aligned with crop growth stages (vegetative, flowering, fruiting) to enhance nutrient efficiency.
- **Record-Keeping and Monitoring:** Maintain detailed records of fertigation schedules, fertilizer formulations, and crop responses. These records serve as valuable references for improving fertigation strategies in future cycles.

Do's in Fertigation

- Use only high-quality, water-soluble fertilizers suitable for fertigation.
- Test and maintain water pH (5.5–6.5) and EC regularly for optimal nutrient uptake.
- Calibrate fertigation units frequently to ensure accurate dosing.
- Flush the irrigation system with clean water after fertigation to prevent clogging.
- Apply nutrients in split doses according to crop phenology (vegetative, flowering, fruiting).
- Keep detailed records of fertigation schedules, fertilizer types, and crop responses.
- Conduct soil and water analysis periodically to refine fertigation strategies.

Don'ts in Fertigation

- Don't use low-grade or insoluble fertilizers that may cause precipitation or block emitters.
- Avoid fertigation during peak hot hours to reduce evaporation losses and crop stress.

- ☆ Don't mix incompatible fertilizers (*e.g.*, calcium nitrate with phosphates/ sulphates) in the same tank.
- ☆ Never ignore filter cleaning and maintenance, as clogged filters reduce system efficiency.
- ☆ Avoid giving large, single fertilizer doses, which may lead to root burn or salt buildup.
- ☆ Don't rely on guesswork–fertigation must be based on crop stage, soil status, and water quality.

7.2.10 Suitability in Protected Cultivation

Fertigation is particularly effective in protected environments such as polyhouses, greenhouses, and shade nets, where high-value horticultural crops like tomatoes, cucumbers, capsicum, strawberries, and cut flowers are grown. It allows growers to optimize limited resources, produce superior quality crops, and achieve higher profitability. Comparison of fertigation methods in protected cultivation is summarised in Table 7.3.

Table 7.3: Comparison of Fertigation Methods in Protected Cultivation

Method	*Description*	*Advantages*	*Limitations*	*Suitability*
Tank Mixing Method	Fertilizer dissolved in a tank and mixed into irrigation line.	Simple design; Low cost; Easy to operate.	Limited control over dosage; Uneven concentration if not well mixed.	Small-scale protected units; Farmers with limited resources.
Venturi Injector Method	Uses pressure difference to draw fertilizer into irrigation system.	No moving parts; Moderate cost; Reliable operation.	Requires specific pressure; Not suitable for low-pressure systems.	Medium-scale greenhouses and polyhouses; Crops needing moderate fertigation precision.
Bypass/ Parallel Tank Method	Part of irrigation water passes through fertilizer tank, then rejoins mainline.	Better control than tank method; Simple and effective.	Still less precise than automated systems; Needs regular monitoring.	Commercial nurseries and medium-scale protected cultivation.
Automated/ Computerized System	Computer-controlled injectors regulate fertigation based on crop needs, soil, and climate.	Highly precise; Saves water and fertilizer; Programmable schedules; Uniform distribution.	High cost; Requires skilled operation; Dependence on power and sensors.	Large-scale greenhouses, hydroponics, and high-value horticultural crops (tomato, cucumber, capsicum, cut flowers).

7.3 Scheduling and Management

Effective scheduling and management in fertigation are essential to optimize crop performance and resource use. By supplying nutrients and water in the right quantity, at the right time, and in the correct form, crops receive balanced nourishment throughout their growth cycle. Proper scheduling minimizes nutrient losses through leaching or evaporation, thereby improving fertilizer-use efficiency. It also ensures that each growth stage–vegetative, flowering, fruiting, and ripening–receives the precise nutrient balance required. Well-planned fertigation promotes uniform crop growth, enhances yield quality, and extends shelf life of produce. Moreover, it helps maintain soil health by preventing excess salt accumulation. Ultimately, systematic fertigation management makes protected cultivation more sustainable and profitable. The following aspects must be carefully managed:

7.3.1 Matching Fertilizer Doses with Crop Growth Stages

Nutrient requirements of horticultural crops vary significantly during different growth phases such as vegetative, flowering, fruiting, and ripening. Fertigation schedules should be designed to meet these changing needs:

- ☆ **Vegetative Stage:** Crops demand higher amounts of nitrogen to stimulate vigorous leaf, stem, and canopy development, which lays the foundation for strong photosynthetic activity.
- ☆ **Flowering Stage:** A balanced supply of nitrogen and phosphorus is required to encourage the initiation and development of flowers, ensuring good fruit set and reduced flower drop.
- ☆ **Fruiting Stage:** Higher levels of potassium, supported by moderate nitrogen, enhance fruit enlargement, improve quality parameters such as firmness, and increase resistance to pests and diseases.
- ☆ **Ripening Stage:** At this stage, nitrogen supply is minimized, while adequate potassium and essential micronutrients like boron, zinc, and magnesium are provided to enhance fruit colour, sweetness, flavour, and overall market value.

7.3.2 Frequency and Dose of Application

In fertigation, the principle of "little and often" is emphasized rather than supplying fertilizers in large, infrequent doses. Applying nutrients in smaller, frequent intervals ensures that they remain continuously available in the crop root zone, where absorption is most efficient. This approach not only minimizes leaching and runoff losses but also prevents sudden surges of salts that can damage roots. Daily or alternate-day fertigation is particularly beneficial in protected cultivation, where crops are highly sensitive to nutrient imbalances. Frequent, precise dosing also aligns better with the plant's physiological needs at different growth stages. Moreover, it helps in maintaining optimum soil moisture and nutrient balance simultaneously. By adopting this strategy, growers can maximize fertilizer-use efficiency, promote steady crop growth, and achieve higher yields with improved quality. Comparison between single dose and frequent split dose fertigation in protected cultivation is depicted in Table 7.4.

Table 7.4: Comparison between Single Dose and Frequent Split Dose Fertigation in Protected Cultivation

Aspect	*Single/Large Dose Fertigation*	*Frequent Split Dose Fertigation*
Nutrient Availability	Nutrients available only for a short period, followed by depletion.	Continuous supply of nutrients to the root zone.
Efficiency	Low fertilizer-use efficiency due to leaching and runoff losses.	High efficiency, with maximum uptake by plants.
Risk of Salt Accumulation	High, as concentrated salts can accumulate and cause root injury.	Low, as nutrients are supplied in dilute, manageable amounts.
Crop Growth	Irregular growth with possible stress periods between applications.	Uniform growth with steady nutrient and water availability.
Suitability	Traditional or low-tech systems with limited precision.	Modern, protected cultivation systems requiring precision management.
Impact on Yield and Quality	Lower yields and inconsistent quality of produce.	Higher yields, better quality, and improved market value.

7.3.3 Timing of Fertigation

Fertigation should be carried out during the cooler hours of the day (early morning or late evening). This practice reduces evaporation losses, enhances nutrient absorption, and avoids plant stress caused by high midday temperatures. Effective timing of fertigation is essential to maximize nutrient uptake, minimize losses, and ensure healthy crop growth. Several factors must be considered:

- ✰ **Fertigation During Cooler Hours:** The ideal time for fertigation is early morning or late evening when temperatures are relatively low. During these hours, plant transpiration is moderate, allowing nutrients to be absorbed efficiently without stress on the plant.
- ✰ **Minimizing Evaporation Losses:** Applying fertilizer solutions during midday, when temperatures are high, leads to excessive evaporation of water from the soil and foliage. Fertigation during cooler periods reduces this loss, ensuring that the maximum amount of water and nutrients reaches the root zone.
- ✰ **Reducing Plant Stress:** High midday temperatures can cause rapid water loss, heat stress, and stomatal closure in plants, which reduces nutrient uptake. Timely fertigation prevents these physiological stresses and supports optimal photosynthesis and growth.

- ☆ **Synchronization with Crop Water Requirements:** Fertigation should also align with the crop's daily water demand. By irrigating during the cooler periods, the system provides both water and nutrients when the crop can absorb them most efficiently, enhancing overall water-use and nutrient-use efficiency.
- ☆ **Practical Considerations in Protected Cultivation:** In greenhouses or polyhouses, microclimate conditions may vary, so growers should monitor temperature, humidity, and light intensity to select the most suitable fertigation timing. Automated systems can help adjust application schedules based on real-time environmental conditions.

7.3.4 Monitoring pH and Electrical Conductivity (EC)

- ☆ **pH control:** The pH of irrigation water and fertigation solution should generally be maintained between 5.5 and 6.5 for optimal nutrient solubility and uptake.
- ☆ **EC monitoring:** Electrical conductivity indicates the total dissolved salts in the solution. Fertigation should be adjusted to maintain EC levels within crop-specific ranges to avoid salt stress or nutrient deficiency. Recommended pH and EC ranges for protected cultivation crops is depicted in Table 7.5.

Table 7.5: Recommended pH and EC Ranges for Protected Cultivation Crops

Crop	*Recommended pH*	*Recommended EC (dS/m)*	*Notes*
Tomato	5.5 – 6.5	2.0 – 3.5	Higher EC during fruiting enhances sweetness and color.
Cucumber	5.5 – 6.5	2.0 – 3.0	Sensitive to high EC; monitor to prevent tip burn.
Capsicum (Bell Pepper)	5.5 – 6.5	1.8 – 3.0	Moderate EC preferred to avoid leaf burn.
Strawberry	5.5 – 6.0	1.5 – 2.5	Low EC ensures better fruit quality and prevents salt stress.
Lettuce and Leafy Vegetables	5.5 – 6.5	1.2 – 2.0	Sensitive to high EC; frequent small doses recommended.
Rose (Cut Flower)	5.5 – 6.5	1.8 – 2.5	Balanced EC promotes flowering and stem quality.
Gerbera	5.5 – 6.5	1.8 – 2.5	Avoid high EC; reduces flower quality.

7.3.5 Water and Nutrient Balance

Both over-irrigation and under-irrigation can negatively affect nutrient uptake. Therefore, irrigation scheduling must ensure that the soil remains at field capacity without waterlogging. Similarly, fertilizer concentrations should not exceed the tolerance level of crops to prevent root injury. Maintaining the proper balance between water and nutrients is essential for healthy plant growth, efficient resource use, and optimal crop yield. Both excess and deficit irrigation can negatively affect nutrient availability and uptake.

- ☆ **Avoiding Over-Irrigation:** Over-irrigation leads to waterlogging, reduces oxygen availability in the root zone, and can cause nutrient leaching. Essential nutrients, particularly nitrogen, may be washed away beyond the root zone, reducing fertilizer-use efficiency and causing environmental contamination.

  Practical Tip

  - ❖ Monitor soil moisture regularly using tensiometers or soil moisture sensors.
  - ❖ Schedule irrigation to maintain soil close to field capacity, avoiding prolonged saturation.

- ☆ **Preventing Under-Irrigation:** Insufficient irrigation limits nutrient mobility and uptake, as roots require water for nutrient transport. Prolonged water deficit can lead to stress, stunted growth, reduced flowering, and poor fruit development.

  Practical Tip

  - ❖ Adjust irrigation frequency and volume based on crop stage, soil type, and climate conditions.
  - ❖ Use drip or micro-irrigation to supply precise amounts of water directly to the root zone.

- ☆ **Managing Fertilizer Concentration:** Excessively concentrated nutrient solutions can injure roots and disrupt osmotic balance, while overly diluted solutions may not meet crop nutrient requirements. Maintaining the proper concentration ensures safe and efficient nutrient absorption.

  Practical Tip

  - ❖ Monitor EC levels of the fertigation solution and adjust fertilizer doses according to crop tolerance.
  - ❖ Apply nutrients in split doses to match crop growth stages and water availability.

- ☆ **Integrated Water-Nutrient Management:** Balancing water and nutrients involve coordinating irrigation schedules with fertigation, soil moisture, and crop needs. Proper integration improves nutrient-use efficiency, prevents stress, and enhances yield and quality in protected cultivation systems.

7.3.6 System Maintenance and Calibration

Regular cleaning of filters, flushing of lateral lines, and calibration of fertigation units are essential for uniform nutrient distribution. Automated fertigation systems should be checked periodically to ensure accurate dosing and flow control.

- ☆ **Cleaning of Filters:** Filters play a critical role in preventing particulate matter from entering irrigation lines and clogging emitters. Regular cleaning of filters is essential to remove debris, sediment, or algae that may accumulate over time. Properly maintained filters ensure consistent water and nutrient flow, which is vital for uniform crop growth.
- ☆ **Flushing of Lateral Lines:** Periodic flushing of lateral lines is necessary to remove accumulated salts, sediments, or microbial growth that could obstruct emitters. Flushing helps maintain uniform distribution of water and nutrients throughout the fertigation system, ensuring all plants receive adequate nourishment.
- ☆ **Calibration of Fertigation Units:** Calibration of injectors, pumps, and automated dosing systems is essential to deliver the correct amount of fertilizer. Accurate calibration prevents over-fertilization, nutrient imbalances, and potential crop stress, ensuring optimal nutrient-use efficiency and plant health.
- ☆ **Maintenance of Automated Systems:** Automated fertigation systems require periodic inspection to verify that sensors, controllers, and valves are functioning correctly. Faulty or malfunctioning components can lead to uneven nutrient distribution, system inefficiencies, or even complete shutdowns, affecting overall crop performance.

7.3.7 Record-Keeping and Crop Monitoring

Maintaining detailed records of fertilizer types, doses, application frequency, pH, EC, and crop responses is crucial for refining fertigation practices. Continuous monitoring of plant growth, leaf colour, and yield trends helps in fine-tuning the schedule.

- ☆ **Importance of Record-Keeping:** Maintaining detailed records of fertigation practices is essential for achieving precision in protected cultivation. Records should include fertilizer types, quantities applied, frequency of application, pH, EC values, and environmental conditions. This documentation allows growers to analyse past performance, identify trends, and make informed adjustments to optimize nutrient management.
- ☆ **Monitoring Crop Growth:** Regular observation of crop growth parameters, such as plant height, leaf size, leaf colour, and overall vigour, provides real-time feedback on the effectiveness of fertigation schedules. Deviations from expected growth patterns may indicate nutrient deficiencies, over-fertilization, or water stress.

- **Tracking Yield and Quality Trends:** Systematic monitoring of flowering, fruit set, yield quantity, and quality attributes helps evaluate the efficiency of fertigation practices. Data on yield trends, fruit size, sweetness, or colour can guide adjustments in nutrient doses and timing for subsequent cycles.
- **Benefits of Record-Keeping and Monitoring:** A combination of precise record-keeping and vigilant crop monitoring enables growers to fine-tune fertigation schedules, enhance nutrient-use efficiency, improve crop health, and maximize yield and quality. Over time, these practices support sustainable and profitable protected horticulture.

7.3.8 Integration with Climate and Soil Factors

Fertigation scheduling should be adapted according to local conditions:

- **Climate Considerations:** Local climate plays a major role in determining the frequency, timing, and amount of fertigation. High temperatures increase plant transpiration, thereby elevating water and nutrient demand. Similarly, high humidity can reduce transpiration rates, affecting nutrient uptake. Fertigation schedules should therefore be adjusted according to daily temperature, humidity, and seasonal variations to ensure optimal nutrient absorption and prevent plant stress.
- **Soil Type Adaptation:** Soil texture and structure influence water-holding capacity, nutrient retention, and drainage. Sandy soils, which drain quickly and retain fewer nutrients, require more frequent and smaller fertigation applications to prevent leaching and nutrient deficiency. In contrast, loamy or clayey soils retain water and nutrients for longer periods, allowing for slightly less frequent fertigation while still meeting crop needs. Understanding soil characteristics is essential for designing an effective fertigation schedule.
- **Protected Cultivation Conditions:** Crops grown in greenhouses, polyhouses, or other protected environments experience a controlled microclimate, including regulated temperature, humidity, and light intensity. These conditions often necessitate more precise and responsive fertigation management compared to open-field cultivation. Automated systems and frequent monitoring help adjust water and nutrient delivery to match the specific needs of crops under protected cultivation, ensuring maximum efficiency and productivity.

Propagation and Production of Quality Planting Material

8.0 Introduction

Propagation and production of quality planting material are fundamental aspects of successful horticultural crop cultivation. High-quality planting material ensures uniform growth, better yield, and disease-free crops, forming the foundation for sustainable production systems. In horticulture, planting material refers to seeds, seedlings, saplings, cuttings, grafts, or tissue-cultured plants that are used to establish new crops. The choice of propagation method and source of planting material significantly influences crop performance, productivity, and profitability. Propagation can be broadly classified into sexual and asexual methods. Sexual propagation involves the use of seeds, which provides genetic variability and is suitable for breeding programs. However, it often results in heterogeneity in growth and fruiting characteristics. Asexual propagation, on the other hand, includes techniques such as cutting, layering, grafting, budding, and tissue culture, which produce true-to-type plants identical to the parent. These methods are preferred for high-value crops where uniformity and early fruiting are desired.

The production of quality planting material requires careful attention to selection of parent plants, propagation techniques, nursery management, and pest and disease control. Parent plants should be healthy, vigorous, and free from pathogens to ensure that progeny inherit desirable traits. Proper propagation techniques, including the use of rooting hormones, grafting methods, or in vitro culture, increase the success rate and improve plant survival. Nursery management plays a critical role in the growth and development of young plants. Factors such as soil or substrate quality, irrigation, fertilization, temperature control, and light management directly affect seedling vigour and overall quality. In protected cultivation systems, these factors can be precisely controlled to produce uniform and robust planting material.

The production of disease-free planting material is particularly important in horticulture, as viral, bacterial, or fungal infections can rapidly spread and severely affect crop yield. Techniques such as seed treatment, sanitation, proper spacing, and use of pathogen-free mother plants are essential to maintain health standards. Advanced methods, including micropropagation and tissue culture, allow for the mass production of elite, pathogen-free plants in a relatively short period. Quality planting material is also important for introducing improved varieties, hybrids, or genetically superior clones into production systems. Uniform planting material ensures consistent crop management, reduces variability in yield and quality, and enhances market acceptability. Furthermore, using superior planting material helps conserve genetic resources, improves resilience against abiotic and biotic stresses, and contributes to sustainable horticultural practices. Overall, the propagation and production of quality planting material combine scientific knowledge, skilled management, and careful monitoring to provide growers with healthy, uniform, and high-performing plants. It is a cornerstone of modern horticulture, enabling efficient crop establishment, higher productivity, and long-term profitability. Ensuring the availability of quality planting material is therefore a critical component of horticultural development programs worldwide.

8.1 Methods of Propagation

Propagation is the process of multiplying plants to produce new individuals for cultivation. In protected horticulture, where environmental conditions such as temperature, humidity, and light can be controlled, propagation techniques can be optimized for higher success rates, uniformity, and production of high-quality planting material. Broadly, propagation methods are classified into sexual and asexual (vegetative) methods.

8.1.1. Sexual Propagation (Seed Propagation)

Sexual propagation involves the use of seeds to grow new plants. It is widely used for vegetables, flowers, and some fruit crops. Seeds are usually selected from healthy, disease-free parent plants to ensure good germination and vigour (Figure 8.1).

Advantages in Protected Cultivation

- ✰ Enables production of large numbers of seedlings in a short period.
- ✰ Promotes genetic variability, which can be useful for breeding programs.
- ✰ Easier to handle and transport.

Figure 8.1: Propagation by Seed
(*Source*: https://cdn.britannica.com)

Limitations

- Offspring may not be uniform, leading to variability in crop performance.
- Some crops have low germination rates or long seedling establishment periods.

Protected environments such as nurseries, polyhouses, and shade nets provide controlled conditions that enhance germination rates and seedling growth, allowing early and uniform crop establishment.

8.1.2. Asexual or Vegetative Propagation

Asexual propagation is the process of producing new plants without the use of seeds, relying instead on vegetative plant parts such as stems, roots, leaves, or specialized structures like bulbs, corms, tubers, and rhizomes. This method ensures that the offspring are genetically identical clones of the parent plant, thereby preserving desirable characteristics such as fruit quality, flower colour, disease resistance, and adaptability. Unlike sexual propagation, which introduces variability, asexual methods guarantee true-to-type plants with uniform growth and performance, which is essential for maintaining the quality of elite varieties. Vegetative propagation is especially beneficial for high-value crops and commercial hybrids where uniformity and predictability are crucial. Crops like sugarcane, banana, potato, and pineapple are commonly multiplied by vegetative means to ensure rapid multiplication and consistency in yield. Similarly, vegetative methods are preferred in fruit crops such as mango, guava, citrus, and grapes, where seed propagation often leads to variability and loss of superior traits. In ornamentals like roses, chrysanthemums, and orchids, vegetative propagation helps in preserving aesthetic qualities such as flower colour, size, and fragrance, which may not remain stable through seed propagation.

Various techniques of vegetative propagation are employed depending on the crop and purpose. Cuttings (as in sugarcane and grapes), layering (as in jasmine and guava), grafting and budding (as in mango, citrus, and rose), and tissue culture (as in banana and orchids) are widely adopted methods. Naturally occurring vegetative structures like tubers (potato), rhizomes (ginger, turmeric), bulbs (onion, garlic), and suckers (banana, pineapple) also serve as propagation units. These methods not only enable large-scale multiplication within a short time but also help in overcoming problems like seed sterility, poor germination, or genetic segregation. Thus, asexual propagation plays a pivotal role in modern horticulture, ensuring the commercial viability and sustainability of many important crops (Figure 8.2).

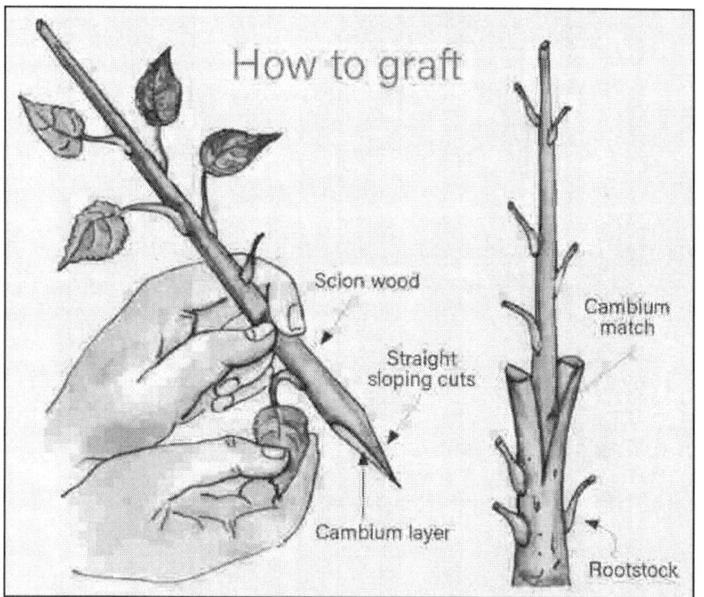

Figure 8.2: Asexual Propagation in Plants
(*Source*: https://www.len.com.)

Common methods include:

- ☆ **Cutting:** Stem, leaf, or root cuttings are used to develop new plants. Cuttings are treated with rooting hormones and grown under controlled moisture and temperature conditions to enhance root initiation (Figure 8.3).
- ☆ **Layering:** In layering, a stem is bent and covered with soil or substrate while still attached to the parent plant. Roots develop at the buried portion, and the new plant can then be separated and transplanted (Figure 4).

Figure 8.3: Propagating Plants from Cutting
(*Source*: https://cimg2.ibsrv.net/cimg/www.davesgarden.com)

Figure 8.4: Layering
(*Source*: www.freepik.com)

- ☆ **Grafting and Budding:** Grafting involves joining a scion (shoot) from a desired variety onto a rootstock with strong root systems. Budding is a form of grafting where a single bud is inserted into the rootstock. These techniques are widely used in fruit crops, ornamental plants, and high-value vegetables for disease resistance, early maturity, and uniformity (Figure 8.5).
- ☆ **Tissue Culture (Micropropagation):** Micropropagation involves the in vitro production of plants under sterile conditions. Small tissue explants from a parent plant are cultured on nutrient media to produce a large number of disease-free, uniform plants. It is highly suitable for elite, high-value crops such as orchids, gerbera, tomato, and capsicum (Figure 8.6).

Figure 8.5: Grafting and Budding.
(*Source*: www.freepik.com)

Figure 8.6: Tissue Culture (Micropropagation)
(*Source*: www.freepik.com)

8.1.3. Advantages of Protected Propagation

- Controlled temperature, humidity, and light enhance rooting and seedling vigour.
- Reduced risk of pests and diseases during early growth stages.
- Higher success rates for difficult-to-propagate crops.
- Enables off-season production and early crop establishment.

In protected cultivation, choosing the appropriate propagation method depends on the crop, desired uniformity, propagation speed, and resource availability. Combining proper propagation techniques with controlled environmental conditions ensures high-quality planting material that forms the foundation for productive and profitable horticultural crops.

8.2 Production of Quality Planting Material

8.2.1 Introduction

Production of quality planting material forms the foundation of successful protected cultivation in horticultural crops. In high-value systems like polyhouses, shade nets, and greenhouses, where every input is precisely managed, the planting material must be of the highest standard. It should be genetically pure, true-to-type, and uniform in growth to ensure that all plants respond consistently under controlled conditions. Healthy and vigorous planting material reduces the risk of pest and disease outbreaks, which can spread rapidly in enclosed structures. Using certified and pathogen-free seedlings not only improves survival rates and early establishment but also shortens the crop cycle, leading to early harvest and higher returns. Since protected cultivation focuses on quality-oriented production, especially for export and domestic premium markets, planting material plays a critical role in achieving desirable traits such as size, colour, flavour, and shelf life. Moreover, uniform planting material allows for efficient resource use, including irrigation, fertigation, and space management, thereby enhancing overall productivity. In fruit crops, ornamentals, and vegetables, superior planting material ensures consistent yield, reduced variability, and better adaptability to microclimatic regulation inside protected structures. The use of hybrids, grafted plants, and tissue culture-derived seedlings has further strengthened the role of quality planting material in achieving commercial success. Thus, ensuring the availability and adoption of reliable, disease-free, and elite planting stock is not just desirable but essential for sustainable and profitable protected cultivation.

8.2.2 Importance of Quality Planting Material in Protected Cultivation

- ✰ **Ensures uniformity in growth and yield:** Quality planting material provides seedlings or propagules that are uniform in vigour, size, and growth habits. In protected structures, where space utilization and precision management are crucial, uniform plants ensure synchronized flowering, fruiting, and harvesting. This uniformity not only simplifies cultural operations such as pruning, training, and fertigation but also leads to predictable yields and consistent quality of produce.
- ✰ **Maintains genetic purity of hybrids and elite varieties:** Protected cultivation often relies on high-performing hybrids and elite clones that possess specific desirable traits such as high yield, resistance to biotic and abiotic stresses, or superior quality attributes. Using genetically pure planting material safeguards these traits, ensuring that farmers obtain

the true-to-type characteristics of the variety or hybrid. This purity is critical for crops like tomato, capsicum, cucumber, orchids, and roses, where even minor genetic variations can affect quality and market value.

- ☆ **Minimizes incidence of pests and diseases inside protected structures:** Since greenhouses and polyhouses create a favourable microclimate for both crops and pathogens, any contamination through infected planting material can spread rapidly, leading to severe economic losses. Disease-free and pathogen-indexed planting material significantly reduces the risk of introducing viruses, fungi, or bacterial infections into the system. This preventive measure is far more cost-effective than managing outbreaks once the crops are established.
- ☆ **Facilitates better resource use efficiency under controlled conditions:** Uniform and healthy planting material makes optimum use of inputs like light, water, nutrients, and space in protected cultivation. Plants with similar growth stages and vigour respond uniformly to fertigation and irrigation schedules, resulting in efficient input utilization. This harmony ensures that energy, water, fertilizers, and other resources invested in the high-cost protected structures are used effectively, reducing wastage and maximizing returns.
- ☆ **Enhances profitability and market acceptance of produce:** High-quality planting material is directly linked to superior crop performance and quality harvest. Uniform size, colour, flavour, and extended shelf life of fruits, vegetables, or flowers enhance their acceptability in both domestic and export markets. Consistency in quality allows growers to build trust with buyers, secure premium prices, and ensure repeat demand. Ultimately, the use of reliable and certified planting material translates into higher profitability and long-term sustainability of protected cultivation enterprises.

8.2.3 Sources of Planting Material

- ☆ **Certified Nurseries:** Certified nurseries, accredited by government agencies or recognized institutions, are the most reliable sources of quality planting material. They supply seedlings, grafts, cuttings, and other propagules that meet prescribed standards of genetic purity, vigour, and health. Such nurseries often follow strict quality control protocols, including pest and disease screening, substrate sterilization, and use of certified mother plants. Farmers who procure planting material from these nurseries can be assured of uniformity, true-to-type plants, and better survival rates under protected conditions.
- ☆ **Research Institutes and State Agricultural Universities (SAUs):** National research institutes, ICAR centres, and state agricultural universities play a pivotal role in developing, maintaining, and distributing breeder and foundation planting material. These institutions serve as

the primary source of elite varieties, hybrids, and disease-resistant lines, ensuring that the genetic potential of newly developed crops is preserved. By supplying nucleus and foundation stock to certified nurseries and seed companies, they act as the backbone of the planting material supply chain.

- **Private Seed Companies:** Private companies have emerged as major contributors in supplying hybrid seeds, tissue-cultured plantlets, and grafted planting material for horticultural crops. Many of these companies invest heavily in research, biotechnology, and hi-tech nursery management to deliver high-performance planting material tailored to market needs. They also ensure year-round availability and offer farmers access to specialized hybrids that cater to domestic as well as export markets. Their role is especially significant in vegetables, flowers, and plantation crops where demand for hybrid and tissue culture plants is increasing rapidly.
- **On-farm Nurseries:** Progressive farmers and grower cooperatives often establish on-farm nurseries under polyhouses or shade-net conditions to meet their own planting material requirements. This practice reduces dependence on external sources, ensures timely availability, and allows growers to propagate elite stock under closely monitored conditions. On-farm nurseries are particularly useful for crops like capsicum, tomato, cucumber, and various flowers where demand is frequent and continuous. With proper training and technical guidance, such nurseries can also become commercial ventures supplying planting material to neighbouring growers.

8.2.4 Methods of Producing Quality Planting Material

Seed-Based Production

- Seeds are selected from genetically pure and disease-free sources.
- Treated with fungicides and bio-inoculants before sowing.
- Raised in pro-trays with cocopeat, vermicompost, and biofertilizers to ensure uniform germination.
- Maintained in controlled conditions (25–28°C, 70–80 per cent RH) with misting and fertigation.
- Hardened under shade before transplanting.

Vegetative Propagation

- **Cuttings:** Used for crops like grapes, chrysanthemum, and geranium. Rooting hormones (IBA, NAA) enhance rooting.
- **Layering and Grafting:** Popular for fruit crops like guava, mango, and citrus where hybrid seed propagation is not possible.

- **Suckers, Bulbs, and Rhizomes:** Banana, pineapple, onion, and ginger are multiplied using these specialized structures.

Tissue Culture and Micropropagation

- Rapid multiplication of elite and disease-free clones under sterile conditions.
- Widely used for banana, potato, orchids, gerbera, carnation, and anthurium.
- Ensures year-round availability of planting material.
- Requires proper acclimatization and hardening in net houses before field planting.

8.2.5 Nursery Management Practices under Protected Cultivation

- **Growing Media:** The choice of growing media plays a vital role in the successful production of quality planting material in protected cultivation. Unlike field soil, which often harbours pathogens, weeds, and nematodes, sterile and porous substrates such as cocopeat, vermiculite, perlite, and peat moss provide a clean and uniform environment for root growth. These media are light in weight, well-aerated, and have excellent water-holding capacity, ensuring proper moisture and oxygen availability to the developing seedlings. Being inert, they reduce the risk of soil-borne diseases and allow precise control over nutrient application through fertigation. Cocopeat, in particular, is widely used due to its renewable nature, buffering capacity, and ability to maintain balanced moisture levels. For crops like capsicum, tomato, cucumber, and flowers, plug-trays filled with such sterile media enable uniform germination and strong seedling growth. In addition, media can be enriched with biofertilizers, beneficial microbes like Trichoderma, or slow-release nutrients to enhance seedling vigour. Proper sterilization of media, either through steam treatment, solarization, or chemical disinfection, is essential to maintain its health. Thus, the use of high-quality, pathogen-free, and porous growing media forms the foundation of robust nursery management in protected cultivation.
- **Irrigation and Fertigation:** Efficient irrigation and fertigation systems are the lifelines of nurseries established under protected cultivation. Since seedlings are delicate and highly sensitive to fluctuations in water and nutrient supply, precise delivery systems such as drip irrigation, micro-sprinklers, and misting units are preferred. These methods ensure that water is applied in small but frequent quantities, preventing both drought stress and waterlogging, which are harmful to root development. Fertigation, the technique of applying nutrients dissolved in irrigation water, allows growers to meet the exact nutrient demands of young plants

during different growth stages. By using water-soluble fertilizers and bio-nutrient solutions, seedlings receive a balanced supply of macro and micronutrients, resulting in vigorous growth and uniformity. In misting systems, the humidity inside the nursery is maintained at optimal levels (70–80 per cent), reducing transplant shock and encouraging better root establishment. Automation of irrigation and fertigation through timers or sensors further enhances efficiency, reduces labour costs, and prevents wastage of inputs. When combined with quality growing media, these systems create ideal root-zone conditions that directly impact seedling survival and future crop performance. Thus, precise water and nutrient management is indispensable for producing healthy, market-ready planting material.

- **Pest and Disease Management:** In protected cultivation, pest and disease management must be approached with utmost care because enclosed structures often provide a favourable microclimate for rapid multiplication of pathogens and insects. The first step is exclusion, achieved by installing insect-proof nets, double-door entry systems, and sticky traps to prevent the entry and spread of pests such as whiteflies, thrips, and aphids. Biological control agents, including predatory insects, beneficial fungi, and bacterial formulations, play an important role in maintaining a natural balance and reducing reliance on chemical pesticides. Soil-borne diseases can be controlled through soil solarization, steam sterilization, and the use of bioagents like Trichoderma and Pseudomonas, which suppress harmful pathogens. Regular monitoring and scouting of the nursery help in early detection, ensuring timely interventions before the problem escalates. Preventive seed treatment with fungicides or biocontrol formulations adds another layer of protection. Maintaining proper spacing and ventilation reduces humidity levels, thereby minimizing the chances of fungal diseases like damping-off, powdery mildew, or downy mildew. Integrated Pest Management (IPM) strategies combining cultural, biological, and mechanical methods are highly recommended in protected cultivation systems. In this way, a preventive, eco-friendly approach ensures pest-free and disease-free planting material, safeguarding both nursery investments and farmer confidence.
- **Hygiene and Sanitation:** Maintaining strict hygiene and sanitation standards in nurseries is critical for producing high-quality planting material. Since even a minor lapse can introduce pests or pathogens that multiply quickly in the enclosed environment, all trays, tools, and nursery equipment must be regularly sterilized using chemical disinfectants or hot water treatment. Structures such as benches, irrigation lines, and misting units should be cleaned periodically to prevent microbial build-up. Workers entering the nursery should follow biosecurity protocols,

including the use of footbaths, clean clothing, and hand disinfection, to minimize the chances of contamination. The disposal of diseased or weak seedlings must be carried out immediately and away from the nursery site to avoid secondary infection. Floor areas should be kept free from weeds, algae, and stagnant water, which often serve as breeding grounds for pests. In tissue culture nurseries, aseptic handling and laminar airflow chambers ensure pathogen-free multiplication. Maintaining records of sterilization schedules, input usage, and pest monitoring further strengthens nursery management. Ultimately, a clean and well-maintained nursery not only protects young seedlings but also builds farmer confidence in the reliability of planting material. Therefore, systematic hygiene and sanitation are non-negotiable pillars of quality planting material production in protected cultivation.

8.2.6 Quality Standards of Planting Material

- **Genetic Purity (True-to-Type Plants):** The foremost quality standard of planting material is maintaining genetic purity, which ensures that the plants are true-to-type and express the exact traits of the parent variety or hybrid. This is particularly important in high-value crops grown under protected cultivation, such as tomato, capsicum, cucumber, and orchids, where even slight genetic variation can affect yield, quality, and market acceptance. True-to-type plants guarantee uniform growth, predictable flowering, synchronized fruiting, and consistent harvest quality. They also allow farmers to achieve the full potential of advanced hybrids and elite clones developed by breeders. The absence of genetic contamination safeguards both economic investment and consumer trust in the produce. Hence, ensuring 100 per cent varietal fidelity is considered a primary benchmark for high-quality planting material.
- **Free from Diseases, Viruses, Nematodes, and Pests:** Another critical standard is that planting material must be completely free from diseases, viruses, nematodes, and insect pests, as these can easily spread in the favourable microclimate of protected structures. Infected material often results in crop failure, reduced productivity, or poor quality of produce, causing significant economic losses to farmers. Virus-indexed, pathogen-free, and nematode-free planting material minimizes these risks and ensures healthy crop establishment. The use of tissue culture, meristem culture, and certified nursery practices helps eliminate hidden infections and guarantees clean stock. Regular diagnostic testing, mother plant indexing, and nursery sanitation are also vital to maintain this standard. By using disease-free planting stock, growers can protect their crops from destructive epidemics and maximize returns.
- **Uniform in Size, Vigour, and Growth Habit:** Uniformity in planting material is essential for efficient management and maximum output

in protected cultivation. Seedlings or propagules should be similar in size, height, and leaf number, with robust root systems and vigorous growth habits. Uniform plants respond equally to irrigation, fertigation, and training practices, ensuring synchronized flowering, fruiting, and harvesting. This homogeneity simplifies crop management operations like pruning, pest control, and harvesting, while also enhancing overall productivity. Uneven planting material leads to irregular growth patterns, competition among plants, and reduced efficiency of resource use. Therefore, nursery managers must carefully grade seedlings before distribution to ensure that only uniform and vigorous stock reaches the growers.

- ☆ **Properly Hardened for Adaptability to Open-Field or Protected Conditions:** Before transplanting, planting material must undergo hardening, a process in which seedlings or plantlets are gradually exposed to conditions outside the nursery to build resilience. Hardening helps plants adjust to fluctuations in temperature, humidity, light intensity, and wind, reducing transplant shock and mortality. For tissue-cultured plants and delicate seedlings raised in pro-trays, hardening under shade nets or low-cost polyhouses is particularly important. This stage ensures that plants develop sturdier stems, stronger roots, and improved water-use efficiency, enabling smooth transition to the main field or protected structures. Properly hardened planting material is better equipped to withstand stress and establish quickly, leading to healthy and uniform crop stands.
- ☆ **Accompanied by a Quality Certification Tag Wherever Applicable:** The presence of a quality certification tag acts as an assurance to farmers that the planting material meets prescribed standards of purity, health, and performance. Certification agencies, government bodies, or recognized institutions issue these tags after verifying genetic fidelity, disease-free status, and nursery management practices. Such labelling builds confidence among growers, reduces the risk of spurious material, and promotes transparency in the supply chain. Certified planting material also fetches premium prices in the market and is often mandatory for export-oriented production systems. For crops like banana, potato, and flowers, certification ensures compliance with national and international quality standards. Thus, a certification tag is not just a label but a guarantee of reliability, authenticity, and quality assurance for planting material.

8.2.7 Hardening and Acclimatization

- ☆ **Tissue-Cultured and Greenhouse-Raised Seedlings are Hardened Gradually Before Planting:** Hardening is a crucial step in preparing delicate seedlings, especially tissue-cultured plantlets and greenhouse-raised seedlings, for successful establishment in the main field or polyhouse. In this process, plants that are initially grown under highly controlled, protected, and humid conditions are slowly exposed to more natural environmental factors. Tissue culture plantlets, for instance, emerge from sterile laboratories with soft tissues and poorly developed cuticles, making them vulnerable to stress and desiccation. By gradually reducing humidity, increasing air circulation, and introducing them to ambient light, the plants develop structural strength and functional adaptability. This transition strengthens the leaves, roots, and stems, ensuring that seedlings can withstand external conditions after transplanting. Without proper hardening, mortality rates can be high, leading to significant economic losses.

- ✰ **Exposed to Lower Humidity and Higher Light Intensity Step by Step:** The process of acclimatization involves systematically exposing seedlings to progressively lower humidity and higher light intensity to help them adapt naturally to open conditions. Initially, seedlings are kept under high humidity to prevent water loss, but over time, humidity levels are reduced to stimulate the development of stronger stomatal control and cuticular layers. Similarly, light exposure is gradually increased to encourage photosynthetic activity, chlorophyll development, and sturdier plant growth. If plants are suddenly shifted from low light and high humidity to field conditions, they may suffer from leaf scorching, wilting, or stunted growth. Controlled, stepwise exposure not only minimizes transplant shock but also prepares seedlings for efficient photosynthesis and resilience to stress. This gradual conditioning ensures a smooth and successful transition to the production environment.
- ✰ **Use of Shade Nets (50–75 per cent) During Hardening Prevents Stress:** Shade nets play an important role in regulating the microclimate during the hardening process. Depending on the crop type, shade nets of 50–75 per cent intensity are commonly used to provide partial protection from direct sunlight while still allowing sufficient light for photosynthesis. This moderated environment prevents sudden stress due to heat, radiation, or strong winds, which could otherwise damage tender seedlings. Shade nets also help maintain a balanced temperature and reduce transpiration losses, creating a more favourable environment for plant acclimatization. For crops like capsicum, tomato, cucumber, banana, and ornamentals such as gerbera and chrysanthemum, the use of shade nets during hardening is highly effective. Gradually reducing shade intensity over time further strengthens seedlings, making them capable of thriving in both open-field and protected conditions.
- ✰ **Application of Mild Fertilizers During This Stage Strengthens Seedlings:** Nutritional support is an essential component of the hardening phase. Application of mild doses of fertilizers, particularly

balanced NPK solutions along with micronutrients, enhances root and shoot development. Foliar sprays of bio-stimulants, seaweed extracts, or water-soluble fertilizers also help strengthen plant tissues and improve stress tolerance. Adequate nutrition during hardening promotes better chlorophyll synthesis, robust root architecture, and enhanced physiological functions, which are crucial for survival after transplanting. Over-fertilization must be avoided, as it can make seedlings tender and more susceptible to stress. Instead, carefully monitored, low-concentration nutrient applications build resilience and vigour. Well-fed and hardened seedlings establish quickly in the field, recover faster from transplant shock, and ultimately ensure higher survival rates and uniform crop growth.

8.2.8 Certification and Regulation

- ☆ **Planting Material Should Conform to Standards Set by Seed Certification Agencies, NHB, and ICAR Institutes:** To ensure uniformity, reliability, and performance, planting material used in protected cultivation must comply with standards laid down by recognized authorities such as Seed Certification Agencies, the National Horticulture Board (NHB), and ICAR institutes. These standards cover parameters such as genetic purity, varietal authenticity, seedling vigour, and absence of pests or diseases. Certification helps in guaranteeing that the planting material is genuine, true-to-type, and capable of delivering expected results under controlled environments. Farmers investing in high-value crops like tomato, capsicum, cucumber, banana, or flowers can rely on such certified material for stable yields and consistent quality. Thus, adherence to official certification standards acts as a safeguard against the risks of spurious or substandard planting material in commercial horticulture.
- ☆ **Accreditation Ensures Quality Assurance and Trust Among Farmers:** Accreditation of nurseries and seed production units by government or authorized agencies provides quality assurance and builds farmer confidence. Accredited nurseries are regularly inspected for their propagation practices, mother plant health, hygiene, pest management, and record-keeping, ensuring transparency and accountability. Farmers who procure planting material from such sources are assured of its performance, which reduces risks and increases trust. Accreditation also helps in streamlining the supply chain, preventing the circulation of counterfeit or low-quality seeds and plantlets in the market. Over time, the system builds brand credibility for accredited suppliers, while also promoting widespread adoption of improved varieties and hybrids. Ultimately, accreditation serves as a guarantee of reliability and professionalism in the horticultural sector.

- **Plant Quarantine Measures Are Followed for Imported Planting Material:** With the increasing globalization of horticulture, many growers depend on imported seeds, bulbs, rootstocks, or tissue culture plantlets. However, importing planting material poses the risk of introducing exotic pests, diseases, and invasive species that could threaten domestic agriculture. To prevent this, strict plant quarantine measures are enforced at entry points by the government. Imported material is thoroughly inspected, tested, and sometimes subjected to treatments like fumigation or disinfection before release. Quarantine procedures ensure that only safe, pest-free, and disease-free material enters the country, thereby protecting local crops and ecosystems. Farmers and nurseries must therefore comply with these regulations and source imported planting material only through authorized channels. This practice not only protects national biosecurity but also sustains the long-term health of the horticultural industry. Certification and regulation of planting material in protected cultivation is summarised in Table 8.1.

Table 8.1: Certification and Regulation of Planting Material in Protected Cultivation

Standard/Measure	*Agency/Authority*	*Purpose/Importance*
Seed Certification Standards	State/National Seed Certification Agencies	To ensure genetic purity, varietal authenticity, and true-to-type planting material.
Quality Standards for Nurseries	NHB (National Horticulture Board)	Accreditation of nurseries for supply of healthy, uniform, and certified planting stock.
Research-based Standards and Foundation Stock	ICAR Institutes and SAUs	Provide breeder and foundation planting material of elite and improved varieties.
Accreditation of Private Nurseries/ Companies	State Departments of Horticulture/NHB	Builds farmer confidence, ensures reliable supply, and prevents circulation of spurious stock.
Plant Quarantine Regulations	Plant Quarantine Organization of India (PQOI)	Prevents entry of exotic pests, diseases, and invasive species through imported material.
Phytosanitary Certification for Export	Directorate of Plant Protection, Quarantine and Storage (DPPQS)	Certifies exported planting material as pest-free, ensuring compliance with global trade standards.

8.2.9 Challenges in Producing Quality Planting Material

- **High Initial Cost of Infrastructure:** One of the foremost challenges in producing quality planting material is the heavy investment required for infrastructure development. Protected cultivation demands modern facilities like polyhouses, shade-net houses, mist chambers, fogging systems, and, in the case of micropropagation, tissue culture laboratories. These facilities involve significant capital costs, which may not be affordable for small and marginal farmers. In addition, the recurring expenses of maintaining environmental control systems, electricity, and specialized equipment further add to the financial burden. Without subsidies or institutional support, many growers find it difficult to establish or sustain such hi-tech nurseries.

- ☆ **Risk of Disease Contamination During Propagation:** Despite maintaining controlled conditions, planting material production remains vulnerable to disease and pest contamination. Pathogens like fungi, bacteria, viruses, and nematodes can enter the system through infected mother plants, contaminated tools, or improper sanitation practices. In tissue culture laboratories, even a minor lapse in aseptic technique can lead to rapid multiplication of infections across thousands of plantlets. Similarly, in nurseries, if preventive measures such as substrate sterilization, insect-proof netting, or regular monitoring are neglected, diseases can spread quickly and compromise the entire batch. Thus, maintaining strict phytosanitary standards is a constant challenge.
- ☆ **Lack of Skilled Manpower in Advanced Propagation Methods:** Producing high-quality planting material requires skilled personnel trained in specialized propagation techniques. Tasks such as grafting, budding, micropropagation, and hardening demand precision and technical expertise, which are often lacking in rural areas. The shortage of trained nursery managers, technicians, and tissue culture operators limits the efficiency and reliability of production systems. Even when skilled manpower is available, retaining them becomes difficult due to limited financial incentives. Continuous training and capacity-building programs are necessary to bridge this skill gap and to ensure that planting material meets the required standards.
- ☆ **Limited Awareness Among Farmers About Certified Planting Material:** Many farmers, particularly in remote and resource-poor regions, are still unaware of the importance of using certified planting material. They often rely on local, uncertified sources due to lower cost or easy availability, which compromises crop performance and increases susceptibility to diseases. Lack of awareness about long-term benefits such as uniformity, higher yields, and improved marketability results in poor adoption of certified material. Extension services, demonstration programs, and awareness campaigns are needed to sensitize farmers about the economic and agronomic advantages of certified planting stock, thereby promoting its wider use in protected cultivation.

8.2.10 Future Prospects

- **Expansion of Hi-tech Nurseries with Automation, Climate Control, and Sensor-based Fertigation:** The future of quality planting material production lies in the expansion of hi-tech nurseries equipped with modern automation and digital technologies. Climate-controlled structures integrated with IoT devices, precision sensors, and automated fertigation systems will ensure optimum growing conditions throughout the year. These technologies not only improve efficiency but also reduce dependency on manual labour, minimize errors, and ensure uniformity in plant growth. By adopting such systems, nurseries will be able to scale up production, maintain consistency, and supply large volumes of high-quality planting material to meet the growing demand in protected cultivation.
- **Increased Adoption of Tissue Culture and Plug-tray Nurseries for Vegetables and Flowers:** Tissue culture has emerged as a reliable method for producing disease-free and genetically uniform planting material, especially in crops like banana, strawberry, ornamentals, and spices. Similarly, plug-tray nurseries are gaining popularity for vegetable and flower seedlings, as they allow efficient space utilization, better root development, and easy transplanting. With increasing demand for hybrid vegetables and high-value flowers in domestic and export markets, the adoption of these techniques is expected to rise rapidly. Both tissue culture and plug-tray systems will play a crucial role in ensuring continuous and timely supply of uniform planting material.
- **Development of Disease-indexed and Virus-free Stock Plants:** Another promising area is the development and widespread distribution of disease-indexed and virus-free stock plants. Using advanced diagnostic tools, molecular markers, and meristem culture techniques, it is now possible to identify and eliminate hidden infections in mother plants. This ensures that the planting material derived from them is completely healthy and free from systemic pathogens. Such clean stock programs will significantly reduce the risk of disease outbreaks in nurseries and protected structures, leading to better crop performance and higher productivity. They will also enhance the export potential of horticultural crops by meeting stringent international phytosanitary standards.
- **Public-Private Partnerships for Large-scale Production and Distribution of Elite Planting Material:** Public-private partnerships (PPPs) will play a vital role in strengthening the planting material supply chain. While research institutes and universities develop elite varieties and maintain breeder stock, private companies and entrepreneurs have the infrastructure and networks to multiply and distribute material on a large scale. Collaborative efforts between these sectors will ensure rapid dissemination of improved varieties, wider farmer access, and

better-quality assurance. Such partnerships can also attract investments in nursery infrastructure, training programs, and certification systems, thereby creating a robust ecosystem for the sustainable production of elite planting material in protected cultivation.

8.3 Nursery Management under Protected Conditions

Protected cultivation nurseries provide an ideal environment for raising high-quality seedlings, grafts, cuttings, and tissue-cultured plants. Effective nursery management under these structures involves a combination of scientific practices, infrastructure, resource management, and plant protection measures to ensure healthy, uniform, and vigorous planting material.

The following sections describe the essential components in detail:

8.3.1 Site Selection and Layout

- Nurseries should be located in areas with good road connectivity, reliable water supply, and access to electricity.
- The site should be well-drained, away from industrial pollution or flood-prone zones.
- Layout should ensure proper division into working areas such as seed sowing beds, hardening units, mist chambers, grafting areas, and storage facilities.
- Orientation of polyhouses and shade-net structures should be east–west to maximize light interception.
- Adequate pathways and space for machinery movement should be planned for efficient operations.

8.3.2 Nursery Infrastructure

- **Polyhouses and Shade-net Houses:** Polyhouses and shade-net houses provide a controlled environment by protecting young plants from harsh climatic factors like excessive heat, rain, and wind. They regulate light intensity and temperature, creating favourable conditions for uniform seedling growth. By minimizing environmental stress, these structures improve survival rates and productivity. Shade-nets also reduce water loss through transpiration. Together, they ensure healthy and vigorous planting material year-round.
- **Mist Chambers and Fogging Units:** Mist chambers and fogging units are essential for maintaining high humidity levels required for seed germination and rooting of cuttings. They create a fine mist that keeps the surrounding air moist without waterlogging the media. This prevents wilting and ensures faster root initiation in delicate propagules. Controlled humidity also reduces transplanting shock. Such units are vital for crops that require precise microclimate management.

- **Benches and Trays:** Raised benches improve aeration around seedlings and reduce the risk of soil-borne infections. Using UV-stabilized plug trays ensures uniform germination, healthy root growth, and easy handling during transplanting. These trays also allow efficient space utilization and better water management. Segregation in trays minimizes disease spread from one plant to another. Thus, benches and trays play a key role in systematic nursery management.
- **Water Storage and Irrigation Units:** Overhead water storage tanks ensure uninterrupted water availability for nursery operations. Precision irrigation through drip lines, misters, and micro-sprinklers provides uniform moisture distribution to seedlings. This prevents both waterlogging and drought stress. Such systems also allow fertigation, ensuring efficient nutrient delivery. Controlled irrigation saves water and enhances seedling quality under protected conditions.
- **Cold Storage Units:** Cold storage units are useful for preserving seeds, bulbs, and other planting material under optimum conditions. They maintain low temperatures and controlled humidity to extend shelf life and viability. This is especially important for crops with seasonal demand where staggered sowing is required. Cold storage prevents premature sprouting and reduces pest or fungal damage. It ensures timely availability of planting stock for nurseries.
- **Hygiene and Biosecurity Measures:** Maintaining strict hygiene is critical to prevent contamination and disease outbreaks in protected nurseries. Double-door entry systems, insect-proof nets, and foot baths act as physical barriers against pests and pathogens. Regular sterilization of tools, trays, and working areas further ensures a clean environment. These measures safeguard against the spread of viruses and soil-borne diseases. A strong biosecurity system enhances farmer confidence in nursery stock.

8.3.3 Growing Media Preparation

- Media should be sterile, well-drained, and porous to promote aeration and root growth.
- Commonly used media include cocopeat, vermiculite, perlite, peat moss, and sterilized sand.
- Nutrient-enriched media ensure better seedling vigour; pH should generally be maintained between 5.8–6.5.
- Media sterilization can be done by steam sterilization, solarization, or use of formalin and fungicidal treatments.
- Recycling of media should be avoided to reduce disease risk.

8.3.4 Irrigation and Fertigation

- ☆ Precision irrigation systems like drip, micro-sprinklers, and misters are essential for uniform water distribution.
- ☆ Mist chambers ensure optimal humidity for germination and rooting.
- ☆ Fertigation with water-soluble fertilizers helps in balanced nutrient supply during different growth stages.
- ☆ pH and EC (Electrical Conductivity) of irrigation water should be regularly monitored to prevent salt accumulation.
- ☆ Automation with timers and sensors ensures efficiency and reduces manual errors.

8.3.5 Seed Sowing and Germination Practices

- ☆ Seeds should be procured from certified sources to ensure genetic purity.
- ☆ Pre-treatment of seeds with fungicides, bio-priming, or growth regulators enhances germination.
- ☆ Sowing in plug trays ensures uniform germination and better transplanting success.
- ☆ Proper labelling of varieties prevents mixing and ensures traceability.
- ☆ Adequate moisture, temperature, and light conditions should be maintained for optimal germination.

8.3.6 Grafting, Budding, and Cutting Management

- ☆ Grafting and budding techniques are widely practiced in vegetable and fruit crops for higher yield and disease resistance.
- ☆ Rootstock and scion material should be healthy, disease-free, and properly hardened.
- ☆ Success of grafting depends on timing, humidity control, and proper healing chambers.
- ☆ Softwood, semi-hardwood, and herbaceous cuttings are propagated under mist chambers for rapid rooting.
- ☆ Hormone treatments (IBA, NAA) improve rooting success in cuttings.

8.3.7 Pest and Disease Management

- ☆ **Susceptibility of Protected Nurseries:** Protected nurseries, while offering favourable conditions for plant growth, also create an ideal microclimate for pests and pathogens. Warm, humid, and sheltered environments encourage rapid multiplication of insects like whiteflies, thrips, and aphids, as well as fungal diseases. If not managed properly, these outbreaks can spread quickly and affect the entire batch of seedlings.

Hence, pest and disease management is one of the most critical aspects of nursery operations under protected cultivation.

- **Preventive Measures and Barriers:** Prevention is always better than cure in nursery management. Insect-proof nets around polyhouses and shade-nets serve as the first line of defence against pest entry. Sticky traps (yellow for sucking pests, blue for thrips) are widely used for monitoring and reducing insect populations. Additionally, biological control agents such as *Trichoderma*, *Pseudomonas fluorescens*, ladybird beetles, and parasitoids help in natural suppression of pathogens and insect pests. Such eco-friendly approaches reduce chemical dependency and ensure safer planting material.
- **Soil-borne Pathogen Management:** Soil-borne pathogens such as *Pythium*, *Fusarium*, and root-knot nematodes pose major challenges in nursery beds and trays. Using sterilized growing media like cocopeat, vermiculite, or perlite helps in preventing infections. Soil solarization, involving the use of transparent polythene sheets over moist soil during hot months, is an effective and low-cost method to kill harmful fungi, nematodes, and weed seeds. Steam sterilization is another scientific method used in hi-tech nurseries for large-scale operations. These practices ensure a clean and safe rooting medium for seedlings.
- **Scouting and Monitoring:** Regular scouting and monitoring are crucial to detect early signs of pest or disease incidence. Nursery staff should inspect seedlings daily for symptoms such as leaf spots, yellowing, wilting, or insect presence. Early detection allows immediate intervention using bio-controls or targeted sprays, thus preventing spread to unaffected batches. Proper record-keeping of pest and disease incidence also helps in identifying seasonal trends and preparing preventive strategies in advance.
- **Eco-friendly and Sustainable Practices:** While chemical pesticides may provide quick results, their overuse can harm beneficial organisms, reduce seedling quality, and leave harmful residues. In protected nurseries, avoiding chemical overload is essential to maintain sustainability and farmer trust. Emphasis should be placed on Integrated Pest Management (IPM), which combines cultural, mechanical, biological, and need-based chemical interventions. This approach ensures eco-friendly production of seedlings, reduces input costs, and enhances long-term sustainability of nursery operations.

Common pests and diseases in protected nurseries and their eco-friendly management is depicted in Table 8.2.

Table 8.2: Common Pests and Diseases in Protected Nurseries and their Eco-friendly Management

Pest/Disease	*Symptoms/Damage*	*Eco-friendly Management Measures*
Whitefly (*Bemisia tabaci*)	Yellowing, curling of leaves, sooty mold growth due to honeydew secretion	Use yellow sticky traps, insect-proof nets, release parasitoids (*Encarsia formosa*), neem oil sprays
Thrips (*Thrips tabaci*)	Silver streaks on leaves, distorted seedlings, stunted growth	Use blue sticky traps, maintain humidity, introduce predatory mites (*Amblyseius* spp.), neem-based biopesticides
Aphids (*Aphis gossypii*)	Clusters on tender shoots, leaf curling, reduced vigour, virus transmission	Spray neem seed kernel extract (NSKE), encourage ladybird beetles and lacewings, use reflective mulches
Damping-off (caused by *Pythium, Rhizoctonia, Fusarium*)	Seed and seedling rot, sudden collapse of young seedlings	Use sterilized media, apply *Trichoderma harzianum* in soil/ media, avoid overwatering, solarize nursery beds
Leaf spot and blight (Fungal pathogens)	Brown or black spots on leaves, premature leaf fall	Remove infected leaves, apply *Pseudomonas fluorescens*, ensure proper ventilation, avoid overhead irrigation
Root-knot Nematodes (*Meloidogyne* spp.)	Root galls, stunted growth, yellowing of seedlings	Soil solarization, use nematode-resistant rootstocks, apply neem cake or bio-nematicides
Powdery Mildew	White powdery patches on leaves and stems	Maintain low humidity, apply sulphur dust or spray potassium bicarbonate, use resistant varieties
Virus Diseases (*e.g.*, Tomato leaf curl, Mosaic virus)	Leaf curling, mottling, stunted seedlings	Use virus-free seeds, eliminate weed hosts, control vectors (whitefly/ aphids) with nets and sticky traps

8.3.8 Hardening and Acclimatization

- ✰ **Need for Hardening:** Seedlings grown under polyhouses, mist chambers, or tissue culture labs are accustomed to controlled environments with optimal humidity, temperature, and light. When directly transplanted into the open field or greenhouse beds, they face sudden stress due to higher temperature fluctuations, stronger light intensity, and varying humidity levels. Hardening is therefore an essential step before transplanting, as it gradually conditions seedlings to withstand external stress, reducing transplanting shock and ensuring better establishment.
- ✰ **Gradual Exposure to External Conditions:** Hardening involves systematically exposing seedlings to outdoor conditions. The process

starts with reducing irrigation frequency, lowering humidity, and slowly increasing exposure to natural sunlight. Over a period of 7–14 days, seedlings are introduced to fluctuating temperatures and reduced protection, which improves their physiological adaptability. This gradual adjustment strengthens their root and shoots system, enabling them to survive in the field environment without stress-related damage.

- ☆ **Role of Shade Nets in Hardening:** During the initial phase of hardening, seedlings are placed under 50–75 per cent shade nets to protect them from direct sunlight and wind. Shade nets reduce the intensity of solar radiation and moderate the microclimate, preventing sudden wilting and transplanting shock. As the seedlings acclimatize, the shade intensity is progressively reduced, allowing them to adapt to natural light conditions. This step-wise approach ensures that seedlings develop tolerance to field conditions without being subjected to stress abruptly.
- ☆ **Nutrient and Biological Support During Hardening:** To improve seedling vigour and enhance their resilience, mild foliar sprays of water-soluble nutrients, seaweed extracts, and bio-stimulants are applied during the hardening stage. Biological agents like Trichoderma and Pseudomonas fluorescens may also be applied to strengthen the root system and protect against soil-borne pathogens. These treatments not only enhance overall growth but also prepare seedlings for healthy establishment after transplantation.
- ☆ **Benefits of Hardening:** Well-hardened seedlings exhibit better adaptability, stronger root development, and improved resistance to pests, diseases, and abiotic stresses. They are less prone to wilting, transplanting shock, or mortality when shifted to the main field or protected structures. Hardening ensures uniform growth after transplantation, leading to higher survival rates and better crop performance. Ultimately, this practice improves profitability for farmers by reducing losses and maximizing the potential of quality planting material.

8.3.9 Hygiene, Sanitation, and Waste Management

- ☆ **Sterilization of Nursery Trays, Tools, and Working Areas:** Regular sterilization of all nursery equipment such as seedling trays, grafting tools, benches, and working surfaces is essential to prevent the spread of pathogens. Unsterilized tools or reused trays often harbour fungal spores, bacteria, or nematodes that can quickly infect young seedlings. Sterilization can be done using formalin, sodium hypochlorite, or steam treatment, depending on the material. Maintaining clean surfaces ensures a healthy environment and enhances the success rate of nursery operations.
- ☆ **Foot Baths and Double-door Entry Systems:** The entry of workers and visitors into nurseries is one of the most common sources of

contamination. Installing disinfectant foot baths at the entrance helps in preventing the introduction of pathogens through soil particles carried on footwear. Similarly, double-door entry systems act as a protective barrier, reducing the direct inflow of pests and diseases from outside. These biosecurity measures are simple yet highly effective in safeguarding sensitive nursery crops.

- ☆ **Waste Disposal and Management:** Proper management of waste plant material is crucial to prevent the buildup of inoculum sources for pests and diseases. Diseased or dead seedlings should not be left in the nursery area as they can serve as breeding grounds for pathogens. Safe disposal methods include composting healthy waste material or burning heavily infected residues. A systematic waste management system ensures a clean working environment and minimizes the chances of reinfection in future nursery cycles.
- ☆ **Worker Hygiene and Sanitation Protocols:** The role of workers in maintaining nursery hygiene cannot be overlooked. Workers should follow strict protocols such as wearing clean uniforms, washing hands before handling seedlings, and disinfecting tools before and after use. Training programs can help in sensitizing staff to the importance of personal hygiene in plant health management. Consistent adherence to these practices ensures that contamination risks are minimized at every operational step.
- ☆ **Importance of High Biosecurity Standards:** Nurseries serve as the foundation for crop production, and any lapse in hygiene can have serious consequences for large-scale farming operations. Maintaining high biosecurity standards prevents disease outbreaks, enhances farmer confidence, and ensures uniformity in seedling quality. With rising demand for certified planting material, hygiene and sanitation are not optional but essential practices. A well-managed nursery with strong waste management protocols sets a benchmark for sustainable and reliable planting material production.

8.3.10 Record Keeping and Certification

- ☆ **Detailed Documentation of Nursery Activities:** Maintaining accurate records is the backbone of scientific nursery management. Every detail, such as the source of seed or planting material, date of sowing, germination percentage, treatments applied, and dispatch date, should be systematically documented. These records not only provide transparency but also serve as a reference for identifying bottlenecks and improving practices. In the event of crop failure or poor establishment, such documentation helps in identifying the root cause and taking corrective action.

- **Labelling and Tagging for Traceability:** Proper labelling of seedling trays, beds, or batches ensures that each lot can be traced back to its source and management history. Information such as variety, date of sowing, and special treatments should be clearly displayed on durable tags. Traceability safeguards farmers from receiving mixed or inferior planting material and assures them of varietal purity. This systematic practice also simplifies inventory management and reduces confusion during large-scale nursery operations.
- **Role of Certification Agencies:** Certification from recognized authorities such as the National Horticulture Board (NHB), Indian Council of Agricultural Research (ICAR), and State Seed Certification Agencies adds credibility to nursery operations. Certified planting material guarantees genetic purity, disease-free status, and adherence to prescribed quality standards. Accreditation not only builds farmer trust but also opens opportunities for premium pricing and wider market reach. It also helps nurseries integrate into government schemes and subsidy programs.
- **Use of Records for Evaluation and Improvement:** Maintaining a systematic record-keeping system provides a valuable database for evaluating nursery performance. By comparing germination percentages, survival rates, and success of different treatments across batches, nursery managers can identify the most effective techniques. Such analysis guides future planning and helps in optimizing resources. Over time, this contributes to consistent improvement in quality, efficiency, and profitability of nursery operations.

Greenhouse Cultivation of Important Horticultural Crops

9.0 Major Flower Crops and their Varieties Cultivated under Greenhouses

It is the technique of providing beneficial environmental or developmental circumstances to plants. In greenhouses, the cultivation environment is adjusted to satisfy the specific requirements of plants. It is mostly utilized to protect plants from adverse climatic circumstances by managing optimal levels of light, temperature, humidity, CO2, and air circulation to foster growth, hence improving yield and quality (Figure 9.1).

Advantages of Growing Flower Crops in Greenhouses

- ☆ **Year-round production** – Flowers can be grown and supplied irrespective of external seasonal variations.
- ☆ **Controlled environment** – Temperature, humidity, light, and CO_2 can be regulated for optimum growth.
- ☆ **Higher yield and quality** – Flowers are uniform, healthier, and have better color, size, and fragrance.
- ☆ **Protection from adverse weather** – Crops are safeguarded from frost, heavy rains, wind, and heat stress.

Figure 9.1: Flower Crops their Varieties Cultivated under Greenhouses
(*Source*: https://www.hillagric.ac.in).

- ✰ **Reduced pest and disease incidence** – Controlled entry, insect-proof nets, and sanitation reduce infestations.
- ✰ **Efficient use of resources** – Water, fertilizers, and land are utilized more effectively through drip and fertigation.
- ✰ **Extended vase life** – Flowers grown under controlled conditions have longer post-harvest shelf life.
- ✰ **Production of exotic and off-season flowers** – High-value flowers not possible in open fields can be cultivated.
- ✰ **Better market price** – Off-season and premium quality flowers fetch higher returns in domestic and export markets.
- ✰ **Support for modern techniques** – Compatible with hydroponics, vertical farming, and automated irrigation systems.

Commercially important greenhouse flowers with their cultivation and market advantages is given in Table 9.1.

Essential Factors for Growing Flowers in Greenhouses

- ✰ **Ground Beds** – Used for large-scale cultivation, allowing efficient use of space and easier management of soil-based crops.
- ✰ **Raised Beds** – Improve drainage, root aeration, and soil temperature; suitable for cut-flower production.
- ✰ **Benches** – Provide better accessibility, hygiene, and control over growing medium; widely used for potted and ornamental plants.
- ✰ **Pots/Containers** – Ideal for high-value flowers, potted plants, and nursery stock; allow mobility and precise control of growing media.

Table 9.1 : Commercially Important Greenhouse Flowers with their Cultivation and Market Advantages

Flower Crop	*Greenhouse Advantage*	*Market Benefit*
Roses	Controlled temperature and humidity → longer stems, bigger blooms, year-round supply	High demand in domestic and export markets; premium price
Carnations	Protection from temperature extremes → uniform size, strong stems, long vase life	Preferred cut flower for events and exports
Gerbera	High light and humidity control → large flower heads, continuous blooming	Popular in bouquet and floral decoration industry
Orchids	Specific humidity and temperature control → exotic flowers of export quality	Premium export flower; high value in hotels, events, gifting
Chrysanthe-mums	Photoperiod manipulation → off-season and year-round flowering	Strong demand during festivals, weddings, and export markets
Tulips	Temperature regulation → uniform flowering and bulb forcing	Export potential; high-value seasonal demand
Lilies	Climate control → uniform flowering, longer stalks, larger buds	Popular as cut flowers and in high-end floral arrangements
Anthuriums	High humidity and shade management → glossy bracts, vibrant colours	Premium indoor and export markets; long vase life
Gladiolus	Protected cultivation → staggered planting, off-season availability	High seasonal demand; used in ceremonies, decorations
Geraniums and Petunias	Bulk production in controlled environment → uniform, disease-free seedlings	Landscaping, nurseries, and potted plant markets

Characteristic Features of a Good Growing Medium

- ✰ Adequate Nutrient Supply – The medium should provide essential macro- and micronutrients for healthy growth and flowering.
- ✰ Proper Support/Anchorage – It must firmly hold the plants and provide stability to the root system.
- ✰ Good Moisture-Holding Capacity – Sufficient water retention is necessary to meet crop requirements between irrigations.
- ✰ High Porosity – Adequate pore space ensures proper aeration and facilitates root respiration.
- ✰ Non-Saline Nature – The medium should be free from excessive salts to prevent toxicity and growth inhibition.

- ✰ Resistance to Pasteurization – It should withstand treatments like steam sterilization or solarization without degrading.
- ✰ Freedom from Contaminants – The medium must be free from weed seeds, nematodes, pathogens, and other harmful organisms.

Decontamination of Growing Medium

To ensure healthy crop growth, the growing medium must be free from harmful pathogens, weed seeds, and pests. Common methods of decontamination include:

- ✰ Chemical Drenching/Fumigation – Application of chemicals such as formaldehyde, chloropicrin, hydrogen peroxide, captan, and vapam helps in eliminating soil-borne diseases and pests.
- ✰ Steaming – Exposure of the medium to high-temperature steam effectively destroys pathogens, nematodes, and weed seeds.
- ✰ Pasteurization – Controlled heating of the medium at moderate temperatures reduces harmful organisms while retaining beneficial microbes.
- ✰ Solarization – Covering the medium with transparent polyethylene sheets and exposing it to solar radiation raises the temperature sufficiently to kill weeds and pathogens.

Management of High Temperature in Greenhouse Flower Cultivation

- ✰ **Naturally Ventilated Polyhouses** – Side vents (1.5 m) and top vents (1 m) facilitate natural air circulation, reducing heat buildup.
- ✰ **Fan and Pad Cooling System** – Ensures effective temperature control by pulling air through wet pads, providing evaporative cooling.
- ✰ **Shade Nets** – Installation of 50 per cent intensity shade nets inside polyhouses, especially between 12 noon and 4 pm, helps in reducing heat stress.
- ✰ **Foggers and Misters** – Operated once or twice during afternoons in the vegetative phase to enhance humidity and lower temperature.
- ✰ **Photo-Selective Shading Paints** – Application of shading paints that reduce excessive radiation while maintaining 69 per cent light transmission; particularly effective in late spring and summer when applied at 450 kg/ha in a 1:3 ratio (paint: water), as recommended by Evans (2009).

9.1 Flowers

9.1.1 Rose

Roses (Rosa spp.) are among the most significant cut flowers worldwide, and greenhouse culture facilitates year-round production under regulated circumstances. The rose is the predominant cut flower in the global floriculture market and the most significant cut flower shipped from India. Over 90 per cent of greenhouses in the country are dedicated to rose cultivation for cut flower supply. The principal export cultivars of rose comprise Cora, Corvetti, Diplomat, Femma, First Red, Grand Gala, Kiss, Konfetti, Lambada, Laser, Nicole, Noblesse, Osiana, Papillon, Parea, Passion, Pavrotte, Rodeo, Rossini, Sacha, Samura, Sandy, Sangaria, Soledo, Susanne, Texas, Tineke, and Vivaldi. (Figure 9.2).

- ✰ Possess a large number of petals for a full, attractive bloom.
- ✰ Exhibit slow-opening petals, ensuring a prolonged flowering period.
- ✰ Have enhanced vase life and longevity, making them suitable for transport and display.
- ✰ Display an attractive and diverse range of colours, catering to various market preferences.
- ✰ Produce long, straight, and sturdy stems, ideal for export and floral arrangements.

Figure 9.2: Rose Cultivation in Greenhouse.
(*Source*: www.freepik.com)

Environmental Factors for Rose Cultivation

- ✰ Temperature: Day temperature 18–28°C and night temperature 15–18°C for optimal growth and flowering.
- ✰ Light: Photoperiod over 12 hours with light intensity of 6,000–8,000 foot-candles to ensure proper photosynthesis and bud formation.
- ✰ Relative Humidity: Maintain 50–60 per cent to reduce disease incidence and promote healthy growth.
- ✰ Aeration: Ensure good air circulation inside the greenhouse and well-aerated soil for root health.

- Carbon Dioxide (CO_2): Concentration of 1,000–3,000 ppm can enhance photosynthesis and flower quality.

Soil and Growing Medium

- Texture and Drainage: Light, well-drained soil to prevent waterlogging and root diseases.
- pH: Around 6.5, which is optimal for nutrient availability.
- Electrical Conductivity (EC): Less than 1.0 dS/m to avoid salinity stress.
- Organic Matter: 10–12 per cent humus content to improve soil fertility, structure, and water-holding capacity.

Climatic and cultural requirements for roses in greenhouses is summarised in Table 9.2

Table 9.2: Climatic and Cultural Requirements for Roses in Greenhouses

Factor	*Requirement*	*Impact on Flower Quality*
Temperature	25–28°C (day), 15–18°C (night)	Promotes uniform growth, bud initiation, and vase life
Light	4,000–6,000 foot-candles (fc)	Enhances stem length, bud size, and petal coloration
Humidity	60–70 per cent RH	Prevents desiccation and bud drying; excess favors disease
Soil/Growing Medium	pH 5.8–6.5, EC < 1.5 dS/m, porous and well-drained	Ensures good root health and nutrient uptake
Water Quality	pH 6.0–6.5, EC < 0.5 dS/m	Prevents salt buildup and nutrient imbalance
Irrigation and Fertigation	Drip system with balanced NPK + Ca, Mg, micros	Produces sturdy stems, large blooms, and high yield
Planting Material	Grafted plants on Rosa indica var. odorata	Increases vigor, disease resistance, and productivity
Planting Density	10–12 plants/m^2	Optimal space utilization and better-quality blooms
Training and Pruning	Regular pruning and upright training	Maintains shape, enhances light interception, new flushes
Harvesting Stage	Tight bud stage, outer petals just unfolding	Ensures longer vase life and better market acceptability

Soilless Growing Medium for Roses

Roses can be effectively cultivated in soilless media, which provide excellent aeration, drainage, and disease control:

- **Sand:** Ensures good drainage and root support; often combined with organic materials to improve nutrient retention.

- **Coco-peat:** Retains moisture efficiently, offers good aeration, and is rich in organic matter; widely used in hydroponics and greenhouse cultivation.
- **Rock-wool:** Lightweight and inert, providing optimal water and air balance; commonly used in commercial greenhouse soilless culture.

Optimum Spacing and Planting Density

- **Planting System:** In greenhouses, single, double, or four-row planting systems are commonly used within beds.
- **Spacing:** Recommended spacing is 30 × 25 cm between plants to ensure adequate light, air circulation, and growth.
- **Planting Density:** Optimum planting density is 60,000–70,000 plants per hectare, balancing high yield with healthy plant development.

Propagation

Roses can be propagated using various methods, depending on whether the goal is commercial production, rootstock development, or disease-free plants:

- Shield or T-Budding (Commercial Method): Widely used in commercial rose production; a bud from the desired cultivar is inserted into a compatible rootstock
- Stenting Method: Budding is performed on cuttings, which are then planted in a mist chamber for rooting and early growth.
- Cuttings: Used primarily for propagating rootstocks. Stem cuttings 15–20 cm long, pencil-thick, with at least three buds, are planted to develop roots.

Micropropagation

Micropropagation is a tissue culture technique used to produce disease-free, uniform plants on a large scale under controlled laboratory conditions. This method ensures rapid multiplication of elite cultivars and maintains genetic stability.

Important Rootstocks

Commonly used rootstocks provide vigour, disease resistance, and adaptability:

- *Rosa bourboniana* – Widely used for its strong growth and compatibility with many cultivars.
- *Rosa indica* var. *Odorata* – Offers good disease resistance and adaptability to diverse soil *conditions.*
- *Rosa multiflora* – Preferred for its hardy nature and robust root system.

Budding and Pruning Time

The timing of budding and pruning varies according to the region and altitude to ensure optimal growth in Northern Indian Plains: November–December.

Optimum Nutrients (ppm) Status

Optimum Nutrient Status for Greenhouse Roses for healthy growth and high-quality flowering, greenhouse roses require a pH of 6.2–6.8 and an electrical conductivity (EC) of 0.7 dS/m. The optimum levels of macronutrients are nitrate 180 ppm, ammonium 2 ppm, phosphate 4 ppm, potassium 45 ppm, calcium 55 ppm, and magnesium 20 ppm. Essential micronutrients should be maintained at iron 0.39 ppm, manganese 0.04 ppm, zinc 0.07 ppm, boron 0.08 ppm, and copper 0.05 ppm, while other elements include sodium 25 ppm, chloride 35 ppm, sulphate 105 ppm, and bicarbonate 30 ppm. Maintaining these nutrient levels ensures vigorous vegetative growth, strong stems, and long-lasting, high-quality blooms suitable for commercial greenhouse production.

Pruning

Pruning in roses depends on several factors, including the cultivar, the class or category of the rose, the health and vigour of the plant, the spacing between plants, the soil fertility, and the desired quality of flowers. Proper pruning helps remove old or weak stems, encourages new shoot development, improves air circulation, maintains plant shape, and ensures production of long, sturdy stems with high-quality blooms suitable for cut flower markets.

Irrigation

Irrigation of greenhouse roses is mainly done through drip lines placed about 15 cm below the soil to maintain optimal root-zone moisture, with daily water needs of 2–5 litres/m^2. Fertigation provides nutrients efficiently, and supplemental misting or sprinklers during vegetative growth helps maintain humidity and plant vigor.

Diseases

Greenhouse roses are prone to several diseases that can significantly affect plant health and flower quality. Dieback is caused by *Diplodia rosarum* and Colletotrichum species, leading to the drying of stems and shoots. Black spot (*Diplocarpon rosae*) produces dark lesions on leaves, causing defoliation and weakened plants. Powdery mildew (*Sphaerotheca pannosa*) appears as a white powdery growth on leaves and buds, while downy mildew (Peronospora sparsa) results in yellowing and browning of foliage. Gray mould (Botrytis cinerea) affects flowers and buds, causing rot, and rust (*Phragmidium* species) forms orange pustules on leaves. Crown gall (Agrobacterium tumefaciens) produces tumor-like growths at the base of stems, and wilt (*Marmor flaccumfasciens*) leads to sudden drooping and death of shoots. Timely monitoring, sanitation, and appropriate fungicidal treatments are essential to manage these diseases effectively.

Insect Pests of Roses

Greenhouse roses are attacked by various insect pests such as red scales, red spider mites, white ants, aphids, thrips, jassids, chafer beetles, mealybugs, stem girdlers, and sawflies, which damage leaves, stems, buds, and flowers. These pests cause yellowing, stunted growth, deformities, and defoliation. Effective integrated pest management, combining biological control, cultural practices, and careful insecticide use, is essential to maintain healthy, high-quality plants.

Physiological Disorders

Greenhouse roses may develop physiological disorders like bull head (misshapen petals), bent or limp neck (drooping stems), blind shoots (no flower buds), and vascular plugging (wilting from blocked xylem), usually caused by water stress, nutrient imbalance, or improper handling. Proper cultural practices, balanced nutrition, and careful irrigation help prevent these issues and maintain high-quality blooms.

Management

- ✰ **Disease Management:** Maintain good sanitation, proper ventilation, and timely pruning. Apply fungicides like Mancozeb, Carbendazim, or Copper-based formulations to control dieback, black spot, powdery/downy mildew, grey mould, rust, crown gall, and wilt.
- ✰ **Pest Management:** Use integrated pest management (IPM): encourage natural predators, maintain cultural hygiene, and apply selective insecticides for aphids, spider mites, scales, thrips, jassids, mealybugs, and stem pests.
- ✰ **Physiological Disorder Management:** Prevent issues like bull head, bent neck, blind shoots, and vascular plugging by ensuring balanced nutrition, proper irrigation, and careful handling of plants and flowers.

Stage of Flower Harvesting

Roses are harvested based on market and flower colour, for the local market, when the outer one or two petals start unfurling, and for distant or export markets, as fully coloured tight buds. Whitc, pink, and yellow cultivars are picked slightly earlier than red, ensuring proper opening, maximum vase life, and high-quality flowers.

Popular Grades

Large-flower cultivars have stem lengths of 60–90 cm and bud sizes of 3–3.5 cm, while small-flower cultivars have stems 40–50 cm long and bud sizes of 2–2.5 cm, catering to different market and arrangement requirements.

Packaging

Cut roses are typically bundled in groups of 10, 12, 20, or 25 and packed in corrugated cardboard boxes measuring 100 × 32.5 × 20 cm, which hold about 80–100 stems of 60–65 cm length. Flower bunches are wrapped in cellophane sleeves, and the boxes are lined with tissue paper or newspaper to protect blooms during transport

Average Yield

In Indian greenhouses, the average yield is 150–200 stems per m^2 per year, whereas European greenhouses achieve 350–400 stems per m^2 per year due to advanced technology and management practices.

9.1.2 Carnation

Carnation (*Dianthus caryophyllus*) is one of the leading cut flowers after rose in the international floriculture trade. It is highly demanded due to its wide colour range, fragrance, long vase life, and suitability for bouquets and floral arrangements. In India, carnation is commercially grown under protected cultivation in states like Himachal Pradesh, Uttarakhand, Karnataka, and Maharashtra for both domestic markets and export (Figure 9.3).

Figure 9.3: Carnation (*Dianthus caryophyllus*) in Greenhouse
(*Source*: https://thumb.photo-ac.com)

Florist Carnations

- ☆ **Standard Carnations**: Single large bloom per stem. Used mainly in bouquets and arrangements. Example: Master, Domingo, White Liberty.
- ☆ **Spray Carnations**: Multiple medium-sized flowers on a single stem. Popular for vase decoration and mixed floral arrangements. Example: Diana, Rony, Euro.
- ☆ **Mini Carnations**: Smaller flowers and shorter stems than spray type. Suitable for small bouquets, floral baskets, and decorations.
- ☆ **Micro Carnations**: Very small-sized blooms, compact growth. Used in floral designs, garlands, corsages, and boutonnieres.

Popular Cultivars

Common cultivars of carnation cultivated in greenhouse environments comprise Dona, Pink Dona, Malaga, White Dona, Rony, Rhodos, Lipstick, Empire, Romana, White Tendra, Corleone, Design, Natila, Dark Tempo, Bagatel Silvery Pink, Solar, Cobra, Pendy, Lorella, Cabaret, Tanga, Sonsara, Dakar, Liberty, Green Lady, Tempo, Varna, Sun Shine, and Charment. These kinds are favored for their varied colors, appealing shapes, extended vase life, and compatibility with both domestic and international markets.

Environmental Factors

- Daytime Temperature: 18–23°C
- Nocturnal Temperature: Winter: 10–11°C; Spring: 12.7°C; Summer: 13–15.4°C
- Relative Humidity: 50–60 per cent
- Ventilation: Unimpeded air circulation to prevent disease and sustain a consistent climate.
- CO_2 Enrichment: 500–1500 ppm to enhance photosynthesis and improve floral quality.

The climatic and cultural requirements for carnation in greenhouses are summarised in Table 9.3.

Table 9.3: Climatic and Cultural Requirements for Carnation in Greenhouses

Factor	*Requirement*	*Impact on Quality*
Temperature	20–25°C (day), 10–15°C (night)	Strong stems, better colour
Light	Long days, bright sunlight	Stimulates stem elongation, flowering
Humidity	50–60 per cent	Prevents disease, supports healthy growth
Soil/Medium	pH 6.5–7.5, EC <1.2 dS/m	Good nutrient uptake
Irrigation/ Fertigation	Balanced NPK + Ca, B	Prevents splitting, ensures quality blooms
Density	30–40 plants/m^2	Optimal space use
Training	Netting support	Prevents bending of stems
Harvest Stage	Paint brush stage	Long vase life

Propagation

Carnation is propagated by terminal stem cuttings (8–10 cm with 4–6 leaf pairs, treated with IBA) under mist conditions, and by micro-propagation for large-scale, disease-free, uniform planting material.

Soil and Growing Medium

Carnation grows best in light textured loam or sandy loam soils that are well-drained and well-aerated, with an optimum soil pH of 6.0–7.0.

Planting Time

In India, the best planting time for carnation in greenhouses is September–October, as the cool season ensures proper vegetative growth and high-quality flower production. In hilly regions, planting may also be done during March–April.

Planting Density and Spacing

Carnation planting density varies with type and production system. Under ordinary conditions, about 25–32 plants/m^2 are accommodated, while under high-density planting, up to 40 plants/m^2 can be grown. The recommended spacing is 20 × 20 cm for standard carnations and 30 × 30 cm for spray carnations to ensure proper growth, branching, and flower quality.

Optimum Nutrition

The optimum nutrition for carnation includes leaf tissue levels of N: 5.4 per cent, P: 0.31 per cent, and K: 3.8 per cent. Traditionally, for standards, about 5 kg FYM, 30 g N, 20 g P, and 10 g K per m^2 is applied, while for spray types, the dose is 5 kg FYM, 40 g N, 20 g P, and 10 g K per m^2. In modern commercial production, nutrients are supplied more efficiently through fertigation, ensuring balanced uptake and improved flower quality.

Fertigation

In carnation cultivation, fertigation is the most efficient method of nutrient application under greenhouse conditions. A balanced supply of N, P, and K along with secondary nutrients (Ca, Mg, S) and micronutrients (Fe, Mn, Zn, B, Cu, Mo) is provided through drip irrigation. The recommended solution maintains an EC of 1.5–2.0 mS/cm and a pH of 5.8–6.2 for optimum nutrient uptake. Fertigation ensures uniform distribution of nutrients, promotes healthy vegetative growth, improves stem strength, and enhances flower yield and quality.

Irrigation

In greenhouse carnation cultivation, irrigation is mainly done through drip system to ensure uniform water supply and prevent excess humidity. The crop generally requires 2–3 irrigations per week depending on season, soil type, and growth stage. Overhead irrigation is avoided as it increases humidity and disease incidence. Maintaining optimum soil moisture without waterlogging is crucial for strong stems, healthy root development, and high-quality flowers.

Pinching

Pinching is the practice of removing the terminal growing shoot (about 2–3 cm long) at the 6–8 leaf pair stage to break apical dominance and encourage side branching, which results in more flowering shoots and higher yield.

Types of Pinching

- Single Pinching: Terminal bud removed once to induce branching.
- Pinch and Half: Initial pinch followed by partial removal of side shoots.
- Pinch plus Pull Pinch: Combination of pinching with removal of weak side shoots.
- Double Pinching: Two successive pinchings to maximize branching and flower production.

De-shooting

Deshooting is the removal of unwanted side shoots to focus the plant's energy on producing fewer, stronger stems with high-quality flowers, improving uniformity and marketable yield.

Staking

Staking is done to support tall carnation stems and prevent lodging, especially in greenhouse conditions. Plants are guided through plastic nets or bamboo/ wooden stakes, ensuring upright growth and better air circulation. Proper staking improves stem quality, flower uniformity, and ease of harvesting, particularly for long-stemmed standard varieties.

Disbudding

Disbudding is the removal of unwanted side buds to focus the plant's energy on producing a single, strong, high-quality flower, improving stem length and marketable yield.

Weeding

Carnation crops require three to four hand weedings during the growing season to keep beds free from competition. In greenhouse cultivation, chemical weed control can be done using oxidiazon (2.2 kg a.i./ha) and napropamide (4.5 kg a.i./ ha). For open-field cultivation, pre-emergence herbicides like fluchloralin (Basalin) and pendimethalin (Stomp) at 1.0 kg a.i./ha each are effective in controlling weeds.

Stages of Flower Harvesting

Carnation flowers are harvested at different stages depending on type and market requirements:

- **Bud Stage:** Flowers are harvested when the petals are tightly closed but showing slight color; ideal for long-distance transport.
- **Half-Open Stage:** Petals begin to unfurl; suitable for domestic markets and immediate sale.
- **Fully Open Stage:** Flowers are completely open; mainly for local markets or on-site use.

Harvesting at the correct stage ensures maximum vase life, quality, and marketability of carnation flowers.

Common Grades in India

Carnation stems are graded based on stem length:

- ☆ **Grade A:** Over 45 cm – premium quality, ideal for export and bouquets.
- ☆ **Grade B:** 30–45 cm – suitable for domestic markets and mixed arrangements.
- ☆ **Grade C:** Less than 30 cm – generally used for local sale or secondary purposes.

Packaging

Carnation stems are bundled and packed carefully to maintain quality and prevent damage during transport:

- ☆ **Bundling:** Typically, in 10, 12, 20, or 25 stems per bundle depending on stem size and market demand.
- ☆ **Boxes:** Corrugated fibreboard boxes of 100 × 32.5 × 20 cm are commonly used, accommodating 80–100 stems of 60–65 cm length.
- ☆ **Protection:** Stems may be wrapped in moist tissue or polyethylene sheets to reduce dehydration.
- ☆ **Transport:** Pre-cooling at 2–4°C before shipment extends vase life.

Diseases

- ☆ **Wilt** – Caused by *Fusarium oxysporum* f. sp. *dianthi*, leading to yellowing and wilting.
- ☆ **Foot rot** – Caused by *Phytophthora*, *Pythium*, *Rhizoctonia solani* or *Sclerotinia sclerotiorum*, resulting root and collar decay.
- ☆ **Stem rot** – Caused by *Fusarium roseum*, producing stem cankers and collapse.
- ☆ **Flower bud rot** – Caused by *Alternaria dianthi*, leading to blackening and rotting of buds.
- ☆ **Bacterial wilt** – Caused by *Pseudomonas caryophylli*, inducing vascular browning and plant death.
- ☆ **Rust** – Caused by *Uromyces caryophyllinus* or *U. dianthi*, showing reddish-brown pustules on leaves.
- ☆ **Flower blight** – Caused by *Botrytis cinerea*, producing grey mould on flowers.
- ☆ **Fairy ring spot** – Caused by *Heterosporium echinulatum*, giving circular leaf spots with concentric rings.

- **Viral diseases** – Caused by viruses like carnation mottle and *etc.*hed ring, resulting in mosaic, mottling and streaks.

Insect-Pests

- **Red spider mites** – Suck sap from leaves causing yellowing, bronzing, and webbing.
- **Thrips** – Lacerate and suck sap from leaves and petals, leading to streaks and flower distortion.
- **Nematodes** – Attack roots causing galls, stunted growth, and wilting.
- **Aphids** – Suck plant sap causing curling, yellowing, and transmitting viral diseases.
- **Helicoverpa/Heliothis/Spodoptera caterpillars** – Chew leaves, buds, and flowers causing severe defoliation and flower damage.

Physiological Disorders

- **Calyx splitting** – Splitting of calyx due to rapid growth, genetic factors, or uneven nutrition.
- **Grassiness** – Production of narrow, grassy shoots caused by hormonal imbalance or virus.
- **Sleepiness** – Premature flower wilting in vases due to vascular blockage.
- **Splitting at nodes and bushiness** – Excessive branching from nodes due to imbalance of growth regulators.
- **Small narrow leaves and tied tips** – Leaves remain small, narrow, and stuck together due to micronutrient deficiency or hormonal imbalance.

Management

- **Cultural Control:** Maintain well-drained and sterilized soil or potting media, ensure adequate spacing and ventilation, and avoid waterlogging. Remove and destroy infected plant parts and use disease-free planting material. Balanced fertilization and proper irrigation help reduce physiological disorders and improve plant vigor.
- **Biological Control:** Use beneficial organisms such as Trichoderma to suppress root and stem rots. Introduce natural predators like ladybird beetles, predatory mites, and lacewings to manage aphids, thrips, and red spider mites. Beauveria bassiana can be applied against caterpillars and thrips for effective pest management.
- **Chemical Control:** For fungal diseases like wilt, stem rot, rust, and blights, apply fungicides such as Mancozeb, Carbendazim, or Copper oxychloride. Use insecticides like Imidacloprid or Thiamethoxam for aphids, Abamectin for mites, and Spinosad or Emamectin for caterpillars. Apply nematicides if root nematode infestations are severe.

- ☆ **Management of Physiological Disorders:** Prevent calyx splitting, grassiness, small or narrow leaves, and excessive bushiness by maintaining balanced nutrition, proper water management, adequate light, and regulated growth hormones. Careful handling and timely harvesting also reduce "sleepiness" or premature flower wilting.

Tinting in Carnation

Tinting in carnations is a floral colouring technique in which a concentrated liquid or powdered dye is dissolved in warm water (≈ 37°C), and cut stems are immersed in the solution. Over 10–24 hours, the dye travels upward through the stem into the petals, creating vibrant and varied colour patterns that enhance the flowers' decorative value and market appeal.

Average Yield

The average yield of carnations under greenhouse cultivation ranges from 200 to 300 stems per m^2 per year, varying with the cultivar, crop management practices, and prevailing environmental conditions.

9.1.3 Chrysanthemum

Chrysanthemum (*Dendranthema grandiflora*) originates from the Northern Hemisphere, specifically China. Chrysanthemum is cultivated for cut flowers, loose flowers, potted varieties, hanging baskets, and as bedding and border plants. It has gained popularity due to its exceptional vase life, diverse array of flower colours and shapes, and reduced floral weight. It is categorized according on inflorescence, photoperiod, and temperature (Figure 9.4).

Figure 9.4: Chrysanthemum (*Dendranthema grandiflora*) in Greenhouse.
(*Source*: www.freepik.com)

The prominent chrysanthemum cultivars for export are Snow Ball, Snow Don White, Mountaineer, Sonar Bangla, Bright Golden, Anne, Chandrama, Ajay, Birbal Sahni, Lehmans, Nanako, and Sonali Tara. Chrysanthemum is a short-day plant that blooms when the critical photoperiod is less than 9.5 hours. Nevertheless, cultivars have been developed that may flower in any season.

The notable off-season flowering varieties of chrysanthemum are enumerated in Table 9.4.

Table 9.4: Season-Wise Carnation Cultivars with Harvesting Period and Flower Traits"

Season/ Months	*Cultivars*	*Harvesting Period*	*Flower Colour/Special Trait*
April – June	Himanshu, Jawala, Jyoti	Harvest in summer	Likely shades of pink, red or variegated tones (popular spring colours)
July – August	Phuhar	Harvest in monsoon	Possibly lighter or pastel shades/heat-tolerant blooms
September – October	Ajay, Sharda	Harvest in early autumn	Strong, vivid colours to suit cooler autumn light (red, deep pink)
October – November	Makhmal, Megami, Mohini, Sharad	Harvest in late autumn	Deep, rich hues; likely with strong petal form and vase life

Soil

Chrysanthemum grows best in well-drained sandy loam soil with a slightly acidic to neutral reaction. An ideal soil pH of 6.2–6.7 ensures optimum growth, nutrient uptake, and flower quality.

Climate

- ☆ **Light:** Chrysanthemum requires moderate light intensity (1.2–1.6 $MJ/m^2/day$) with a wavelength range of 600–800 nm, and it performs best under short-day conditions of less than 9.5 hours photoperiod.
- ☆ **Temperature:** Ideal temperature for growth is 18–21°C during the day and 10–16°C at night, which supports healthy vegetative growth and quality flowering.
- ☆ **CO_2:** An enriched CO_2 concentration of 500–1000 ppm enhances photosynthesis, growth, and flower yield under protected cultivation.

Climatic and cultural requirements for chrysanthemum in greenhouses is summarised in Table 9.5.

Table 9.5: Climatic and Cultural Requirements for Chrysanthemum in Greenhouses

Factor	*Requirement*	*Impact on Quality*
Temperature	20–28°C (day), 15–18°C (night)	Uniform growth, large blooms
Light	Short-day response	Controls flowering time
Humidity	70–80 per cent	Promotes healthy foliage and bloom
Soil/Medium	pH 6.0–6.5, EC <1.5	Better nutrient absorption
Irrigation/ Fertigation	Balanced NPK, high K in bloom stage	Improves color, size, vase life
Density	32–40 plants/m^2	Maximizes yield per unit area
Pinching	Regular	Encourages branching, uniform flowers
Harvest Stage	Outer petals open, disc florets compact	Better shelf life

Propagation

Chrysanthemum can be propagated efficiently using terminal stem cuttings of 4–5 cm taken during June–July, which root quickly and produce healthy plants. Additionally, propagation through suckers is recommended from February to April to maintain varietal purity and vigorous growth. Both methods ensure rapid multiplication and uniform flowering in the crop. Proper care of cuttings and suckers during the initial stages is essential for successful establishment.

Planting Density and Spacing

- ☆ **Greenhouse cut flowers:** Plant at a density of 40–54 plants per m^2 for optimal growth and flowering.
- ☆ **Loose flowers:** Space plants 30 × 20 cm apart, accommodating 20–25 plants per m^2.
- ☆ **Standards:** Maintain a spacing of 20 × 20 cm for standard varieties.
- ☆ **Sprays:** Space 30 × 30 cm between plants to allow multiple spray production.
- ☆ **Pot mums:** Plant 3–5 cuttings per 15 cm pot for compact, flowering potted chrysanthemums.

Nutrition

- ☆ **FYM:** Apply 3–5 kg per m^2 to improve soil fertility and plant vigour.
- ☆ **N:P:K:** Provide 30:10:15 g/m^2 monthly to support balanced growth and flowering.
- ☆ **Loose flowers:** Use 10–15 tons FYM per hectare along with 150 kg N, 100 kg P, and 120 kg K/ha for optimal yield.

- ☆ **Bud stage spray:** Spraying a light solution of cake + SSP during bud development enhances flowering.
- ☆ **Nitrogen source:** Apply nitrogen through CAN, as urea may cause phytotoxicity.

Pinching

Pinching should be done twice, first at 4 weeks and then at 8 weeks after transplanting, to promote bushy growth. This practice encourages more lateral branches, leading to increased flower production and better plant shape.

De-shooting

Retain 4–5 main shoots in standard cultivars and 8–12 shoots in spray cultivars to ensure optimal flowering and uniform plant structure.

Disbudding

In standard cultivars, lateral buds should be removed to allow the main flower to develop fully, while in spray cultivars, the terminal bud is removed to encourage multiple well-formed blooms. This practice improves flower size, quality, and uniformity across the plant.

Irrigation

Maintain soil moisture at 60–70 per cent for optimal growth, applying 8–10 irrigations of 2.5–5 cm depth based on prevailing weather conditions. Proper irrigation ensures healthy vegetative growth and high-quality flowering.

Staking

Plants are supported using wire mesh, plastic nets, string, or bamboo canes. Wire mesh or plastic nets with 10–15 cm square openings are placed on the ground in three layers and elevated to 20, 35, and 50 cm as plants grow. Alternatively, strings or ropes are set up in three rows along the planting rows. For potted chrysanthemums, a single bamboo stick supports standard cultivars, while 3–4 sticks are positioned around spray cultivars to provide adequate support and protection.

Weeding

Conduct three to four hand weedings during the growing season to keep the crop free of competition. Pre-transplant application of Atrazine at 1.0 kg a.i. per hectare effectively controls weeds and ensures healthy plant establishment.

Diseases

- ☆ **Wilt:** Caused by *Fusarium oxysporum* f. sp. *chrysanthemi*, leading to yellowing, wilting, and plant death.
- ☆ **Stem and foot rot:** *Rhizoctonia solani* infects stems and collars, causing rot and plant collapse.

- **Root rot:** *Pythium* and *Phytophthora* spp. attack roots, resulting in decay and poor plant growth.
- **Bacterial rot:** *Erwinia chrysanthemi* causes soft rot of stems and leaves with foul odour.
- **Powdery mildew:** *Oidium chrysanthemi* produces white powdery growth on leaves and stems.
- **Leaf spot and flower blight:** *Alternaria* and *Septoria* spp. cause dark lesions on leaves and flowers.
- **Gray mould:** *Botrytis cinerea* leads to grey fuzzy mould on flowers, stems, and leaves.
- **Viral diseases:** Chrysanthemum viruses like stunt, mosaic, rosette, tomato spotted wilt, and tomato aspermy cause stunted growth, flower distortion, and mosaic patterns.

Insect-Pests

- **Aphids:** Suck sap from leaves and shoots, causing curling, yellowing, and virus transmission.
- **Red spider mites:** Feed on leaf sap, leading to bronzing, yellowing, and webbing.
- **Hairy caterpillars:** Chew on leaves and flowers, causing defoliation and flower damage.
- **Thrips:** Damage flowers and leaves by lacerating tissues and sucking sap, resulting in streaks and deformities.
- **Grubs:** Feed on roots and underground parts, causing plant weakening and stunted growth.
- **Leaf miners:** Larvae tunnel inside leaves, producing winding trails and reducing photosynthetic area.
- **Nematodes:** Attack roots, forming galls and causing stunted growth and wilting.

Disorders

- **Premature budding:** Flowers develop too early, reducing overall yield and uniformity.
- **Quilling of florets:** Florets roll or curl abnormally, affecting flower appearance and market value.
- **Crown bud formation:** Multiple buds form at the crown, leading to uneven flowering.
- **Heat delay:** High temperatures slow down bud initiation and flower development.
- **Petal burn:** Petals get scorched or discoloured due to excessive sunlight or high temperature.

Management

- **Cultural Control:** Use well-drained soil or growing media and avoid waterlogging to prevent root and stem rots. Maintain proper spacing and ventilation to reduce humidity and disease incidence. Regularly remove and destroy infected plant parts, and ensure the use of disease-free planting material. Proper management of temperature and light also helps prevent physiological disorders.
- **Biological Control:** Apply beneficial microorganisms such as Trichoderma or Pseudomonas to suppress root and stem rots. Encourage natural predators like ladybird beetles, lacewings, and predatory mites to control aphids, thrips, and spider mites. Beauveria bassiana can be used effectively against caterpillars, thrips, and whiteflies.
- **Chemical Control:** For fungal diseases like rots, blights, and powdery mildew, use fungicides such as Mancozeb, Carbendazim, or Copper oxychloride. Insect pests like aphids and whiteflies can be managed with Imidacloprid or Thiamethoxam, mites with Abamectin, and caterpillars with Spinosad. Apply nematicides only if root nematode infestation is severe.
- **Management of Physiological Disorders:** Prevent premature budding, quilling of florets, crown bud formation, and other disorders by adjusting temperature, light, and irrigation, and ensuring balanced nutrition. Proper cultural care maintains uniform growth and high-quality flowering.

Harvesting Stages

- **Standards:** For distant markets, harvest when the outer row of florets begins to unfurl to ensure longer vase life. For local markets, flowers can be picked at the half-open stage for immediate sale.
- **Sprays:** For local markets, harvest when at least two flowers are fully open and others have started showing colour. For distant markets, pick when around 50 per cent of the flowers display colour to maintain quality during transport.
- **Loose flowers:** Harvest when the flowers are fully open to provide maximum visual appeal and freshness for sale.
- **Pot mums:** Pick when approximately 50 per cent of the buds have developed colour, ensuring a balanced display and prolonged flowering in pots.

Packing

Chrysanthemum flowers are bundled in groups of 10, 20, or 25 and placed in corrugated cardboard boxes measuring 91 × 43 × 15 cm, accommodating approximately 80–100 stems. Flower bunches are wrapped in cellophane sleeves to protect them during transport and maintain freshness.

Yield

Standard chrysanthemums produce 2.5–4.5 lakh flowers per hectare, while spray varieties yield 1.5–1.75 lakh per hectare. Under greenhouse conditions, chrysanthemums can yield 150–250 stems per m^2 annually.

9.1.4 Gerbera

Gerbera, popularly known as Transvaal daisy, Barberton daisy, or African daisy, is a vibrant ornamental flower valued worldwide for its beauty. The genus name was given in honour of the German naturalist Traugott Gerber. It traces its origin to the Natal and Transvaal regions of South Africa, as well as parts of Asia. Gerbera is widely cultivated for cut flowers and decorative purposes due to its wide range of attractive colours (Figure 9.5).

Figure 9.5: Gerbera in Greenhouse.
(*Source*: www.freepik.com)

Important Species of the Genus Gerbera

The important species of the genus Gerbera include *G. asplenifolia*, *G. aurantiaca*, *G. jamesonii*, *G. kunzeana*, and *G. viridifolia*. Among these, *G. jamesonii* is the most commercially important species used in hybridization for cut flower production. *G. aurantiaca* is known for its brilliant orange flowers, while others contribute valuable traits like foliage and adaptability. These species form the genetic base for modern cultivated gerbera varieties.

Important Cultivars of *Gerbera*

Some of the important cultivars of Gerbera are Diana, Thalsa, Sonsara, Paganini, Anneke, Nette, Rosetta, Gloria, Ginna, Ingrid, Pricilla, Alexias, Intense, Sunway, Zingaro, Balance, and Monique. These cultivars are popular in commercial cultivation and cut flower trade due to their wide range of attractive colours, large flower size, long vase life, and suitability for greenhouse as well as open field conditions.

Factors Affecting Growth and Flowering of Gerbera

- **Light:** Gerbera performs best under long day conditions which promote growth and flowering.
- **Temperature:** Optimum temperature is 16–22°C during the day and 12–15°C at night for healthy growth.
- **Growth Regulators:** Application of GA_3 promotes flowering, while CCC helps in controlling excessive vegetative growth.

Climatic and cultural requirements for gerbera in greenhouses is summarised in Table 9.6.

Table 9.6: Climatic and Cultural Requirements for Gerbera in Greenhouses

Factor	*Requirement*	*Impact on Quality*
Temperature	18–24°C	Continuous flowering, quality blooms
Light	400–600 μmol/m^2/s	Promotes stem length and color
Humidity	70–75 per cent	Enhances leaf health, reduces stress
Soil/Medium	pH 5.5–6.5, EC <1.0	Good root growth
Irrigation/ Fertigation	Balanced NPK + Mg, Fe	Prevents chlorosis, large flowers
Density	8–10 plants/m^2	Proper spacing for flower size
Harvest Stage	Disc florets at right angle	Ensures longer vase life

Propagation

Gerbera can be propagated by different methods such as seeds, which are mainly used for breeding purposes; division of plants, where clumps are separated for multiplication; cuttings, taken from suckers or side shoots; and micro-propagation, which ensures large-scale production of disease-free, uniform plants.

Ideal Planting Density and Spacing

The ideal planting density of gerbera is 8–10 plants per m^2, with recommended spacing of 30 × 30 cm or 40 × 25 cm, ensuring proper aeration, light penetration, and healthy flower production.

Nutrient Status in Gerbera

The optimum nutrient status in gerbera leaves is:

- Nitrogen (N): 2.7–3.1 per cent
- Phosphorus (P): 0.19–0.35 per cent
- Potassium (K): 3.06–3.64 per cent

- Calcium (Ca): 1.66–2.18 per cent
- Magnesium (Mg): 0.30–0.48 per cent

These ranges indicate the balanced nutrient requirement for healthy growth, optimum flowering, and prolonged vase life in gerbera.

Ideal Fertilizer Application Schedule

The ideal fertilizer application schedule in gerbera is:

- Vegetative stage: Apply N:P:K:Ca:Mg in the ratio 3:2:3:1:1 at the rate of 75 g/m^2 to promote healthy leaf and root development.
- Flowering stage: Apply N:P:K:Ca:Mg in the ratio 3:2:4:1:1 at the rate of 75 g/m^2 to enhance bud initiation, flower quality, and yield.

Diseases

- Root rot (*Pythium irregularae, Rhizoctonia solani*): Causes rotting of roots leading to wilting and stunted growth.
- Foot rot (*Phytophthora cryptogea*): Affects the collar region, causing rotting and plant collapse.
- Sclerotium rot (*Sclerotium rolfsii*): Produces white mycelial growth with sclerotia at the base of plants.
- Blight (*Botrytis cinerea*): Leads to grey mold on flowers and leaves under high humidity.
- Powdery mildew (*Erysiphe cichoracearum, Oidium crysiphoides*): Appears as white powdery patches on leaves and stems.
- Leaf spots (*Phyllosticta gerberae, Alternaria* spp.): Characterized by brown to black spots causing leaf damage.
- Viral disease (Cucumber mosaic virus and Tobacco rattle virus): Results in leaf mottling, yellowing, and flower deformation.

Insect-Pests

- **White fly:** Sucks sap from leaves, causing yellowing and transmitting viral diseases.
- **Red spider mites:** Feed on leaf undersides, leading to speckling, bronzing, and leaf drop.
- **Nematodes:** Attack roots, causing galls, poor growth, and reduced flower yield.
- **Aphids:** Sap-sucking insects that distort young leaves and act as virus vectors.
- **Leaf miner:** Larvae tunnel into leaves, creating winding mines and reducing photosynthesis.

- **Caterpillars:** Chew leaves, buds, and flowers, resulting in severe feeding damage.

Management

- **Cultural Control:** Maintain well-drained soil or potting media and avoid waterlogging to prevent root, foot, and sclerotium rots. Ensure proper spacing and ventilation to reduce humidity and disease incidence. Remove and destroy infected leaves, flowers, or plants, and use virus-free planting material.
- **Biological Control:** Apply beneficial microorganisms like Trichoderma or Pseudomonas to suppress soil-borne rots. Encourage natural predators such as ladybird beetles and predatory mites to control aphids, whiteflies, and spider mites. Beauveria bassiana is effective against caterpillars and thrips.
- **Chemical Control:** Use fungicides like Mancozeb, Carbendazim, or Copper oxychloride to manage rots, blights, powdery mildew, and leaf spots. Apply insecticides such as Imidacloprid or Thiamethoxam for whiteflies and aphids, Abamectin for mites, and Spinosad for caterpillars. Nematicides can be applied in case of severe nematode infestations.
- **Viral Disease Management:** Control vectors like aphids, whiteflies, and leaf miners to reduce virus spread, and use virus-free planting material. Remove and destroy infected plants promptly to prevent disease outbreaks.

Harvesting Stages of Cut Flowers

Gerbera cut flowers are harvested just before the outer row of ray florets start showing pollen to ensure longer vase life. The ideal stage is when the outer row of petals stands perpendicular to the stalk, indicating proper maturity and quality.

Packaging of Cut Flowers

Gerbera cut flowers are packed in insulated boxes to protect them from freezing injury during transport. They can also be placed on plastic-coated metal grids (50 × 70 cm) with 2 × 2 cm mesh for safe handling and ventilation.

Average Yield of Gerbera Cut Flowers

The average yield of gerbera cut flowers is 200–250 flowers/m^2/year under greenhouse conditions and 120–150 flowers/m^2/year in open field cultivation.

9.1.5 Orchid

Dendrobium Orchid (*Dendrobium* sp.)

Dendrobium is a highly esteemed tropical orchid, renowned for its graceful blossoms and versatility. This species, which prefers warm and humid environments, flourishes particularly in coastal areas like Chennai, where elevated humidity and moderate temperatures provide optimal growth conditions. Dendrobium orchids are esteemed for their enduring blossoms, diverse colour palette, and elegant form, making them highly sought after as decorative plants and cut flowers in the floral business (Figure 9.8).

Figure 9.8: Dendrobium in Greenhouse.
(*Source*: www.freepik.com)

Prominent Commercial Varieties of Dendrobium Orchid

- ✰ **Sonia 17** – A highly popular hybrid with vibrant violet-purple flowers and white centres, well-suited for cut flower production due to its long vase life and year-round blooming ability.
- ✰ **Sonia 28** – Similar in appearance to Sonia 17 but produces slightly larger and brighter flowers with better spike quality, making it preferred in export markets.
- ✰ **Emma White** – A striking pure white variety symbolizing elegance and purity, widely demanded for weddings, decorations, and premium floral arrangements.
- ✰ **Sakura Pink** – Known for its delicate pink blossoms with soft hues resembling cherry blossoms, this variety is especially valued for ornamental displays and gifting purposes.

Climate Requirements

Dendrobium orchids thrive best under 75 per cent green shade net, which provides filtered light essential for healthy growth and flowering. An ideal relative humidity of 70–80 per cent and a temperature range of 18–28°C create favourable conditions for year-round cultivation. Optimal light intensity of 1,500–2,000 foot-candles ensure proper photosynthesis without causing leaf scorch, making this tropical orchid well-suited to warm and humid regions.

Climatic and cultural requirements for orchids in greenhouses is summarised in Table 9.8

Table 9.8: Climatic and Cultural Requirements for Orchids in Greenhouses

Factor	*Requirement*	*Impact on Quality*
Temperature	18–30°C; 10–15°C for Cymbidium nights	Species-specific flowering
Light	30–40 per cent shade	Prevents leaf scorch, ensures bud development
Humidity	70–80 per cent	Supports root health and flower initiation
Growing Medium	Bark, moss, perlite, charcoal	Good aeration, root anchorage
Irrigation/ Fertigation	Low EC 0.5–0.8	Healthy growth, long-lasting spikes
Density	15–20 plants/m^2	Optimal growth and spike length
Harvest Stage	80 per cent flowers open on spike	Long shelf life and market value

Propagation

Dendrobium orchids can be propagated through several methods:

- ☆ **Division of clumps** – Mature plants are divided into smaller units with healthy pseudobulbs and roots for replanting.
- ☆ **Keikis (plantlets)** – Small offshoots that develop on the mother plant are carefully detached and grown independently.
- ☆ **Back bulbs** – Old but viable pseudobulbs are used to regenerate new shoots.
- ☆ **Tissue culture plants** – Large-scale commercial propagation is carried out through micropropagation, ensuring uniform, disease-free planting material.

Containers and Support

Perforated earthen pots or plastic pots with good drainage are ideal for healthy root growth. The growing medium is kept loose and aerated, usually with a mix of charcoal, brick pieces, and coconut husk chips. Plants are supported with bamboo stakes or sticks to maintain upright spikes and prevent damage during flowering.

Growing Media

The most prevalent potting mixture comprises charcoal, fragmented bricks and tiles, as well as coconut husk and fiber.

Irrigation

Utilization of mist or overhead sprinklers to supply water and regulate humidity.

Nutrients

For optimum growth and flowering, Dendrobium orchids respond well to foliar feeding. A foliar spray of NPK (20:10:10) @ 0.2 per cent is recommended at weekly intervals, starting 30 days after planting. This balanced fertilizer promotes healthy vegetative growth, strong pseudobulbs, and robust spikes. As plants mature, the nutrient ratio may be adjusted towards higher potassium to enhance bud initiation, flower colour, and longevity.

Growth Regulators

Foliar spray of GA_3 @ 50 ppm at bimonthly intervals starting 30 days after planting promotes vigorous growth, uniform shoots, and strong pseudobulbs. This practice enhances spike quality and increases flower yield.

Repotting

Orchids generally require repotting every two to three years to maintain healthy growth. Repotting becomes necessary when:

- ☆ The plant outgrows its container.
- ☆ The potting medium decomposes or loses its structure.
- ☆ The plant needs to be split or divided.
- ☆ Epiphytic orchids, in particular, benefit from annual repotting.

The ideal time for repotting is when fresh roots begin to emerge at the base of the previous year's growth. In monopodial orchids, repotting or division should be carried out when new leaves appear at the top along with fresh root growth.

Splitting or Division of Plants

When Dendrobium orchids grow into large clumps with 2–3 old canes and new shoots, they should be divided before repotting. Each division must include at least one mature cane (about two years old), one healthy new shoot, and a portion of fresh roots to ensure successful establishment.

Pests of Dendrobium Orchids

- ☆ Aphids – Suck sap from tender shoots and flower buds, causing curling and distortion.
- ☆ Mealybugs – Found on roots, leaves, and sheaths, secreting honeydew that leads to sooty mold.
- ☆ Scales – Attach to leaves and stems, weakening plants by sap feeding.
- ☆ Thrips – Damage flowers and buds, leading to scarring, discoloration, and poor quality blooms.

- ☆ Mites (Spider mites) – Cause yellow speckling and bronzing of leaves under dry conditions.
- ☆ Snails and slugs – Feed on tender roots, shoots, and flowers, leaving irregular holes.

Diseases of Dendrobium Orchids

- ☆ Bacterial soft rot (*Erwinia* spp.) – Causes water-soaked, foul-smelling rot of pseudobulbs and leaves.
- ☆ Bacterial brown spot (*Pseudomonas* spp.) – Brown, sunken spots on leaves, spreading rapidly under high humidity.
- ☆ Leaf spot (*Cercospora* spp., *Phyllosticta* spp.) – Circular yellow to brown spots on foliage.
- ☆ Anthracnose (*Colletotrichum* spp.) – Produces black or brown necrotic lesions on leaves and flowers.
- ☆ Botrytis blight (*Botrytis cinerea*) – Small brown spots on petals, often appearing in cool, damp conditions.
- ☆ Fusarium wilt (*Fusarium oxysporum*) – Affects roots and pseudobulbs, causing yellowing, wilting, and plant death.
- ☆ Rhizoctonia root rot – Roots turn brown and decayed, reducing plant vigor.

Management

- ☆ **Cultural Control:** Maintain good air circulation and 70–80 per cent humidity to reduce disease incidence. Avoid overhead irrigation to prevent leaf wetness. Remove and destroy infected leaves, flowers, or pseudobulbs, and use well-drained, sterilized potting media. Regular repotting helps maintain plant health and prevent root rots.
- ☆ **Biological Control:** Release natural predators such as ladybird beetles and lacewings to manage aphids and mealybugs. Apply Trichoderma harzianum in potting media to suppress root rot, and use Beauveria bassiana or Verticillium lecanii against thrips and scale insects.
- ☆ **Chemical Control:** For sucking pests, spray Imidacloprid (0.3 ml/L) or Acetamiprid. Control mites and thrips with Abamectin (0.5 ml/L). Manage fungal leaf spots using Copper oxychloride (2 g/L) or Mancozeb, and treat bacterial diseases with Streptocycline (0.5 g/L) in rotation with copper sprays.

Harvest

Dendrobium flowers take about 3–4 days to reach full maturity after opening. Harvesting is done when the flowers are fully open, as cutting them earlier causes premature wilting before reaching the market. After harvest, about 0.75 cm of the peduncle is trimmed, and the spike is immediately placed in a tube of fresh water with preservative. The ideal stage for harvest is when around 75 per cent of the florets are open, while the remaining buds are still unopened (Figure 9.9).

Figure 9.9: Spike Ready for Harvest.
(Source: www.freepik.com)

Postharvest Handling

- Yield: 8 - 10 spikes/plant/year
- Pulsing: 8-HQC 500 ppm + Sucrose 5 per cent for 12 hrs
- Holding solution: $AgNO_3$ 25 ppm + 8-HQC 400 ppm + Sucrose 5 per cent
- Wrapping material: 50 gauge polythene with base of spikes dipped in 8-HQC 25 ppm

9.1.6 Anthurium (*Anthurium andreanum*)

Anthurium (*Anthurium andreanum*), also referred to as the flamingo flower or painter's palette, is a highly esteemed tropical ornamental plant recognized for its vivid spathes and extended vase life. In greenhouses, it is grown under regulated conditions that guarantee optimal temperature, humidity, and light, which are essential for its growth and flowering. Protected cultivation facilitates continuous production throughout the year, enhances flower yield, and improves quality relative to open-field conditions. Anthurium flourishes optimally in shade net rooms with 70–80 per cent relative humidity and diffused light that accentuates the vibrancy of its spathes. The crop necessitates well-drained, organic-rich substrate and meticulous irrigation by misting or sprinklers to sustain humidity levels. Through effective fertigation, cultural methods, and pest-disease management, greenhouse-cultivated Anthurium has become a lucrative enterprise in floriculture (Figure 9.10).

Figure 9.10:Anthurium Under Shade Net.
(*Source*: https://agritech.tnau.ac.in)

Climate

Anthurium necessitates porous, well-drained, aerated soil abundant in organic matter. The soil pH must range from 5.5 to 6.5. It functions effectively under a green shade net providing 70-80 per cent shade, with humidity levels of 80-90 per cent, daytime temperatures of 24-28°C, nighttime temperatures of 15-22°C, and light intensity ranging from 1500 to 2000 foot candles.

Growing Environment

Anthurium flourishes in 75 per cent shade with 70–80 per cent relative humidity. The ideal temperatures for good growth and flowering are 24–28°C during the day and 15–22°C at night. Climatic and cultural requirements for anthurium in greenhouses is summarised in Table 9.9.

Table 9.9: Climatic and Cultural Requirements for Anthurium in Greenhouses

Factor	*Requirement*	*Impact on Crop Performance*
Temperature	Day: 24–28 °C; Night: 15–22 °C	Ensures optimum growth, flowering, and prevents stress due to heat or cold.
Relative Humidity	70–80 per cent RH	Maintains turgidity, prevents desiccation, and improves flower quality.
Light/Shade	75 per cent shade net; diffused light preferred	Provides ideal photosynthesis without leaf scorching; improves spathe color and size.

Factor	*Requirement*	*Impact on Crop Performance*
Growing Media	Leaf mould : Coco peat (1:1), pH 5.5–6.5; good drainage and water retention	Supports healthy root growth, prevents root rot, and ensures nutrient uptake.
Propagation	Tissue culture (commercial), suckers, stem cuttings, seeds	Ensures uniform, disease-free plants with faster establishment.
Planting Material	Tissue culture plants, 15 cm height with 4–6 leaves	Ensures quick establishment and early flowering.
Irrigation	Mist or overhead sprinkler system	Maintains required humidity, prevents water stress, and enhances flower longevity.
Fertilization	Foliar spray or fertigation with NPK and micronutrients; cow dung + DAP early	Promotes vigorous vegetative growth, higher yield, and improved spathe size and color.
Growth Regulators	GA_3 (200 ppm) for growth; Planofix (NAA) for flower induction	Enhances vegetative vigor, increases flower production, and improves quality.
Cultural Practices	Raised bed or pot cultivation; leaf pruning (retain 4–6 leaves); burying roots	Reduces disease incidence, promotes continuous flowering, and maintains plant vigor.

Growing Media

Anthurium thrives in 75 per cent shade with 70–80 per cent relative humidity. The optimal temperatures for robust development and flowering are 24–28°C during the day and 15–22°C at night.

Propagation

Anthurium can be propagated via tissue culture, suckers, stem cuttings, or seeds. Tissue culture plants are frequently utilized in commercial production because of their homogeneity and disease-free condition.

Seed Propagation

Seeds germinate in 10 days and are transplanted after 4 to 6 months, requiring 2 to 3 years to flower. Seeds are planted on finely shredded substrate under 75 per cent shade or aseptically in Nitsch/MS media enriched with BAP and Adenine.

Varieties

- **Big-Flowered Varieties:** Popular varieties with large spathes include Red Champion, Red Glory, Ardens, Andra, and Linden. These are widely grown for cut-flower production due to their vibrant colors and long vase life.

- **Medium-Flowered Varieties:** Varieties like Marriot, Calypsa, and Alice offer medium-sized spathes with good color intensity and are suitable for both pot and cut-flower cultivation.
- **Miniature Varieties:** Compact varieties such as Ace, Cirque, and Pride are ideal for pot cultivation and decorative purposes.
- **Color Variants:** Protected cultivation allows growing multiple colors including red, pink, white, green, and bi-colored spathes, catering to market demand and ornamental value.

Suckers

Suckers are generated at the plant's base during the 4–5 leaf stage, with 2–3 roots detached for transplantation. The monthly injection of 57 ppm BAP to plants over one year old promotes increased sucker output.

Stem Cutting

Stem tips from 3–4-year-old plants with a few roots are removed for propagation. Each cutting should have a single eye or bud, and treatment with IBA 500 ppm promotes good root formation.

Tissue Culture

Tissue culture is increasingly popular for commercial cultivation. Explants used include leaf segments, root segments, stem sections, vegetative buds, flower stalks, spathes, and spadices, grown on MS medium.

Planting

Plants can be grown in pots or raised beds. Tissue culture plants of 15 cm height with 4–6 leaves are ideal for transplanting.

Irrigation

Water is supplied using misting or overhead sprinklers, which also help maintain and improve relative humidity around the plants.

Pot Cultivation

In pot cultivation, foliar sprays of NPK are applied at 0.2 per cent concentration–30:10:10 during the vegetative stage and 10:20:20 during the flowering stage. Fertigation practices similar to raised bed cultivation can also be adopted to ensure balanced nutrient supply.

Raised Bed Cultivation

For the first six months, Anthurium plants are sprayed with a solution of cow dung and DAP (250 ml per plant) for initial nutrition. After six months, weekly fertigation is applied to supply essential nutrients. Schedule 'A' includes calcium nitrate, potassium nitrate, and micronutrients, while Schedule 'B' provides mono ammonium phosphate, potassium nitrate, and magnesium sulphate. This system ensures balanced nutrition, supporting healthy plant growth and continuous flowering. Regular fertigation also maintains soil fertility and promotes optimal development.

Fertilizer Requirement

Apply NPK at 30:10:10 @ 0.2 per cent as a foliar spray starting 30 days after planting, repeated weekly.

Growth Regulators

Spray GA_3 at 200 ppm as a foliar application every 2 months to enhance growth and flower quality.

Pest and Disease Management

Pest

- ☆ **Aphid (*Myzus circumflexus*):** Yellowing of leaves and poor plant growth; nymphs and adults suck sap from the plant.

  Management: Spray Dicofol @ 2 ml/L, Malathion @ 2 ml/L, or Phosphomidon @ 2 ml/L.
- ☆ **Thrips (Chaetanaphothrips orchid):** Brown streaks on flowers and leaves; mottled appearance on foliage and flowers.

  Management: Spray Vertimac @ 0.4 ml/L or Dimethoate 30 EC @ 1 ml/L.
- ☆ **Whitefly (*Bemisia tabaci*):** Nymphs and adults suck sap from leaves, causing weakening of the plant.

  Management: Spray Methomyl @ 1.8 ml/L.
- ☆ **Red Spider Mite (*Tetranychus cinnabarinus*)**: Attacks lower leaves, causing bronzing of both upper and lower leaf surfaces.

  Management: Spray Kelthane @ 1 ml/L or Vertimac @ 0.4 ml/L.

Diseases

- ☆ **Anthracnose:** Dark, sunken lesions on leaves, stems, or flowers; can lead to rotting of affected parts.

  Management: Spray Bavistin @ 0.1 per cent.
- ☆ **Leaf Spot:** Small, brown or black spots on leaves that may enlarge and coalesce, causing leaf blight.

  Management: Spray Dithane M-45 @ 0.2 per cent.
- ☆ **Root Rot:** Roots become soft, brown, and decayed; plant shows wilting and yellowing of leaves.

  Management: Apply Captan @ 2 g/L as a soil drench.
- ☆ **Bacterial Wilt:** Sudden wilting of leaves and entire plant; vascular tissues show brown discoloration.

  Management: Treat with Streptocyclin @ 200 ppm.

Harvest

Suckers are produced at the base of the plant during the 4–5 leaf stage, with 2–3 roots severed for transplantation. The monthly administration of 57 ppm BAP to plants over one year old enhances sucker production.

Yield

Suckers are generated at the plant's base during the 4–5 leaf stage, with 2–3 roots cut for transplantation. The monthly application of 57 ppm BAP to plants over one year old increases sucker production.

9.1.7 Lilium

The Lilium species are indigenous to Asia (49 species), Europe (12 species), and North America (24 species). The genus is categorically divided into two groups: Oriental and Asiatic. Among the cultivars, those that blossom not discolour clothing or table surfaces with pollen–such as Tiara (pink single), Aphrodite (pink double), and Sphinx (red double)–are particularly favoured by floral aficionados (Figure 9.11).

Figure 9.11: Lilium in Greenhouse.
(*Source*: www.freepik.com)

- ☆ **Characteristics of Oriental lilies:** The Lilium genus has 49 species native to Asia, 12 species to Europe, and 24 species to North America. The genus is unequivocally classified into two categories: Oriental and Asiatic. Among the cultivars, those that do not stain clothing or table surfaces with pollen–such as Tiara (pink single), Aphrodite (pink double), and Sphinx (red double)–are especially preferred by floral enthusiasts.
- ☆ **Characteristics of Asiatic lilies:** Asiatic lilies, developed through hybridization of species such as Lilium amabile, L. bulbiferum, L. concolor, L. dauricum, L. davidii, L. hollandicum, L. maculatum, L. leichtlinii, L.

pumilum, and L. tigrinum, have an extended flowering period with flowers in a wide range of colors including orange, red, and yellow. They are mostly odorless, with narrow, upward-growing leaves and small bulbs measuring 10–16 cm. Key export cultivars include Connecticut King, Gran, Paradiso, Elite, Pollyana, Prato, and Solemio, which are highly valued in the floral industry.

Soil or Growing Medium

Lilies grow best in well-drained soil that is rich in organic matter with a pH of 6.5–7.5. Suitable potting mixtures include soil and peat moss in a 1:1 ratio, or a blend of soil, peat moss, and perlite in equal parts, providing good aeration, drainage, and moisture retention for healthy bulb and root development.

Climate

Lilies grow best under day temperatures of 18–25°C and night temperatures of 12–18°C. They require partial shade (40–50 per cent light) and good aeration/ventilation to ensure healthy growth, prevent diseases, and promote optimal flowering.

Climatic and cultural requirements for lilium in greenhouses is summarise in Table 9.11.

Table 9.11: Climatic and Cultural Requirements for Lilium in Greenhouses

Factor	*Requirement*	*Impact on Flower Quality*
Temperature	20–25°C (day), 12–15°C (night); bulbs pre-chilled	Uniform sprouting, strong stems, better buds
Light	Full light with 30–40 per cent shade in summer	Prevents scorching, enhances flower colour
Humidity	60–70 per cent RH	Reduces stress; excess leads to Botrytis
Growing Medium	pH 6.0–6.5, EC <1.0, sandy loam/cocopeat mix	Good root health, prevents bulb rot
Planting Material	Large, virus-free bulbs	Improves stem length, bud size, vase life
Density	35–45 bulbs/m^2	Optimum yield and flower quality
Irrigation/ Fertigation	Balanced NPK, high K at bud stage	Enhances colour, bud size, stem strength
Support	Netting/staking	Prevents stem bending and lodging
Harvest Stage	1–2 buds showing colour	Ensures vase life and market acceptability

Propagation

Lilies can be propagated through bulbs and bulblets. Additionally, scales and bulbils are used in certain species such as *Lilium bulbiferum, L. sargentiae, L.*

tigrinum, and *L. wallichianum*, allowing multiplication of plants for commercial and ornamental purposes.

Planting Time

The optimal planting time for lilies varies with altitude. In low hills, planting is recommended during October–November. In mid hills, the best periods are February–March and July–September, while in high hills, lilies should be planted between March and April. These timings ensure proper bulb establishment and healthy growth.

Treatment of Bulbs

Bulbs should be dipped for approximately 20 minutes in a fungicidal solution such as Emisan (0.2 per cent), Thiram (0.3 per cent), Captan (0.2 per cent), Bavistin (0.2 per cent), or Benlate (0.2 per cent). After treatment, bulbs must be dried in the shade before planting or storage. Prior to planting, bulbs should be treated with a systemic fungicide, while for storage, a contact fungicide is recommended. It is essential that the bulbs are thoroughly dried after treatment to ensure effective protection and prevent rotting.

Planting Density and Spacing

The planting density and spacing of lilies depend on the group of *Lilium*, bulb size, and the cultivation site. For open-field cultivation, the recommended density is 20–40 bulbs per square meter, with a spacing of 40 × 15 cm between bulbs.

Nutrition

For lilies, apply NPK fertilizers at a rate of 30:20:20 g/m^2. For liquid feeding, use an NPK ratio of 14:10:14. Care should be taken to ensure that fertilizers do not come into direct contact with the bulbs, as this may cause rotting.

Weeding

Perform three to four hand weedings during the growing season. Additionally, pre-emergence herbicides such as Propyzamide (2.25 kg/ha) and Chloropham (3.5 L/ha) can be applied to control weeds effectively.

Irrigation

Maintain soil moisture at 60–90 per cent, but do not water until bulbs sprout. Depending on weather conditions, provide 8–12 irrigations of 2.5–5 cm each. Ensure uniform moisture throughout the growing period.

Staking

Support lilies using wire mesh or plastic nets with 15–20 cm square openings, placed on the ground in three layers at heights of 20, 35, and 50 cm. Additionally, strings or ropes can be erected in three rows along the plant rows at the same heights to provide extra support.

Diseases

- ✰ **Gray mould (*Botrytis elliptica, B. cinerea*):** Causes grayish-brown spots on leaves, stems, and flowers, leading to tissue decay.
- ✰ **Soft bulb rot (*Rhizopus stolonifer*):** Bulbs become soft, water-soaked, and prone to rapid rotting.
- ✰ **Fusarium bulb rot (*Fusarium oxysporum* f. *lilii*):** Leads to yellowing leaves and rotting of bulbs, often causing plant death.
- ✰ **Brown scale (*Colletotrichum lilii*):** Causes brown lesions on bulb scales, reducing bulb quality and growth.
- ✰ **Root rot (*Pythium splendens*):** Roots become blackened and mushy, affecting nutrient and water uptake.
- ✰ **Bacterial soft rot (*Pseudomonas* spp.):** Bulbs turn soft, foul-smelling, and decay rapidly under humid conditions.
- ✰ **Viral diseases (mosaic):** Causes mottled or streaked leaves, stunted growth, and reduced flowering.

Insect-Pests

- ✰ **Aphids:** Small sap-sucking insects that cause leaf curling, yellowing, and transmit viral diseases.
- ✰ **Thrips:** Tiny insects that feed on flowers and leaves, causing silvering, browning, and distortion.
- ✰ **Whiteflies:** Sap-sucking pests that weaken plants, cause yellowing, and excrete honeydew leading to sooty mold.
- ✰ **Mites:** Microscopic pests that damage leaves and buds, causing speckling, bronzing, and stunted growth.

Disorders

- ✰ **Leaf scorch:** Caused by Mn and Al deficiency from excessive nitrates, corrected by applying lime @ 10 t/ha.
- ✰ **Bud blast:** Results from water accumulation at the plant top, nutrient competition, low light high nitrates, and fluctuating carbohydrates.
- ✰ **Puffy foliage:** Occurs due to frost injury, leading to stunted and swollen leaves.

Management

- ✰ Diseases: Use disease-free bulbs, treat bulbs with systemic and contact fungicides before planting and storage, ensure proper drainage, and avoid overcrowding.
- ✰ Insect-pests: Monitor regularly and control aphids, thrips, whiteflies, and mites using recommended insecticides or biocontrol agents.

- ☆ Disorders: Apply lime to correct nutrient deficiencies, avoid waterlogging, maintain proper spacing, regulate fertilization, and protect plants from frost.

Harvesting Stages

- ☆ **Local market:** Harvest when 1–2 florets are open.
- ☆ **Distant market:** Harvest when 1–2 florets just start showing color.
- ☆ **Stem cutting:** Cut stems a few centimeters above ground to promote bulb growth.

Storage of Flowers

Store lilies at a low temperature of 1–2/°C to maintain freshness and prolong vase life.

Harvesting and Storage of Lilium Bulbs

Harvest bulbs 40–50 days after flowering or when the foliage begins to yellow. Initially, store bulbs in moist sand at –2/°C for three weeks, then at 0–2/°C until two weeks before planting. In hilly areas, bulbs can alternatively be stored in a mixture of moist moss and sand.

9.1.8 Tulip (*Tulipa* spp.)

Tulips are among the most popular bulbous ornamental plants, valued for their vibrant colours, elegant form, and high market demand. Cultivating tulips in greenhouses allows precise control of temperature, light, and humidity, supporting off-season production, uniform quality, and higher profitability (Figure 9.12).

Figure 9.12: Tulips in Greenhouse
(*Source*: https://pixabay.com)

Climate and Environment

- **Temperature:** Tulips need a cool period for bulb sprouting and flower initiation. Bulbs are pre-cooled at 2–9/°C for 10–14 weeks depending on the cultivar. In greenhouses, optimal day temperatures are 15–18/°C and night temperatures 12–14/°C.
- **Light:** Tulips grow best under full sunlight; supplemental lighting can be used in low-light areas.
- **Humidity:** Moderate relative humidity (60–70 per cent) is ideal; excessive humidity may promote fungal diseases.
- **Ventilation:** Proper ventilation is essential to regulate temperature and reduce the risk of fungal infections.

Climatic and cultural requirements for tulip in greenhouses is summarised in Table 9.12.

Table 9.12: Climatic and Cultural Requirements for Tulip in Greenhouses

Factor	*Requirement*	*Impact on Flower Quality*
Temperature	2–9°C chilling (12–16 weeks); 12–18°C growth	Ensures sprouting, uniform flowering, stem elongation
Light	Bright light, avoid excessive heat	Improves stem quality and bud colour
Humidity	60–70 per cent RH	Prevents drying and reduces disease incidence
Growing Medium	pH 6.0–7.0, EC <1.0; well-drained sandy loam/ cocopeat mix	Prevents bulb rot and supports root health
Bulbs	Large, healthy (10–12 cm circumference)	Produces bigger flowers and longer stems
Density	40–50 bulbs/m^2	Maximizes production and quality per unit area
Irrigation/ Fertigation	Moderate irrigation; balanced NPK, high K	Improves bud colour, stem strength, vase life
Forcing	Pre-cooled bulbs forced under greenhouse	Enables off-season, high-value cut flowers
Harvest Stage	Bud shows full colour but remains closed	Ensures longer vase life and transport quality

Soil and Media

Tulips thrive in well-drained sandy loam soil rich in organic matter, with an ideal pH of 6.0–7.0. In greenhouse cultivation, they are commonly grown in raised beds, pots, or crates using a growing media composed of cocopeat, perlite, and sand, which ensures proper aeration and drainage for healthy bulb and root development.

Bulb Selection and Treatment

Select large bulbs (>10/cm circumference) for superior flowering. Break dormancy by providing a chilling period (vernalization) before planting, and treat bulbs with a fungicide dip, such as carbendazim or captan, to prevent rotting.

Planting Practices

Tulips are usually planted in autumn–winter for spring flowering, though greenhouse conditions allow year-round cultivation. Maintain a spacing of 10–15/ cm between bulbs, planting at a depth 2–3 times the bulb height. Bulbs can be arranged in rows or blocks depending on the bed or pot system used.

Irrigation

Provide light, frequent watering to keep the soil consistently moist without waterlogging; drip irrigation is ideal for greenhouse cultivation.

Fertilization

Equitable fertilizer management is essential for optimal tulip growth and superior flowering quality. Prior to planting, integrate basal organic matter, such as well-decomposed farmyard manure or compost, into the soil to enhance fertility and soil structure. During the growth phase, administer NPK fertilizers in a balanced ratio (10:10:10 or 12:12:18) to facilitate vegetative growth, flower initiation, and bulb development. Furthermore, foliar applications of micronutrients including magnesium (Mg), zinc (Zn), and boron (B) may be administered at consistent intervals to improve bloom size, coloration, and overall quality. The appropriate timing and dose of fertilizers are crucial to prevent nutrient imbalances and enhance tulip performance.

Cultural Practices

- ✰ **Weeding:** Regular weeding, either manual or with appropriate herbicides, is essential to prevent competition for nutrients, water, and light, ensuring healthy tulip growth. Maintaining weed-free beds also reduces pest and disease pressure.
- ✰ **Staking/Support:** While tulips generally do not require staking, support becomes important in cut-flower cultivation with long stems to prevent bending or breakage and maintain an attractive appearance.
- ✰ **Disbudding:** Removing malformed or secondary buds enhances the size, shape, and overall quality of the primary flowers, resulting in more marketable blooms.

Flowering and Harvesting

Tulips bloom within 8–10 weeks after planting under greenhouse conditions. For cut flowers, harvest buds when they are still tight but beginning to show colour to ensure longer vase life. Potted tulips are typically sold at full bloom for ornamental display.

Post-Harvest Management

For cut tulip flowers, maintain storage at 0–2/°C with high relative humidity to prolong vase life. For bulb production, allow leaves to remain after flowering to photosynthesize and nourish bulbs; mature bulbs are harvested once foliage yellows, then cured and stored at 15–20/°C before undergoing chilling treatment.

Pests and Diseases

- ☆ **Pests:** Tulips are commonly affected by aphids, thrips, and bulb mites, which feed on leaves, flowers, and bulbs, causing stunted growth and transmitting viral diseases. Regular monitoring and timely control measures help minimize damage.
- ☆ **Diseases:** Major diseases include Fusarium basal rot, Botrytis blight, tulip fire (caused by *Botrytis tulipae*), and viral infections, which lead to bulb decay, leaf spotting, and reduced flowering. Proper sanitation, disease-free bulbs, and fungicidal treatments help manage these problems.

Management

The health of tulips can be preserved by utilizing disease-free bulbs, providing enough greenhouse ventilation, administering suitable fungicide applications, and deploying biological control agents to reduce pest and disease prevalence.

Ideal Temperatures for Tulip Cultivation

Tulips flourish optimally at a relative humidity of 70–75 per cent, with a vapor pressure deficit (VPD) between 0.2 and 0.55 kPa, promoting sufficient transpiration and reducing the risk of fungal diseases.

Ideal Relative Humidity and VPD for Tulip Cultivation

Tulips thrive well in a relative humidity of 70–75 per cent, with a vapor pressure deficit (VPD) ranging from 0.2 to 0.55 kPa, facilitating adequate transpiration and mitigating the likelihood of fungal illnesses.

9.2 Vegetables

9.2.1 Tomato

Tomatoes are one of the most widely grown vegetables worldwide, valued for their high nutritional content and market demand. Greenhouse cultivation allows year-round production by providing controlled temperature, humidity, and light conditions. It protects plants from pests, diseases, and adverse weather, ensuring better fruit quality. With proper management, greenhouse-grown tomatoes achieve higher yields and superior market value compared to open-field cultivation. Progressive farmers increasingly adopt protected cultivation of tomatoes for its multiple advantages, including better quality produce, higher productivity, and improved nursery raising and hardening of plants. It allows effective insect and disease control with reduced pesticide use, enables off-season cultivation, and ensures efficient utilization of resources. Additionally, precise application of fertilizers under protected conditions further enhances plant growth and yield (Figures 9.13 and 9.14).

Figure 9.13: Protected Cultivation of Tomato in Open Ventilated Oolyhouse
(*Source*: Image by Filmbetrachter from Pixabay)

Figure 9.14: Cherry Tomato grown under Protected Cultivation
(*Source*: https://pixabay.com/)

Climate

Tomato is a warm-season crop that can tolerate mild frost, with seed germination occurring at 8–10°C. Night temperatures of 16–22°C are critical for fruit setting, while fruit fails to set below 12°C or when day temperatures exceed 32°C and night temperatures exceed 22°C. Greenhouse cultivation allows year-round production for 10–12 months by cooling in summer and heating in winter. This controlled environment ensures optimal growth and consistent fruit yield throughout the year.

Varieties

Tomato varieties suitable for greenhouse cultivation include big-fruited, indeterminate types such as R 144, FA 574, FA 180, GS 600, Delphi, Astona, and Cronos. Other indeterminate varieties like Avtar, Indam Hybrid, All Rounder, Arka Meghali, and Arka Surabhi are also widely grown. For cherry-type tomatoes, popular varieties include HA818, NS Cherry 1, NS Cherry 2, and T56, offering diversity in fruit size and market preference.

Season

- Spring: Tomato seedlings are transplanted in the greenhouse during late February to early March.
- Autumn: Transplanting in the greenhouse occurs around mid-August.

Bed Preparation

For tomato cultivation, raised beds are recommended. Well-decomposed organic manure (about one bamboo basket per sq.m) is mixed thoroughly into the soil, and the beds are levelled. In heavy soils, up to 25 per cent sand is added to improve aeration in the root zone.

Fertilizer Application

Apply commercial fertilizers supplying 50:50:50 NPK (kg/ha) to the growing beds before soil fumigation.

Mulching

Cover the planting bed with a 100-micron (400-gauge), 1.2 m wide polyethylene mulch film. Make 5 cm diameter holes in the film according to the recommended spacing of 60 × 45 cm. Secure the edges of the mulch by burying them in the soil to keep the film in place over the bed.

Transplanting

Five- to six-week-old seedlings, about 16 cm tall, are transplanted under suitable temperature and light conditions, with around 3,000 seedlings needed for 1,000 sq.m. Spacing of 90 × 50 × 40 cm is maintained, and beds are watered to field capacity before transplanting. Seedlings are sprayed with imidacloprid (0.3 ml) one day prior and drenched with carbendazim (0.1 per cent) on the planting day, preferably in the early morning or late evening. Seedlings are placed in holes on the mulch without touching the film, irrigated daily until established, and foggers are used if humidity is low. If seedling mortality occurs, beds are drenched with 0.3 per cent copper oxychloride.

Irrigation

Drip irrigation is started 10 days after transplanting and is applied daily, supplying 2–3 L of water per sq.m per day, depending on local weather conditions.

Fertigation

Water-soluble fertilizers are applied in a ratio of 200:200:200 kg NPK/ha throughout the crop growth period. From the 3rd week after transplanting, 3.75 g/sq.m of each NPK is supplied twice weekly through fertigation. Micronutrients (3 g/L) are sprayed 2–3 times at 30-day intervals starting 60 days after transplanting. Additionally, calcium nitrate is applied 2–3 times at 15-day intervals.

Growth Regulators

Planofix is sprayed four times at 2.5 ml per 10 L of water at 15-day intervals, starting from the initiation of first flowering to enhance flowering and fruit set.

Application of Fertilizers

A total of 250:250:250 kg NPK/ha is applied in 10 equal splits through the drip irrigation system at 10-day intervals from the date of transplanting.

Special Intercultural Operations

- **Staking:** Staking provides support to tomato plants using bamboo poles, wooden sticks, or other sturdy materials, keeping foliage and fruits off the ground. This practice improves fruit yield and size, reduces fruit rot, facilitates spraying, and makes harvesting easier.
- **Pruning:** Pruning begins 20–30 days after transplanting and is done weekly. After the first flower cluster, the main stem branches into two, which are retained, while all other branches–including those at the base–are removed. Removing excess fruits also results in larger tomatoes and higher market value.
- **Cluster Thinning:** Each tomato cluster produces 1–19 flowers, but only 6–8 fruits typically develop under good pollination. Excess fruits in a cluster reduce size, shape, quality, and uniformity. Thinning is done to maintain the optimal number of fruits per cluster.
- **Suckering:** Indeterminate tomato plants develop suckers between compound leaves and the stem. These suckers are removed to leave only the main stem as the growing point, ensuring better plant structure and fruit development.
- **Topping:** Small fruits developing at the top growing point are removed six weeks before the crop's termination. This allows fruits set lower on the plant enough time (7–8 weeks from anthesis) to reach full size and maturity, improving overall fruit quality.

Harvesting

The first tomato harvest occurs 75–85 days after transplanting. Large tomatoes are picked individually with the calyx attached. During summer, harvesting is best done in the early morning or late evening to minimize postharvest losses. Cherry tomatoes are harvested either with the stem attached or individually with the calyx.

Yield

Under protected cultivation, tomatoes can yield 6–8 kg/m^2 with proper care. On average, each plant produces 3–4 kg of fruit, with 40–60 fruits per plant, each weighing 80–90 g.

Climatic and cultural requirements for tulip in greenhouses is summarised in Table 9.13

Table 9.13: Climatic and Cultural Requirements for Tulip in Greenhouses

Factor	*Requirement*	*Impact on Crop Performance*
Temperature	20–25°C (day), 16–18°C (night)	Ensures proper fruit set and color development
Light	High light intensity; partial shade in summer	Improves photosynthesis and yield
Humidity	60–70 per cent RH	Reduces flower drop and disease incidence
Growing Medium	Cocopeat/perlite mix; pH 6.0–6.8; EC 2–3 dS/m	Promotes root growth and nutrient uptake
Varieties/ Hybrids	Indeterminate hybrids (Arka Rakshak, Shanty, NS 501)	High yield and disease resistance
Density	2.5–3 plants/m^2	Maximizes yield without overcrowding
Irrigation/ Fertigation	Drip with balanced NPK and micronutrients	Improves yield, fruit quality, and uniformity
Training/ Pruning	Vertical trellis, single/double stem pruning	Enhances ventilation and fruit size
Pollination	Bumblebees/manual vibration	Ensures high fruit set and uniform shape
Harvest Stage	Mature green, breaker, or red ripe depending on market	Increases shelf life and marketability

9.2.2 Bell Pepper

Capsicum is the second most grown vegetable under protected cultivation after tomato. Capsicum farming in polyhouse is a very profitable and popular agribusiness in India. Capsicum grown under protected cultivation has better quality and round the year production. Capsicum is commonly known as Bell pepper, Paprika, Sweet pepper, or Shimla mirch, it is valued for its taste, colour, and nutritional content. The crop is popular in both fresh vegetable markets and processing industries (Figure 9.15).

Figure 9.15: Capsicum Farming in Polyhouse
(*Source*: https://www.farmatma.in)

Soil

Capsicum thrives in well-drained, porous loamy or sandy loamy soils rich in organic matter. A slightly acidic to neutral soil with pH 6.0–7.0 is ideal for optimal growth. Proper soil preparation ensures good root development and high fruit yield.

Climate

The crop requires moderate temperatures with 25–30°C during the day and 18–20°C at night. Relative humidity of 50–60 per cent supports healthy flowering and fruit set. Extreme temperatures or humidity can reduce fruit quality and yield.

Climatic and cultural requirements for bell pepper in greenhouses is summarised in Table 9.14.

Table 9.14: Climatic and Cultural Requirements for Bell Pepper in Greenhouses

Factor	*Requirement*	*Impact on Crop Performance*
Temperature	22–28°C (day), 16–20°C (night)	Ensures proper flowering, fruit set, and growth
Light	Bright sunlight; 20–30 per cent shading in summer	Prevents sunscald; promotes fruit development
Humidity	60–70 per cent RH	Reduces fungal infection; improves pollination
Growing Medium	Sandy loam/cocopeat-perlite mix; pH 5.8–6.5; EC 1.5–2.0	Good root growth and nutrient uptake
Varieties/ Hybrids	California Wonder, Arka Lohit, Nusantara	High yield, uniform fruit size and quality

Factor	*Requirement*	*Impact on Crop Performance*
Density	2–3 plants/m^2	Optimal space utilization and yield
Irrigation/ Fertigation	Drip with balanced NPK + Mg, Ca	Prevents blossom-end rot; improves quality
Training/ Pruning	Vertical trellis, remove lower leaves/lateral shoots	Improves air circulation and fruit size
Pollination	Hand vibration or bumblebee-assisted	Ensures high fruit set and uniformity
Harvest Stage	Mature green or fully coloured stage	Long shelf life and marketability

Capsicum Variety

Most capsicum hybrids initially grow green and later turn red, yellow, or orange at maturity, depending on the crossbreed. Farmers should select varieties that offer high yield, fast growth, self-protective features, and strong local market demand. Popular commercial hybrids grown in India include Bomby, Indra, Swarna, Orobelle, Bachata, Mohini, Arka Gaurav, and Natasha.

Plant Population

In polyhouse cultivation, 2.5–3.5 plants per square meter is recommended to balance growth and light exposure. This density accommodates around 10,000–14,000 plants per acre. Proper spacing prevents overcrowding and promotes uniform fruit development.

Harvesting and Yield

Harvesting Time

- ☆ Green capsicum: 55–60 days after transplanting
- ☆ Yellow capsicum: 70–80 days
- ☆ Red capsicum: 80–90 days

Polyhouse Yield: 80–100 tonnes/acre

Open Field Yield: 20–40 tonnes/acre

Winter Cultivation: Green capsicum in open field yields 8–15 tonnes/acre over 4–5 months

Shade Net/Polyhouse Cultivation

Green and coloured capsicum yield 32–40 tonnes/acre in 8–10 months; polyhouse yields ~20 per cent more, but shade nets are cheaper to build and maintain (Figure 9.16).

Figure 9.16: Green Capsicum.
(*Source*: https://www.farmatma.in)

Nursery Raising

Farmers can buy seedlings or raise their own in 98-cell pro-trays (1.5 × 2 cm) inside a greenhouse. Seeds are sown 0.5 cm deep, covered with coco peat, and watered lightly; germination occurs in 5–7 days. Seedlings are sprayed with Acephate (0.75 ml/L) and supplied nutrients with mono-ammonium phosphate, ready for transplanting in 30–35 days.

Bed Preparation/Sterilisation

Greenhouse soil is ploughed and mixed with 20 kg/sq.m vermicompost, and raised beds of 80 × 40 cm with 50 cm spacing are prepared. Beds are sterilized with 0.4 per cent formaldehyde (4 L/sq.m) covered with polythene for 5 days. Two weeks before transplanting, neem cake with bio-agents is applied, and seedlings are sprayed with imidacloprid (0.2 ml/L) and chlorothalonil (1 g/L) to prevent pests.

Fertigation

Fertigation delivers nutrients through irrigation, supplying 150–180 kg N:P_2O_5 per hectare over the 6–8 month crop cycle. It starts from the 3rd week after transplanting, with 2.5–4 g/sq.m applied regularly. This ensures steady growth, proper fruit development, and higher yields.

Mulching/Spacing and Planting

Mulching with 30–100-micron polyethylene conserves moisture, enhances soil fertility, and suppresses weeds. Seedlings aged 30–35 days are transplanted into 6 cm diameter holes spaced 45 × 35 cm and treated with copper oxychloride/copper hydroxide (25–30 ml/plant). Watering is continued for 7 days to help plants adapt to greenhouse conditions.

Pruning/Training/Irrigation

Pruning starts three weeks after planting and is done weekly to remove weaker branches while retaining stronger stems. Each node produces one strong and one weak branch, and pruning ensures better growth of the stronger branch. Sharp cutting tools are recommended to achieve clean cuts and promote superior fruit quality. Proper pruning increases productivity and fruit size. Regular irrigation supports plant health and maximizes yield throughout the crop cycle.

Pest and Disease and Management

- ✰ **Mites:** Infestation by larvae that suck sap from leaves, buds, and fruits, causing leaf curling, reduced fruit size, and stunted plant growth.

  Management: Remove and discard affected plant parts. Spray Neem soap (8–10 g/L), Pongamia oil (5–8 ml/L), Wettable sulphur (2 ml/L), or chlorfenapyr (1 ml/L).

- ✰ **Aphids:** Adults and nymphs feed on leaf veins, damaging leaves, reducing plant growth, and lowering yield.

  Management: Monitor regularly and spray Imidacloprid (0.5 ml/L), Neem soap (8–10 g/L), or dimethoate (2 ml/L). Maintain plant hygiene.

- ✰ **Thrips:** Sucking of leaf juice causes leaf curling, stunted growth, and poor nutrient absorption, reducing yield.

  Management: Remove affected leaves, buds, flowers, and fruits. Keep areas clean with sterilized water. Spray Imidacloprid (0.5 ml/L), Neem seed kernel extract 4 per cent, fipronil (1 ml/L), or acephate (1.5 g/L).

- ✰ **Fruit Borer:** Eggs laid on leaves, fruits, and flowers hatch into larvae that feed on them, severely damaging produce quality.

  Management: Destroy eggs and larvae. Spray thiodicarb (1 ml/L), chlorfenfur (1.5 ml/L), carbaryl (3 g/L), fipronil (1 ml/L), or indoxacarb.

- ✰ **Nematodes:** Root infestation causes yellowing leaves, reduced leaf and fruit size, and stunted plant growth. Large nodes on roots often indicate heavy infestation.

  Management: Rotate with non-solanaceous crops (sweet corn, marigold, cabbage). Apply Neem cake @ 900 kg/acre 5 days before planting and carbofuran granules @ 25 kg/acre. Monitor for reinfestation after 2–3 cropping cycles.

- ✰ **Bacterial Wilting:** Attacks seedling base causing drooping, drying, and plant death; often introduced from nurseries or during transplanting.

  Management: Maintain strict hygiene in nurseries and greenhouses. Drench plant bases with carbendazim (1 g/L) or copper oxychloride @ 30–50 ml/plant.

- ✰ **Powdery Mildew:** Yellow spots on leaves resembling powder, leading to leaf drop and low-quality fruits.

Management: Spray Neem oil (7 ml/L) + sulphur WDG-80 (2 g/L), or drench leaves with hexaconazole (0.5 ml/L), azoxystrobin (0.5 ml/L), or flusilazole (0.5 ml/L).

- ☆ **Phytophthora:** Tiny oily spots on leaves during flowering/fruiting; blackening and rapid death of plants, especially in high humidity.

  Management: Prefer polyhouse over net houses. Spray azoxystrobin (0.5 ml/L), copper hydroxychloride (3 g/L), or metalaxyl MZ (2 g/L) upon detection.

Harvesting and Yield of Capsicum

Harvesting starts at different stages for different coloured capsicums. Typical harvesting turnover time from their transplantations are as follows-

1. **Green capsicum:** 55 to 60 days
2. **Yellow capsicum:** 70 to 80 days
3. **Red capsicum:** 80 to 90 days

Harvesting should be done with due care without causing any harm to the plant, flowers and budding fruits. Morning is the best harvesting time and can be continued for 3-4 days.

Average yield is found to be 80-100 tons per acre in polyhouse opposed to 20-40 tons under the traditional open field system. Capsicum is also known as bell pepper, paprika, sweet pepper or Shimla mirch. In the open field cultivation, green capsicum can be grown in the winter season with proper climatic conditions. In its 4-5 months crop duration a yield of 8-15 tonnes/acre can be achieved. Green and coloured capsicum can be grown under shade net houses and polyhouses. 32-40 tonnes/acre of yield can be achieved with a crop duration of 8-10 months. Yields from polyhouses are 20 per cent higher than shade net houses but shade net houses are cheaper to build and maintain.

Post Harvest Management

Harvested capsicums should be kept under shade for necessary grading and packaging. Good quality marketable fruits should be assorted and to be cleaned with soft dry clothes to discard powdery residues of pesticides, droplets of water or derbies, if any.

1. Grading

Harvested capsicums are sorted as per grades for better marketing and profitability. Selected fruits should have uniformity in terms of their colour, maturity, size, and shape and are free from all kinds of decay, defective spots, or fungicide residues while packing as per market demand and transportation need.

Standard Grades for Capsicums

Grade	*Specification*
A	Fruits having 3-4 lobes and weighing 150+ grams
B	Fruits with 2-3 lobes and weighing less than 150 grams

2. *Packaging and Storing*

Consider maintaining a 7-8 degree temperature (not more or less) and a relative humidity of 90-95 per cent in the storage place to keep up the optimum standard of fruits. This should be carefully noted that capsicum must not be stored together with fruits like tomato, papayas, or mangos, papayas to protect them from ripening that tax on your revenue.

9.2.3 Cucumber

Cucumber is an important greenhouse crop, highly demanded in five-star hotels. It is commonly used in salads, raita, and pickles. Compared to many other crops, cucumbers are easier to cultivate, making their farming popular among growers. They are generally grown throughout the year. For cucumber plants to bear fruit, pollination is essential, usually carried out by bees and other insects. However, since it is difficult for such insects to enter a polyhouse, it is recommended to cultivate parthenocarpic cucumber varieties in polyhouse conditions. Cucumber can be grown in almost all types of soils, but well-drained soil is considered most suitable. The ideal soil pH for cultivation ranges between 5.5 and 6.7 (Figure 9.17).

Figure 9.17: Cucumber in Greenhouse.
(*Source*: https://cdn.pixabay.com)

Climatic and cultural requirements for cucumber in greenhouses is summarised in Table 9.15.

Table 9.15: Climatic and Cultural Requirements for Cucumber in Greenhouses

Factor	*Requirement*	*Impact on Crop Performance*
Temperature	22–28°C (day), 16–20°C (night)	Ensures proper flowering, fruit set, and growth
Light	12–14 hrs bright light; shading during peak summer	Promotes uniform fruit development and size
Humidity	70–80 per cent RH	Reduces stress and disease incidence
Growing Medium	Cocopeat/perlite/vermiculite or sandy loam; pH 6.0–6.8; EC 1.5–2.0	Good root growth and nutrient uptake
Varieties/ Hybrids	Parthenocarpic hybrids: Poinsett, Solan, Green Wonder	High yield, uniform fruits, disease resistance
Density	2–3 plants/m^2	Optimal growth and maximum yield
Irrigation/ Fertigation	Drip with balanced NPK + Ca	Prevents blossom-end rot, improves fruit quality
Training/ Pruning	Vertical trellis, single stem, remove laterals	Enhances airflow, fruit uniformity, quality
Pollination	Parthenocarpic: not required; others: bee-assisted	Ensures fruit set and uniformity
Harvest Stage	Marketable size (18–25 cm slicing)	Best taste, quality, and shelf life

Preparation of Seedlings

Cucumber seeds have a high germination capacity, so they are either sown directly on raised beds or transplanted using 5–6-week-old seedlings. In polyhouse cultivation, seedlings are raised in pro trays filled with a mixture of coco peat and vermicompost. One seed is placed in each cell, and germination usually occurs within 3–4 days of sowing. The seedlings are ready for transplanting after 20–25 days. Cucumber is cultivated on raised beds that are carefully prepared to ensure proper growth and management. The beds should have a surface width of about 90 cm, with a walkway of 50 cm between two beds to allow easy movement and operations. The height of each bed should be maintained at around 40 cm for better drainage and root development.

Use of Mulching

The use of mulching in cucumber cultivation under polyhouse conditions is highly beneficial, as it helps suppress weed growth and retain soil moisture in the beds. However, mulching is not recommended when planting a new crop during the summer season, as it may adversely affect crop establishment.

Planting Spacing

For cucumber planting, proper spacing is essential to ensure healthy growth and maximum yield. The recommended spacing is 60 cm between two plants and 50 cm between two rows, which provides adequate room for development and ease of intercultural operations.

Irrigation

In cucumber cultivation, drip irrigation is commonly used to supply water and fertilizers in a balanced manner according to the crop's growth stage. Farmers must ensure that adequate moisture is consistently maintained in the root zone of the plants.

Special Cultivation Methods in Cucumber

In greenhouse cultivation, cucumber plants develop large leaves and grow rapidly, which creates a high demand for sunlight. To ensure proper light penetration and healthy growth, special practices such as trellising, pruning, and training is essential. Trellising supports vertical growth, pruning removes excess foliage for better airflow, and training helps guide the plant structure. These methods together enhance yield and fruit quality in greenhouse cucumbers.

- **Cucumber Trellising, Pruning, and Training:** In cucumber cultivation, trellising, pruning, and training are key practices for healthy growth and higher yield. The crop is trained vertically on wires fixed at a height of 2.5–3 meters. Once the vine reaches the top wire, its tip is pruned to form an umbrella-like structure. At the stage when one or two leaves appear on the stem, the growing tip of the main stem is removed to encourage side branches. These lateral branches are extended along the wire and allowed to grow downward, and when they reach near the ground, their tips are pruned to maintain balance and productivity.

Fruit Thinning

Fruit thinning in cucumber depends on the plant's vigour and the weight of the fruits. When too many fruits develop at once, their growth slows, leading to small and undersized cucumbers that have little market value. To maintain quality and uniformity, it is important to remove small, deformed, and low-grade fruits, ensuring that the remaining ones grow well and fetch better prices.

Harvesting

Harvesting of cucumbers generally begins 50 to 65 days after planting, depending on the variety and growing conditions. Since cucumbers grow rapidly, they must be picked at regular intervals of every 2 to 4 days to ensure quality and continuous production.

Major Diseases of Cucumber Crop

- **Damping Off:** This disease occurs mainly during germination, causing seedlings to lose vigour and stop growing. Symptoms include drying of leaves from the apex and browning of the stem. Poor drainage and humid or extreme temperatures increase its severity.
- **Fusarium Wilt:** A fungal disease that attacks the vascular tissues of roots and stems, blocking water and nutrient transport. This leads to stunted growth, wilting, and eventually plant death. It is one of the major yield-reducing diseases in cucumbers.
- **Downy Mildew:** Symptoms begin as pale green to yellow angular spots on older leaves, later spreading quickly in cloudy weather. The disease reduces flower and fruit formation, lowering yield and quality. Infected plants produce fewer and inferior cucumbers.
- **Powdery Mildew:** This fungal disease appears as white powdery patches on leaf surfaces and fungus growth underneath. Affected leaves become deformed, fall prematurely, and reduce photosynthesis, leading to poor fruit size, yield, and taste.
- **Cucumber Yellow Stunted Disorder Virus (CYSDV):** Spread by whiteflies, this viral disease causes bright yellowing of leaves with green veins. It reduces chlorophyll and photosynthetic activity, leading to poor plant health and a significant drop in cucumber production.

Major Pests of Cucumber Crop

- **Mites:** Mite infestation causes small yellow or white spots on cucumber leaves, which eventually turn dull and dry. In severe cases, webs form on the leaves, leading to premature leaf fall. Mites are most problematic in hot weather conditions.
- **Thrips:** Thrips are minor pests but still harmful, as they cause white to brown spots and slight curling of cucumber leaves. More importantly, they act as carriers of viral diseases. Proper monitoring and control are necessary to avoid secondary damage.
- **Whiteflies:** Whiteflies are one of the most destructive greenhouse pests of cucumber. They multiply rapidly, colonizing the underside of leaves, causing yellowing, stunted growth, and black sooty mold due to honeydew secretion. They also spread viral diseases, making early control essential.
- **Leaf Miners**: Leaf miner larvae feed on the sap of shoots and leaves, creating tunnels within the leaf tissue. This damages chlorophyll, reduces photosynthesis, and weakens plant growth. Infestation lowers the overall productivity of the crop.
- **Nematodes:** Nematodes attack the roots, forming nodules that restrict the absorption of water and nutrients. This severely hampers plant growth

and reduces yield. Infested plants show poor vigour and often fail to reach their full production potential.

Management

Disease Management

- **Damping Off:** Use well-drained soil, avoid overwatering, and treat seeds with fungicides or biocontrol agents before sowing.
- **Fusarium Wilt:** Grow resistant varieties, practice crop rotation, and apply Trichoderma or suitable fungicides.
- **Downy Mildew:** Ensure good ventilation, avoid overhead irrigation, and spray recommended fungicides at early symptoms.
- **Powdery Mildew:** Maintain proper plant spacing, remove infected leaves, and apply sulphur-based fungicides or bio-controls.
- **CYSDV (Cucumber Yellow Stunted Disorder Virus):** Control whiteflies with insect-proof nets, sticky traps, and recommended insecticides; remove infected plants.

Pest Management

- **Mites:** Maintain humidity, use neem oil or miticides, and encourage natural predators.
- **Thrips:** Install blue sticky traps, use neem-based sprays, and apply insecticides if infestation is severe.
- **Whiteflies:** Use yellow sticky traps, practice sanitation, and apply systemic insecticides; introduce bio-agents like Encarsia formosa.
- **Leaf Miners:** Remove and destroy infested leaves, use neem-based sprays, and introduce parasitoids like Diglyphus isaea.
- **Nematodes:** Practice crop rotation with non-host crops, use resistant rootstocks, apply neem cake or bio-fumigants, and maintain soil health.

9.2.4 Strawberry

Strawberry is a high-value fruit crop that thrives well under controlled conditions. Protected cultivation allows year-round production by safeguarding the plants from extreme weather, pests, and diseases. It helps in achieving uniform growth, better flowering, and higher fruit quality compared to open-field cultivation. Greenhouses or polyhouses provide an ideal environment with regulated temperature, humidity, and light, which enhances fruit size, colour, and taste. This method also allows efficient water and nutrient management, reducing wastage and improving yield. Overall, protected cultivation of strawberries ensures better returns and consistent market supply (Figure 9.18).

Figure 9.18: Strawberry in Greenhouse
(*Source*: https://pixabay.com/)

Soil and Climate

Strawberries grow well in sub-tropical and temperate climates, with optimal day temperatures of 22–25°C and night temperatures of 7–13°C. Sandy loam soils with a pH of 5–6.5 are ideal, while waterlogged or heavy soils should be avoided. Proper site selection, good air drainage, and plastic mulch help reduce frost damage. Fields previously grown with tomato, potato, chilli, or brinjal are unsuitable for strawberry cultivation.

Climatic and cultural requirements for strawberry in greenhouses is summarised in Table 9.16.

Table 9.16: Climatic and Cultural Requirements for Strawberry in Greenhouses

Factor	*Requirement*	*Impact on Crop Performance*
Temperature	18–25°C (day), 10–15°C (night)	Ensures flowering, fruit set, and quality
Light	Bright light; shading during peak summer	Promotes uniform fruit development and color
Humidity	60–70 per cent RH	Reduces fungal infections; improves fruit quality
Growing Medium	Sandy loam or cocopeat-perlite mix; pH 5.5–6.5; EC <1.0	Good root growth, fruit yield, and quality
Varieties/ Hybrids	Chandler, Festival, Sweet Charlie, Selva, Camarosa	High yield, uniform fruits, export suitability
Density	8–10 plants/m^2	Optimum plant growth and fruit quality

Factor	*Requirement*	*Impact on Crop Performance*
Irrigation/ Fertigation	Drip with balanced NPK + high K and Ca	Improves fruit size, firmness, and shelf life
Training/ Support	Raised beds, pots, or vertical towers; remove runners	Maintains vigor and fruit uniformity
Pollination	Bees or hand vibration	Ensures high fruit set and even development
Harvest Stage	Fully red fruits	Maximum taste, color, and market value

Propagation

Strawberries are commercially propagated through runners, which produce true-to-type plants but can transmit viral diseases. To prevent infection, a separate bed should be used for runner production on land where strawberries have not been grown for at least 3–4 years. Planting is done at a spacing of 1.2 × 1.2 m or 1.8 × 1.8 m, with 30–40 runners per plant considered optimal. The production of runners can be enhanced by applying GA3 at 50–100 ppm. Careful management ensures healthy, disease-free planting material for successful cultivation.

Varieties

In India, only a few improved strawberry varieties are grown, selected for regional adaptation and fruit size. Popular cultivars include Pusa Early Dwarf, Premier, Dilpasand, Bangalore, Florida-90, Seascape, Redgauntlet, Sujata, Majestic, Chandler, Camarosa, Sweet Charlie, and Senga. These varieties are developed to enhance yield and fruit quality under protected conditions. They cater to both small- and large-fruited preferences for different markets.

Field Preparation and Planting Method

Strawberry requires friable soil, prepared by ploughing and raised beds about 15 cm high, with 60 cm wide beds on terraced slopes. Healthy runners with only one or two young leaves are planted after thoroughly washing the roots. Using the matted-row method, runners are spaced 45 cm apart with 90 cm between rows, forming a mat 40–45 cm wide, which increases plant numbers but may reduce fruit size. Small pits are made for planting, roots are pressed firmly, and light irrigation is applied immediately after planting.

Manures and Fertilizers

Strawberry plants have high nutrient requirements for optimal growth. In sub-tropical regions, 20–50 tons of well-decomposed cow dung along with one-third of nitrogen (100 kg), full dose of phosphorus (70 kg), and potash (150 kg) per hectare are applied during field preparation before planting. The remaining nitrogen is supplied through drip irrigation in equal doses at intervals of 10–15 days after planting to support continuous growth.

Uses of Plant Growth Regulators (PGRs)

Plant growth regulators (PGRs) act as catalysts for flowering, fruit development, and ripening, and are widely used in fruit crops to control growth sustainably. Foliar application of GA3 at 25 ppm improves fruit weight, TSS, ascorbic acid, and yield, while 50 ppm promotes runner development and delays flowering. These compounds help enhance productivity in an environmentally friendly manner.

Irrigation

Irrigation is essential during the first year for proper rooting of runners, with strawberries needing about 1 inch of water per week. As roots are concentrated in the top 6–10 inches of soil, light irrigation at short intervals is recommended–10–15 days in autumn and 5–7 days in summer. Watering should be avoided during harvest to prevent fruits from becoming too soft, which reduces their marketability for long-distance transport.

Weed Control

Weeds, which thrive especially in wet conditions, are a major threat in strawberry fields, with common species including Bermuda grass, Goose grass, Sorrel, and Mock Strawberry. Weed management is costly but essential, and can be done using herbicides like simazine (3 kg/ha) or pendimethalin (330 g/l). In protected cultivation, black plastic mulch is widely used to suppress weed growth effectively.

Mulching

Mulching helps reduce winter damage, control weeds, and prevent soil erosion. Ideal mulching materials are loose and light, such as white straw, marsh hay, or sawdust. Black polyethylene mulch is particularly effective, promoting early harvest and higher yields. However, it must be removed and properly disposed of after each fruiting season, which is a drawback.

Diseases and Management

- ☆ **Verticillium Wilt (*Verticillium alboatrum*):** Plants wilt in dry weather; older leaves turn brown, shrivel, and eventually die.

  Management: Avoid planting tomatoes, potatoes, peppers, brinjal, or raspberries in the same field. Treat soil with Formalin @ 5000 l/ha or Chloropicrin @ 210 l/ha before planting.

- ☆ **Anthracnose (*Colletotrichum* spp.):** Round, black or grey leaf lesions; dark brown or black circular lesions on stems, petioles, and runners; stunted growth, yellowing, wilting, and eventual plant collapse.

  Management: Use crop rotation with non-host crops, rinse soil around plant crowns before planting, practice regular weeding, plant disease-free runners, and avoid excessive nitrogen application.

- ☆ **Leaf Spot (*Mycosphaerella fragariae*):** Small, circular dark purple to reddish-purple spots on upper leaf surfaces; black spots on fruits during wet weather.

  Management: Remove and burn crop residues, avoid overhead irrigation, and spray Bordeaux mixture 2–3 times at 10–15 day intervals.

- ☆ **Powdery Mildew (*Sphaerotheca macularis*):** Upward folding of leaf edges, erratic purple blotches on upper surfaces along veins, fragile foliage; thick spores are absent.

  Management: Remove and destroy infected plant parts, use sulphur as a preventive, and spray Bayleton or Karathane (0.05 per cent) at 15-day intervals.

- ☆ **Botrytis Rot (*Botrytis cinerea*):** Light brown spots on green and ripening fruits, fruits dry, harden, and are covered with fungus, resulting in complete rotting.

  Management: Protect fruits from soil contact, ensure good field drainage, and avoid excessive nitrogen fertilizer use.

Insect- Pest Management and Management

- ☆ **White Grubs (*Phyllophaga* spp.):** Larvae feed on roots and stems of young plants, visible around roots in the soil, potentially killing heavily infested plants.

  Management: Practice deep ploughing before planting and spray soil with Chlorpyriphos @ 1–2 ml/L of water to control grubs.

- ☆ **Thrips (*Scirtothrips dorsalis*):** Feeding on stigmas and anthers causes flowers to turn brown and wither prematurely; fruit develops russeting, seedy or bronze appearance.

  Management: Use crop rotation with legumes or cowpea, remove old field debris after harvest, renew beds after the first fruiting season, and apply Cypermethrin (6.6 g a.i/ha) or Deltamethrin (2.25 g a.i/ha) to reduce thrips and deformed fruits.

- ☆ **Red Spider Mite (*Tetranychus urticae*):** Causes speckling and mottling on leaf surfaces, stunted growth, reduced yield, and webbing on leaf undersides.

  Management: Apply Aldicarb (10G) @ 4–5 kg/ha, monitor natural predators like ladybird beetles, green lacewings, and phytoseiid mites, and release ladybird beetles and green lacewings at two per plant to maintain biological control.

Harvesting

In the plains, strawberries start ripening from the last week of February to March. Fruits should be harvested when more than half of the fruit turns red, plucking them along with the stem to prevent damage.

Yield

Recommended strawberry varieties produce an average yield of 250–500 g per plant. Maximum yield is generally obtained from plants that are two years old, while the crop remains productive for up to three years in temperate regions.

Storage

Strawberries are packed in plastic punnets and placed in fibre or ventilated cardboard boxes to avoid damage. Proper storage in a controlled atmosphere containing 2 per cent O_2 and 5 per cent CO_2 at 3/°C can extend shelf life to 10–12 days, increasing market availability.

9.3 Cultivation of Pot Plants

Greenhouse cultivation of pot plants in India has gained immense popularity due to the rising demand for ornamental foliage and flowering plants in urban homes, offices, hotels, and landscapes. It provides a controlled environment where temperature, humidity, and light can be regulated for healthy growth and year-round production. Protected cultivation minimizes the impact of adverse weather conditions, pests, and diseases, ensuring higher survival rates and better-quality plants. A wide range of foliage plants like dracaena, pothos, and ficus, as well as flowering plants like anthurium, orchids, and gerbera, are successfully grown in pots under greenhouse conditions. The system allows efficient use of water, fertilizers, and space, making it suitable for commercial nurseries as well as small-scale entrepreneurs. By using improved potting media, scientific irrigation, and fertigation methods, pot plants achieve attractive shape, colour, and longer shelf life. Greenhouses also enable off-season and export-oriented production of ornamentals. With increasing urbanization and interest in interior landscaping, the market potential for pot plants is rapidly expanding. Government support through subsidies and training further promotes greenhouse cultivation. Thus, greenhouse pot plant production is emerging as a profitable and sustainable enterprise in India (Table 9.17).

Why Greenhouse for Pot Plants?

The greenhouse cultivation of potted plants is significantly advantageous, providing numerous benefits compared to open-field circumstances. It safeguards plants against detrimental weather conditions, including extreme heat, cold, torrential rain, hail, and high winds, so assuring their survival and consistent growth. The regulated environment facilitates the management of light, temperature, humidity, and water, leading to superior plant quality, continuous output, and off-season accessibility. It also reduces losses from pests and diseases while enhancing the efficiency of water and nutrient utilization. Furthermore, greenhouse environments facilitate the cultivation of high-value and delicate ornamental plants such as anthurium.

Table 9.17: Selection of Containers based on Size of the Plants

Size of the Plant	*Container Size*
Small annual plants (3-8 inches tall)	4 inches
Medium annual plants (9-18 inches tall)	6 inches
Large annual plants (>18 inches tall)	8 inches
Small perennial plants (3-10 inches tall)	8 inches
Large perennial plants (>10 inches tall)	10 inches

Figure 9.19: Pot Plants
(*Source*: https://pixabay.com)

Environmental Control – Basics for Indian Climates

- **Temperature:** In most parts of India, the ideal range for pot plants is 18–30 °C depending on the species. Day temperatures of 20–28 °C and night temperatures of 15–22 °C favour foliage and flowering plants. Sensitive species should be protected from extremes above 35 °C or below 10 °C.
- **Relative Humidity (RH):** Foliage plants grow best at 50–70 per cent RH, while epiphytic species like orchids and anthuriums prefer 60–80 per cent RH. Though higher humidity reduces transpiration, it can also increase disease incidence, so ventilation must be well managed.
- **Light:** Shade-loving plants such as pothos and ferns need 1000–2000 lux of diffuse light for healthy growth. Flowering plants like gerbera and anthurium require moderate to bright diffused light, usually 50–70 per cent of full sunlight. Shade nets and diffused covers help prevent direct scorching from intense midday sun.
- **Ventilation and Cooling:** Proper ventilation through side vents and ridge vents maintains uniform conditions in moderate climates. In hot and dry

regions, evaporative cooling or pad-and-fan systems are highly effective. Short misting or fogging cycles can increase humidity, but continuous misting should be avoided to reduce disease risks.

- **Heating:** Heating is rarely needed in most Indian climates but becomes important in colder hill areas for tropical ornamentals. Simple portable heaters or hot water pipe systems can maintain optimal conditions. This ensures sensitive plants remain healthy during cold nights or winter months.

Growing Media and Containers

In greenhouse cultivation of pot plants, the choice of containers and growing media plays a vital role in healthy plant growth. Containers may include plastic pots, clay or terracotta pots, fiber pots, and even grow bags, selected according to the species and size of the plant. All containers should have proper drainage holes and, preferably, saucers to collect excess water. The potting medium should be light, porous, well-drained, and capable of retaining some moisture while ensuring adequate aeration. A common mix for foliage pot plants consists of cocopeat or peat moss (40–50 per cent), coco coir or leaf mould (20–30 per cent), perlite or vermiculite (10–20 per cent), and well-decomposed compost (10 per cent). For epiphytic plants like orchids, a coarse mixture of bark, charcoal, sphagnum moss, and perlite is preferred, while succulents require a medium with higher sand and coarse grit content and less organic matter to avoid rotting. To maintain plant health, sterilization and hygiene are essential–media should be pasteurized by steam or solarization, or treated with bio-control agents like *Trichoderma*. Used media should never be reused without proper sanitization to prevent disease spread.

Propagation Methods

- **Cuttings (stem/leaf):** This is the most common method for multiplying foliage plants like pothos, coleus, and ficus. Stem or leaf cuttings are rooted in a sterile medium under mist benches with adequate humidity. Application of rooting hormones such as IBA ensures quick root initiation and higher success rates.
- **Division and offsets:** Clumping and rhizomatous plants such as peace lily and snake plant are propagated by dividing mature clumps or using offsets. Each division should have healthy roots and shoots for quick establishment. This method ensures uniformity and maintains the true-to-type character of the mother plant.
- **Seed sowing:** Some ornamental species that produce viable seeds can be propagated through seed sowing in trays or small pots. A fine, well-drained medium along with humidity domes encourages uniform germination. Seedlings are later transplanted to pots after proper hardening.

- **Tissue culture:** For large-scale production of high-value ornamentals like orchids and anthurium, tissue culture is widely used. This laboratory-based technique provides disease-free, genetically uniform planting material. It ensures rapid multiplication and supply throughout the year.
- **Grafting and layering:** Although less common, certain ornamental cultivars may be propagated by grafting or layering. These methods are employed when cuttings or seeds do not guarantee true-to-type plants. They are mainly useful for specific hybrids or rare species.
- **Propagation environment:** Successful propagation requires controlled conditions of 20–28 °C temperature, high humidity, and low light intensity. Mist chambers or fogging systems help maintain a favourable microclimate for rooting. Hardened plants should gradually be shifted to the main greenhouse for further growth.

Irrigation and Water Management

- **Water quality:** check pH (6.0–6.8 preferred) and EC (electrical conductivity). Avoid saline water; treat if high salts or sodium.
- **Systems:** drip irrigation (emitters per pot), capillary mats for benches, ebb and flow (flood and drain), manual hand watering for small operations. Drip + fertigation is common and efficient.
- **Irrigation scheduling:** based on pot size, media, plant species, season. Monitor substrate moisture (tensiometers, finger test). Allow limited drying between irrigations for many foliage plants; do not keep pots soggy.
- **Overhead vs root watering:** root/soil irrigation preferred; overhead increases disease risk.

Fertilizer and Fertigation

- A balanced supply of nutrients is essential for healthy growth of pot plants. Regular feeding can be done using water-soluble fertilizers such as 20:20:20 or 19:19:19 as a base mix. During flowering stages, the nutrient ratio should shift towards higher potassium, for example 15:15:30, to promote better blooms and stronger plants.
- Nutrients are best supplied in low concentrations, usually around 200–400 ppm nitrogen, through fertigation on a weekly basis. Foliar feeding at a low dose every 10–14 days can also support growth. The exact frequency and strength should be adjusted depending on the plant species and its growth stage.
- In addition to NPK, micronutrients like iron, manganese, zinc, boron, and molybdenum should be applied occasionally to prevent deficiencies. Substrate pH must be maintained between 5.5 and 6.8 for proper nutrient availability, and irrigation water can be acidified if pH levels are too high.

- To prevent salt accumulation in the potting medium, it is important to flush or leach the substrate periodically. This practice is particularly useful where irrigation water has moderate to high electrical conductivity, ensuring root systems remain healthy and active.

Integrated Pest Management (IPM)

- **Prevention and Hygiene:** Strong preventive practices are the foundation of IPM in greenhouses. Always quarantine new plants before introducing them to the main house, sanitize tools regularly, and keep the area free from weeds. Avoid overwatering, as excess moisture encourages root diseases and pest outbreaks.
- **Monitoring:** Regular monitoring helps in early detection and timely action. Yellow sticky traps are effective for catching whiteflies and aphids, while magnifiers can be used to spot tiny pests like mites. Frequent scouting across different crop sections ensures that no infestation goes unnoticed.
- **Biological Controls:** Beneficial organisms provide eco-friendly pest suppression without harming the crop. Predatory mites like Phytoseiulus control spider mites, Encarsia targets whiteflies, and ladybugs feed on aphids. Soil pests can be managed with entomopathogenic nematodes, while Trichoderma suppresses root diseases effectively.
- **Cultural Controls:** Good cultural practices reduce disease pressure and pest habitats. Pruning infected plant parts, maintaining proper plant spacing, and improving air circulation lower the chances of fungal attacks. Controlling nighttime humidity spikes also helps prevent mildew and botrytis.
- **Chemical Controls:** Chemical measures should be used only when other methods fail and pest pressure crosses thresholds. Apply selective insecticides or miticides, always rotating chemical groups to avoid resistance. Prefer systemic treatments for sucking pests and strictly follow label instructions and safety intervals.
- **Record Keeping:** Detailed records are vital for evaluating IPM effectiveness. Keep track of pest monitoring results, action thresholds, control methods applied, and their outcomes. These records help refine future strategies and prevent repeated mistakes.

Common Nutrient/Pest/Disease Management

- **Yellow lower leaves:** normal ageing or nitrogen deficiency → apply balanced fertilizer.
- **Brown leaf tips:** salt build-up or inconsistent watering → flush substrate; adjust irrigation.

- ☆ **Wilting with soft roots:** root rot (Pythium) → remove plant, discard infected media, improve drainage, use clean pots and Trichoderma/appropriate fungicide.
- ☆ **White cottony masses:** mealybugs → treat with soft soap, neem oil, systemic insecticide or introduce Anagyrus/cryptolaemus predators.
- ☆ **Sticky leaves + honeydew:** aphids/whitefly → release parasitoids (Encarsia) or use neem oil/soap.
- ☆ **Fine webbing:** spider mites (dry, hot conditions) → raise humidity briefly, use predatory mites or miticide.

Popular Pot Plants

- ☆ **Foliage houseplants:** pothos (Epipremnum), philodendron, dracaena, sansevieria (snake plant) – low light, moderate fertigation.
- ☆ **Flowering pot plants:** anthurium, peace lily (Spathiphyllum), gerbera, kalanchoe – need brighter diffused light, strict fertigation, humidity control.
- ☆ **Orchids:** special bark/sphagnum mixes, high humidity, strong airflow, indirect light, lower N, higher humidity.
- ☆ **Succulents and cacti:** very porous media, low organic matter, less frequent watering, bright light.
- ☆ **Ferns and mossy pots:** high humidity, shaded conditions, fine porous medium.

Harvesting, Finishing and Post-Production Care

- ☆ **Finishing for Retail:** Before sending pot plants to market, ensure they are uniform in size, healthy, and visually appealing. Clean the pots, remove any damaged leaves or signs of insect attack, and add a decorative mulch or top dressing for better presentation. Foliar shine sprays may be used cautiously, and shade-acclimated plants should be hardened gradually before shifting to outdoor display or brighter retail environments.
- ☆ **Packing and Transport:** Proper packing is essential to maintain plant quality during handling and transit. Use protective sleeves around foliage and secure pots in sturdy boxes or trays to prevent tipping or breakage. Transport should avoid exposure to extreme heat or direct sunlight, and each pot should be labelled with the plant's name along with clear care instructions to enhance customer satisfaction and plant survival.

Cultivation of Economically Important Medicinal and Aromatic Plants

10.1. Introduction

Medicinal and aromatic plants (MAPs) are vital for pharmaceuticals, cosmetics, perfumery, and nutraceuticals. India, being one of the biodiversity hotspots, has rich resources of MAPs, but productivity and quality are often limited under open field cultivation. Protected cultivation (greenhouses, shade nets, polyhouses, hydroponics) ensures controlled environment, enhanced yield, better quality, and round-the-year availability. Economically important medicinal and aromatic plants in protected cultivation is summarized in Table 10.1.

10.2. Importance of Protected Cultivation of MAPs

10.2.1 Controlled Environment

Protected cultivation provides a controlled temperature, humidity, and light intensity, creating an ideal microclimate for the growth of medicinal and aromatic plants (MAPs). This ensures optimum physiological functions and better crop performance throughout the year.

Reduction of Biotic Stresses: Growing plants under protection significantly reduces biotic stresses such as pest attacks, disease incidence, and weed infestation.

Table 10.1: Economically Important Medicinal and Aromatic Plants in Protected Cultivation

Crop	*Type*	*Suitable Structure*	*Cultivation Practices*	*Yield/Quality Advantage*	*Major Uses*
Ashwagandha (Withania somnifera)	Medicinal	Polyhouse/Shade net	Raised beds, organic manures, controlled irrigation	Higher root biomass, uniform alkaloid content	Roots used in Ayurveda, immunity booster
Stevia (Stevia rebaudiana)	Medicinal	Polyhouse/ Hydroponics	Seedlings in trays, drip fertigation, 3–4 harvests/year	Leaves richer in steviosides, year-round production	Natural sweetener, diabetic-friendly product
Safed Musli (Chlorophytum borivilianum)	Medicinal	Shade net house	Rhizome planting, mulching, high humidity maintained	Higher root yield, disease-free	Aphrodisiac, tonic in Ayurvedic formulations
Aloe vera	Medicinal	Polyhouse	Suckers planted on raised beds, minimal irrigation	Higher gel content, cleaner leaves	Pharmaceutical, cosmetics, nutraceuticals
Kalmegh (Andrographis paniculata)	Medicinal	Shade net	Direct sowing/ transplants, organic manures	Higher andrographolide content	Liver tonic, herbal medicines
Tulsi (Ocimum sanctum)	Medicinal/Aromatic	Shade net/ Polyhouse	Spacing 30x30 cm, organic fertigation	Enhanced essential oil yield	Ayurveda, perfumery, herbal teas
Geranium (Pelargonium graveolens)	Aromatic	Polyhouse/ Greenhouse	Stem cuttings, pruning for bushy growth, fertigation	Higher oil yield (citronellol, geraniol)	Perfumery, cosmetics
Patchouli (Pogostemon cablin)	Aromatic	Greenhouse/Shade net	Vegetative propagation, 2–3 cuttings per year	Quality oil rich in patchoulol	Perfume, incense, cosmetics

Crop	*Type*	*Suitable Structure*	*Cultivation Practices*	*Yield/Quality Advantage*	*Major Uses*
Rose (Rosa damascena)	Aromatic	Polyhouse/ Greenhouse	High-value cuttings, fertigation, humidity control	Higher attar yield, better oil content	Rose oil, gulkand, perfumery
Vetiver (Vetiveria zizanioides)	Aromatic	Shade net	Root propagation, minimal irrigation	Roots richer in vetiver oil	Perfumes, aromatherapy
Mint (Mentha spp.)	Aromatic	Polyhouse/ Hydroponics	Stolons planting, frequent irrigation, fertigation	High menthol content, quick turnover	Menthol oil, pharma, confectionery
Vanilla (Vanilla planifolia)	Aromatic	Greenhouse with humidity and support poles	Stem cuttings, hand pollination, shade 50–60 per cent	High-quality cured pods	Flavoring agent, perfumery, confectione

This lowers dependency on chemical pesticides, thereby producing cleaner and safer products.

10.2.2 Enhanced Germination and Growth

The uniformity of germination, plant growth, and flowering is much higher under controlled conditions. This results in consistent crop quality and improved yield.

10.2.3 Support for Organic and GAP Production

Protected structures make it easier to adopt organic farming and Good Agricultural Practices (GAP). By minimizing contamination and allowing better input management, they promote sustainable and eco-friendly cultivation.

10.2.4 Higher Active Principles and Essential Oils

Medicinal and aromatic plants cultivated under protection accumulate higher concentrations of active principles, alkaloids, and essential oils. This improves the medicinal and commercial value of the produce.

10.2.5 Off-season Cultivation and Market Advantage

Protected cultivation allows farmers to grow crops even in the off-season. This ensures a continuous supply to the market, better price realization, and enhanced profitability.

10.3. Suitable Structures for MAP Cultivation

10.3.1 Polyhouses

Polyhouses are highly suitable for cultivating temperature-sensitive medicinal and aromatic plants (MAPs) such as Ashwagandha, Aloe vera, and Stevia. They help maintain optimum warmth and protection from external fluctuations, ensuring better growth and uniform quality.

10.3.2 Greenhouses

Greenhouses are ideal for high-value aromatic crops like Geranium, Patchouli, and Vanilla. These structures provide a stable microclimate, which enhances essential oil content, flowering, and overall commercial quality of the produce.

10.3.3 Shade Net Houses

Shade net houses are best suited for shade-loving MAPs such as Safed Musli, Kalmegh, and Tulsi. By filtering excess light and reducing heat stress, they promote healthy vegetative growth and higher biomass production.

10.3.4 Mist Chambers and Fogger-Equipped Greenhouses

Mist chambers or greenhouses fitted with foggers are primarily used for raising elite planting material, rooting cuttings, and hardening tissue culture plantlets.

They maintain high humidity and prevent desiccation, thus improving survival rates of propagules.

10.3.5 Hydroponics and Aeroponics

Advanced techniques like hydroponics and aeroponics are gaining popularity for MAPs such as Stevia, Mint, and other leafy herbs. These systems ensure high purity, rapid growth, and contamination-free production, making them suitable for pharmaceutical and nutraceutical applications.

10.4. Propagation Techniques under Protected Conditions

10.4.1 Seed Propagation

Seed propagation under protected conditions ensures uniform germination by maintaining controlled humidity and optimum temperature. The regulated environment prevents seedling mortality caused by sudden weather fluctuations. It also reduces fungal infections and damping-off diseases, which are common in open fields. With the use of seed trays and sterilized media, high-quality planting material can be raised efficiently. This method is especially useful for crops like Ashwagandha and Kalmegh. Overall, it ensures vigorous seedlings with better field establishment.

10.4.2 Vegetative Propagation

Vegetative propagation methods such as stem cuttings, suckers, rhizomes, and stolons perform best under mist chambers or fogging systems. These controlled structures maintain the right level of humidity to promote rooting and prevent desiccation. Cuttings of Tulsi, Mint, and Geranium show high survival when propagated under protection. Similarly, rhizome-based crops like Ginger and Turmeric benefit from disease-free conditions. This method shortens the crop cycle and produces true-to-type plants. As a result, farmers get uniform plantations with assured yield quality.

10.4.3 Tissue Culture

Tissue culture or micropropagation is highly effective for multiplying rare, endangered, and high-value MAPs like Vanilla, Orchids, and Safed Musli. Protected laboratories and greenhouses provide aseptic conditions for culture initiation and growth. This technique ensures disease-free, genetically uniform plants that can be produced year-round. Tissue-cultured plantlets are later hardened in mist chambers to improve survival rates. It is especially important for species where conventional propagation is slow or unreliable. Thus, tissue culture bridges the gap between demand and supply in commercial MAP cultivation.

10.4.4 Grafting and Layering

Grafting and layering are specialized propagation techniques adopted for selected MAPs like Sandalwood and Patchouli. In protected structures, these methods achieve higher success due to stable humidity and temperature. Grafting improves plant vigour, while layering allows quick multiplication of elite genotypes. Controlled conditions minimize desiccation and infection at grafting or rooting sites. These methods help in producing robust planting stock with desirable traits. Hence, they are widely used for commercial scale production of high-value MAP species. Comparison of propagation techniques for MAPs is summarised in Table 10.2.

Table 10.2: Comparison of Propagation Techniques for MAPs

Method	*Examples of MAPs*	*Advantages*	*Limitations*
Seed Propagation	Ashwagandha, Kalmegh, Tulsi	✰ Cost-effective and simple ✰ Produces vigorous seedlings ✰ Suitable for large-scale production	✰ Genetic variability ✰ Some seeds show ✰ Slower multiplication rate
Vegetative Propagation	Tulsi, Mint, Geranium, Ginger, Turmeric	✰ True-to-type plants ✰ Faster multiplication ✰ High survival in mist chambers	✰ Limited to species with vegetative parts ✰ Risk of disease transfer
Tissue Culture	Vanilla, Orchids, Safed Musli, Aloe vera	✰ Disease-free plants ✰ Rapid, large-scale multiplication ✰ Conservation of rare/ endangered species	✰ Expensive setup ✰ Requires skilled labor ✰ Needs hardening stage
Grafting and Layering	Sandalwood, Patchouli, Rosewood	✰ Combines vigor and desired traits ✰ Higher success under protected conditions ✰ Produces robust planting stock	✰ Labor-intensive ✰ Species-specific ✰ Limited scale multiplication

10.5. Growing Media and Soil Management

Soilless Media

Soilless media such as cocopeat, vermiculite, and perlite are increasingly used in protected cultivation of MAPs for disease-free and uniform plant growth. These substrates are lightweight, well-aerated, and have high water-holding capacity. They reduce the risk of soil-borne pests and pathogens, which ensures healthy seedlings. Cocopeat also provides a balanced root environment, while perlite and vermiculite improve porosity and drainage. By using soilless media, farmers can precisely control nutrients through fertigation. This method is especially suitable for high-value herbs like Stevia, Mint, and Aloe vera.

Raised Beds with Sterilized Soil

Raised beds are prepared using sterilized soil to minimize disease incidence and improve root development. Techniques like soil solarization (using polythene sheets under sunlight) and **chemical fumigation** are applied to kill harmful fungi, nematodes, and weeds. Raised beds also enhance aeration, water infiltration, and root penetration. In MAP cultivation, this ensures better establishment and uniform plant stands. Proper sterilization reduces the need for heavy pesticide use, thereby promoting eco-friendly production. This approach is widely practiced in greenhouse and shade net cultivation systems.

pH and EC Management

Maintaining the right **pH and electrical conductivity (EC)** of soil or substrate is crucial for optimum nutrient uptake. Most MAPs grow well in slightly acidic to neutral pH (6.0–7.0), though specific crops may have unique preferences. For instance, Geranium requires slightly acidic media, while Aloe vera tolerates near neutral conditions. EC must be monitored regularly to avoid salt accumulation, which hampers growth. Proper water quality and fertigation schedules help maintain balance. Continuous monitoring ensures higher biomass and better quality of active principles in MAPs.

Organic Amendments

The addition of **farmyard manure (FYM), compost, and biofertilizers** enriches soil health and supports organic MAP cultivation. Organic matter improves soil texture, microbial activity, and nutrient availability. Biofertilizers like Azotobacter, PSB, and Trichoderma enhance nutrient cycling and disease resistance. This reduces the dependence on chemical fertilizers while maintaining productivity. In protected systems, organic amendments improve the quality of medicinal raw material, making it more acceptable for herbal and pharmaceutical industries. Thus, soil enriched with organics ensures sustainability and eco-friendly production. Recommended growing media, pH and EC for important MAPs is summarised in Table 10.3.

Table 10.3: Recommended Growing Media, pH and EC for Important MAPs

Crop	*Preferred Media*	*pH Range*	*EC Range (dS/m)*	*Remarks*
Aloe vera	Sandy loam/ Cocopeat + Perlite mix	6.0 – 7.5	1.0 – 1.5	Tolerates slightly alkaline soils; requires good drainage.
Ashwagandha	Well-drained sandy loam, raised beds	6.0 – 7.5	0.8 – 1.2	Prefers light soils; excessive salts reduce alkaloid content.

Crop	*Preferred Media*	*pH Range*	*EC Range (dS/m)*	*Remarks*
Stevia	Cocopeat + Vermiculite + Perlite (1:1:1)	6.5 – 7.0	1.0 – 1.5	Sensitive to salinity; best in soilless media for uniform sweetness.
Geranium	Sandy loam/ Cocopeat + FYM + Vermiculite	5.8 – 6.5	1.2 – 1.8	Slightly acidic media enhances oil yield and quality.
Safed Musli	Raised beds with sterilized sandy loam soil	6.0 – 7.0	0.8 – 1.2	Requires loose, friable soil; sensitive to waterlogging.
Mint	Cocopeat + Perlite/Sandy loam beds	6.0 – 7.0	1.0 – 1.5	Soilless systems give high-purity leaves suitable for essential oil distn.
Vanilla	Cocopeat + Charcoal + Brick pieces + FYM mix	5.5 – 6.5	1.0 – 1.2	Requires porous, organic-rich substrate and high humidity.
Tulsi	Sandy loam/ Cocopeat + FYM	6.0 – 7.0	1.0 – 1.5	Well-drained beds improve root growth and essential oil concentration.

10.6. Nutrient and Water Management

10.6.1. Drip Irrigation with Fertigation

Drip irrigation combined with fertigation ensures precise and uniform delivery of water and nutrients directly to the root zone. This method minimizes wastage, reduces nutrient leaching, and enhances fertilizer-use efficiency. It helps in maintaining optimum soil moisture, which is critical for sensitive medicinal and aromatic plants. Overall, it improves crop yield and quality with sustainable resource use.

10.6.2. Organic Liquid Manures (Panchagavya, Jeevamrit)

The use of organic liquid manures like panchagavya and jeevamrit is highly beneficial in medicinal crops. These inputs enrich soil microbial activity, improve nutrient availability, and enhance plant immunity. Being chemical-free, they ensure the purity of raw drugs and extracts. Their regular application also promotes sustainable soil health and eco-friendly production.

10.6.3. Hydroponic Nutrient Solutions for Leafy MAPs

Hydroponic systems provide controlled nutrient solutions that support the rapid growth of leafy medicinal and aromatic plants. This method avoids soil-borne diseases and allows for year-round cultivation in limited spaces. Nutrient uptake efficiency is maximized, resulting in higher biomass and active compound concentration. Such practices are ideal for urban farming and commercial herbal production.

10.6.4. Balanced NPK + Micronutrients

A well-balanced supply of nitrogen, phosphorus, potassium, along with essential micronutrients, is crucial for MAPs. These nutrients directly influence the synthesis of essential oils, alkaloids, and other bioactive compounds. Deficiencies can reduce medicinal quality, aroma, and therapeutic value. Proper nutrient management ensures both higher yields and superior pharmacological properties.

10.7. Crop Management Practices

10.7.1. Spacing and Plant Density

Proper spacing and plant density are essential to maximize light interception and air circulation in medicinal and aromatic crops. Optimum plant population ensures uniform growth, reduces competition for nutrients, and minimizes disease incidence. Wider spacing benefits root and canopy development, while closer spacing may increase yield per unit area in some crops. Scientific spacing practices enhance both productivity and quality.

10.7.2 Pruning and Training

Pruning and training are vital crop management practices, especially in aromatic plants like geranium and patchouli. Regular pruning improves canopy shape, stimulates new shoots, and enhances essential oil content. Training helps direct plant growth for better exposure to sunlight and easier harvesting. Together, these practices improve yield and maintain the health of perennial crops.

10.7.3. Mulching

Mulching helps retain soil moisture, suppress weeds, and regulate soil temperature in MAP cultivation. Organic mulches such as straw, dry leaves, or coir pith improve soil fertility upon decomposition. Plastic mulches are also used in commercial cultivation to conserve water and reduce weed pressure. This simple practice significantly improves crop growth and reduces input costs.

10.7.4. Pollination Management

Pollination management is critical in crops requiring assisted fertilization, such as vanilla under greenhouse conditions. Since natural pollinators are absent or inefficient, hand pollination is practiced to ensure fruit set. This precise technique improves pod quality and yield in commercial vanilla cultivation. Efficient pollination practices directly impact productivity and profitability of such crops.

10.8. Pest and Disease Management

10.8. 1 Protected Structures and Disease Incidence

Protected cultivation reduces the risk of many pests and diseases by creating a controlled environment. However, high humidity inside greenhouses often favours fungal pathogens like powdery mildew and root rots. Proper ventilation and climate regulation are crucial to minimize disease outbreaks. Regular monitoring helps in early detection and timely intervention.

10.8.2 Integrated Pest Management (IPM)

IPM is a sustainable approach combining preventive, biological, and eco-friendly measures for MAPs. Sticky traps and pheromone traps help in monitoring and reducing insect populations effectively. Biological control agents such as Trichoderma and Beauveria suppress soil-borne and foliar pathogens naturally. Neem-based biopesticides provide additional protection while avoiding harmful residues.

10.8.3 Residue-Free Production

In medicinal and aromatic crops, maintaining purity is vital, hence chemical pesticides are discouraged. Excessive use of synthetic chemicals may leave residues that compromise the therapeutic quality of raw drugs. Eco-friendly methods ensure compliance with quality standards for herbal and pharmaceutical industries. Thus, pest and disease management should prioritize residue-free cultivation.

10.9. Harvesting and Post-Harvest Handling

10.9.1 Physiological Maturity

Harvesting medicinal and aromatic plants at the correct physiological stage is essential to obtain maximum active principles. Immature harvesting reduces bioactive compound concentration, while delayed harvest may cause losses in quality. Careful timing ensures optimum yield of alkaloids, glycosides, and essential oils. Thus, maturity index is crop-specific and must be strictly followed.

10.9.2 Essential Oil Crops

For aromatic crops, harvesting is done when the essential oil content and composition reach their peak. For instance, lemongrass and palmarosa are harvested at full tillering stage for highest oil recovery. Delayed cutting often leads to reduced oil yield and inferior aroma quality. Scientific scheduling of harvest improves both oil quantity and quality.

10.9.3 Drying Methods

Post-harvest drying is critical to preserve volatile compounds and medicinal value. Shade drying under well-ventilated conditions is preferred for delicate herbs, while controlled dryers maintain uniform quality. Direct sun-drying should be avoided as it causes loss of aroma and degradation of active ingredients. Proper drying prevents microbial contamination and improves shelf-life.

10.9.4 Distillation Techniques

Aromatic oils are extracted through distillation methods like steam, hydro-distillation, or solvent extraction depending on the crop. Steam distillation is commonly used for mint, lemongrass, and eucalyptus. Hydro-distillation suits hardy plant parts like roots and bark, while solvent extraction is preferred for delicate flowers. Choice of method influences oil purity and yield.

10.9.5 Grading, Packaging, and Storage

Post-distillation, proper grading ensures uniformity and market quality of medicinal produce. Packaging in clean, airtight, and inert containers helps prevent contamination and loss of aroma. Storage in cool, dry, and dark conditions prolongs shelf life and preserves active compounds. Adhering to quality standards is crucial for export and pharmaceutical use.

10.10. Economically Important Medicinal Plants in Protected Cultivation

10.10.1 Ashwagandha (*Withania somnifera*)

Ashwagandha is a highly valued medicinal crop cultivated under protected conditions for its roots. The roots are rich in alkaloids like withanolides, which are known for their adaptogenic and rejuvenating properties. Controlled cultivation ensures uniform root development and higher alkaloid concentration. It is widely used in Ayurveda for stress relief, immunity boosting, and general wellness. Protected cultivation also safeguards the crop from soil-borne diseases and climatic stress. This makes Ashwagandha a reliable crop for both domestic and export markets.

10.10.2 Stevia (*Stevia rebaudiana*)

Stevia is an important natural sweetener crop gaining popularity as a sugar substitute. The leaves contain stevioside and rebaudioside, compounds 200–300 times sweeter than sugar but calorie-free. Cultivation in greenhouses improves leaf biomass and stevioside content. It is in high demand for diabetic-friendly products, beverages, and nutraceuticals. Protected conditions also allow year-round harvesting of quality leaves. Stevia thus offers great commercial potential in the health food industry.

10.10.3 Safed Musli (*Chlorophytum borivilianum*)

Safed Musli is a high-value medicinal crop known for its tuberous roots used as an aphrodisiac. The roots contain saponins and alkaloids, which enhance vitality and immunity. Protected cultivation ensures better root size, reduced pest infestation, and uniform quality. Due to over-exploitation in the wild, greenhouse production has become essential for sustainable supply. It is widely used in herbal tonics, nutraceutical formulations, and traditional medicine. The crop fetches premium prices in the herbal drug market.

10.10.4 *Aloe vera*

Aloe vera is one of the most commercially important medicinal plants used in pharmaceuticals, cosmetics, and nutraceuticals. Its fleshy leaves yield gel rich in polysaccharides, vitamins, and bioactive compounds. Protected cultivation enhances leaf succulence and maintains gel quality. The crop has diverse uses in skin care, wound healing, and digestive health remedies. It is also processed into juices, powders, and herbal formulations. Consistent production under greenhouses ensures high-quality raw material for industries.

10.10.5 Kalmegh (*Andrographis paniculata*)

Kalmegh is a medicinal herb cultivated for its bitter leaves rich in andrographolide, a powerful bioactive compound. The plant is known for its hepatoprotective, antipyretic, and immunomodulatory properties. Cultivation under protected structures improves leaf biomass and enhances secondary metabolite accumulation. It is used extensively in Ayurveda, Unani, and modern herbal medicines for treating liver ailments and fevers. Greenhouse cultivation also prevents excessive loss due to pests and heavy rains. This crop has strong market demand in the pharmaceutical sector.

10.10.6 Tulsi (*Ocimum sanctum*)

Tulsi, revered as a sacred herb in India, is cultivated for its aromatic leaves and essential oils. The leaves contain eugenol and other bioactive compounds with antimicrobial, anti-inflammatory, and adaptogenic properties. Protected cultivation ensures better growth, uniform oil content, and extended harvesting cycles. Tulsi leaves and extracts are widely used in Ayurveda, herbal teas, and natural remedies. It also has high demand in essential oil industries for perfumes and therapeutics. Sustainable greenhouse production ensures consistent supply of quality Tulsi throughout the year. Economically important medicinal plants in protected cultivation is summarised in Table 10.4.

Table 10.4: Economically Important Medicinal Plants in Protected Cultivation

Plant	*Botanical Name*	*Part Used*	*Major Active Principles*	*Major Uses*
Ashwagandha	*Withania somnifera*	Roots	Withanolides (alkaloids)	Adaptogen, stress relief, immunity booster, rejuvenator
Stevia	*Stevia rebaudiana*	Leaves	Stevioside, Rebaudioside	Natural sweetener, diabetic-friendly products, nutraceuticals

Plant	*Botanical Name*	*Part Used*	*Major Active Principles*	*Major Uses*
Safed Musli	*Chlorophytum borivilianum*	Roots (tubers)	Saponins, Alkaloids	Aphrodisiac, vitality enhancer, immunity booster
Aloe vera	*Aloe barbadensis/ Aloe vera*	Leaves (gel)	Polysaccharides, Vitamins, Aloin	Cosmetics, skin care, wound healing, digestive health
Kalmegh	*Andrographis paniculata*	Leaves	Andrographolide	Liver protection, antipyretic, immuno-modulator
Tulsi	*Ocimum sanctum*	Leaves	Eugenol, Ursolic acid	Ayurveda, anti-microbial, essential oil, herbal teas

10.11. Economically Important Aromatic Plants in Protected Cultivation

10.11.1 Geranium (*Pelargonium graveolens*)

Geranium is an important aromatic crop cultivated for its fragrant leaves. The essential oil extracted from its foliage is rich in citronellol and geraniol, widely used in perfumery and cosmetics. Protected cultivation ensures high-quality oil with consistent aroma and chemical composition. Geranium performs well in controlled environments, producing multiple harvests annually. The oil is valued for its floral fragrance and therapeutic properties in aromatherapy. This crop offers high returns in the essential oil industry.

10.11.2 Patchouli (*Pogostemon cablin*)

Patchouli is a commercially important aromatic plant cultivated for its strongly scented leaves. The oil contains patchoulol, a key compound in cosmetics, incense, and perfumery. Cultivation under polyhouses or greenhouses protects the delicate leaves from excess rainfall and pests. Protected conditions enhance leaf biomass and oil yield. Patchouli oil is highly stable, making it ideal for fixatives in perfumes. It enjoys strong export demand, especially in the fragrance industry.

10.11.3 Rose (*Rosa damascena*)

The damask rose is cultivated for its highly valued flowers, used in the production of rose oil, rose water, and attar. Protected cultivation ensures year-round flowering, better flower quality, and higher oil content. The essential oil, rich in citronellol and geraniol, is one of the most expensive natural fragrances. Roses grown under controlled conditions are also less prone to diseases. This makes rose a profitable crop in the perfumery and cosmetic sector. The flowers also have therapeutic and culinary applications.

10.11.4 Vetiver (*Vetiveria zizanioides*)

Vetiver is a hardy aromatic grass grown for its roots, which are a rich source of vetiver oil. The oil has a deep, woody fragrance and is used in perfumery, cosmetics, and aromatherapy. Protected cultivation improves root quality and oil yield by maintaining uniform moisture and reducing fungal diseases. Vetiver roots are also valued for cooling mats and traditional crafts. Its essential oil acts as a natural fixative in perfumes. Sustainable cultivation ensures steady supply for domestic and export markets.

10.11.5 Mint (*Mentha spp.*)

Mint is an economically important aromatic herb cultivated for menthol-rich essential oil. Species like *Mentha arvensis* and *Mentha piperita* are widely grown under protected structures. Controlled conditions improve biomass yield, oil recovery, and menthol content. Mint oil and menthol crystals are extensively used in pharmaceuticals, confectionery, and cosmetics. Protected cultivation also helps reduce incidence of pests like rust and aphids. Mint cultivation provides high profitability with assured industrial demand.

10.11.6 Vanilla (*Vanilla planifolia*)

Vanilla is one of the most expensive aromatic crops, cultivated for its pods used as a natural flavouring agent. The cured pods contain vanillin, a highly valued compound in the food and beverage industry. Protected cultivation provides the warm, humid conditions essential for good pod development. Since natural pollinators are absent, hand pollination is practiced inside greenhouses. Proper curing and processing of pods significantly increase their market value. Vanilla cultivation is highly profitable but requires skilled management. Economically important aromatic plants in protected cultivation is summarised in Table 10.5.

Table 10.5: Economically Important Aromatic Plants in Protected Cultivation

Plant	*Botanical Name*	*Part Used*	*Major Active Principles*	*Major Uses*
Geranium	*Pelargonium graveolens*	Leaves	Citronellol, Geraniol	Essential oil, perfumery, cosmetics, aromatherapy
Patchouli	*Pogostemon cablin*	Leaves	Patchoulol	Cosmetics, perfumery, incense, fixatives
Rose	*Rosa damascena*	Flowers	Citronellol, Geraniol	Rose oil, attar, rose water, perfumery
Vetiver	*Vetiveria zizanioides*	Roots	Vetiverol, Vetivone	Perfumery, cosmetics, aromatherapy, cooling mats
Mint	*Mentha spp.*	Leaves	Menthol, Menthone	Pharmaceuticals, confectionery, cosmetics
Vanilla	*Vanilla planifolia*	Pods	Vanillin	Flavouring, food industry, confectionery

10.12. Challenges in Protected MAP Cultivation

10.12.1 High Initial Cost of Structures

One of the major challenges in protected cultivation is the high investment required for greenhouses, polyhouses, and irrigation systems. The cost of construction, maintenance, and climate-control equipment can be prohibitive for small farmers. Although subsidies are available, initial capital remains a barrier. This restricts widespread adoption of protected MAP cultivation.

10.12.2 Skilled Labor Requirement

Protected cultivation of medicinal and aromatic plants requires skilled labour for nursery management, fertigation, pruning, pollination, and post-harvest handling. Lack of technical knowledge often leads to poor crop performance and reduced quality. Farmers need regular training and exposure to modern practices. Human resource development is thus critical for success.

10.12.3 Market Fluctuation

The herbal drug and essential oil markets are highly volatile due to demand–supply gaps and international trade variations. Prices of crops like vanilla, mint, and geranium may fluctuate drastically. Such unpredictability makes it difficult for farmers to maintain consistent profitability. Market linkages and contract farming models can help reduce this risk.

10.12.4 Need for Certification

Medicinal and aromatic plants grown under protected structures often require certification like organic, GAP (Good Agricultural Practices), or fair-trade. These certifications are essential to access premium export markets and ensure consumer trust. However, obtaining certification is costly and time-consuming. Many small farmers find it challenging to meet these stringent requirements.

10.12.5 Post-Harvest Infrastructure

Efficient post-harvest handling is essential to maintain the quality of MAPs, especially for export. Lack of drying chambers, distillation units, cold storage, and value addition facilities limits profitability. Poor infrastructure often leads to contamination, quality deterioration, and lower market prices. Strengthening processing and storage facilities can greatly improve farmer incomes.

Off-Season Production of Flowers and Vegetables

11.1 Off-season Production of Vegetables

Off-season vegetable production under protected cultivation allows farmers to grow high-value, short-duration, and perishable vegetables beyond their regular season. Selection of crops is carefully guided by market demand, climatic suitability, and the type of protective structure, such as polyhouses, greenhouses, or shade nets. This approach ensures a continuous supply of vegetables even during off-peak periods, helping to stabilize market availability. Protected cultivation also enhances crop quality by providing controlled environmental conditions, reducing pest and disease incidence. Farmers can achieve higher yields and earn premium prices by supplying vegetables when they are scarce in the open market. It promotes efficient use of resources like water, fertilizers, and space, making production more sustainable. Off-season farming also opens opportunities for crop diversification and experimentation with exotic or high-demand varieties. Overall, it is a profitable and strategic approach for modern vegetable cultivation. The following is a categorized list of vegetable crops commonly grown off-season under different types of protected structures:

11.1.1. Cucurbits

These vegetables thrive in warm conditions and are highly thermophilic. Their growth and productivity are strongly influenced by controlled environments. They are especially sensitive during late winter and early spring, requiring careful management for off-season cultivation.

- ☆ **Bitter Gourd (*Momordica charantia*)** – This thermophilic vegetable grows well in controlled environments. Cultivation in net houses or low tunnels allows early fruit production. It is highly suited for off-season harvesting in late winter or early spring.
- ☆ **Bottle Gourd (*Lagenaria siceraria*)** – Performs best in low tunnels or walk-in tunnels during late winter and early summer. Controlled conditions enhance growth and reduce pest incidence. Early harvests fetch better market prices.
- ☆ **Cucumber (*Cucumis sativus*)** – European varieties thrive in polyhouses with trellis systems. Such protected cultivation ensures uniform growth and high-quality fruits. Early production meets off-season market demand.
- ☆ **Ridge Gourd (*Luffa acutangula*)** – Walk-in tunnels help prevent humidity-related diseases. Controlled conditions support vigorous growth and quality fruiting. Off-season cultivation ensures steady market availability.
- ☆ **Muskmelon (*Cucumis melo*)** – Low tunnels encourage early fruiting before peak summer production. Controlled conditions reduce stress and improve fruit quality. Early harvests capture higher market prices.

11.1.2. Solanaceous Crops

Solanaceous vegetables thrive under controlled temperature and humidity, which significantly boost their growth, yield, and overall quality. Protected cultivation helps regulate environmental stress, ensuring uniform flowering and fruiting. These conditions also reduce pest and disease incidence, leading to healthier crops. As a result, solanaceous crops like tomato, capsicum, chili, and brinjal become highly profitable for off-season production.

- ☆ **Tomato (*Solanum lycopersicum*)** – One of the most lucrative greenhouse crops, well suited for all-round production throughout the year.
- ☆ **Capsicum/Bell Pepper (*Capsicum annuum*) –** Grows best in polyhouse conditions since it is prone to temperature fluctuations.
- ☆ **Chili (*Capsicum* spp.) –** Can be grown early or pushed through beyond normal seasons in net houses.
- ☆ **Brinjal (Eggplant) (*Solanum melongena*) –** Tolerates insect-proof structures to prevent shoot and fruit borer infestation.

11.1.3. Leafy Vegetables

Leafy vegetables are generally short-duration crops with low light requirements, making them ideal for protected cultivation. They can be grown year-round in covered structures, ensuring steady supply and high-quality produce for both local and urban markets.

- ☆ **Spinach (*Spinacia oleracea*)** – Spinach grows well in hydroponic systems and shade net houses, where temperature, humidity, and light can be carefully managed. Off-season cultivation allows early harvests and continuous supply, while controlled conditions reduce bolting and pest problems.
- ☆ **Lettuce (*Lactuca sativa*)** – Prefers cooler temperatures and performs best in greenhouses or NFT (Nutrient Film Technique) hydroponics. Protected environments ensure tender, crisp leaves, prevent premature flowering, and support year-round production for high-value markets.
- ☆ **Fenugreek (*Trigonella foenum-graecum*)** – A fast-growing leafy vegetable, fenugreek is well-suited for off-season cultivation in shade net houses. Controlled conditions improve leaf quality, allow early harvests, and meet home market demand efficiently.

11.1.4. Leguminous Vegetables

Leguminous vegetables enrich the soil through nitrogen fixation and are valuable for cropping systems. They perform well in off-season cultivation under protected structures, ensuring quality produce and stable yields.

- ☆ **French Bean (*Phaseolus vulgaris*)** – A temperature-sensitive crop that grows best in walk-in tunnels or shade nets during hot months. Protected cultivation ensures uniform flowering and pod development. Off-season production helps meet market demand when open-field yields are low.
- ☆ **Cowpea (*Vigna unguiculata*)** – Heat- and drought-tolerant, cowpea thrives in net houses during off-season periods. Controlled environments improve pod quality and reduce pest incidence. Early or off-season harvests fetch higher market prices and ensure continuous supply.

11.2 Categories of Protected Cultivation Structures in Vegetables

Protected cultivation encompasses several structural systems that promote crop growth under modified environmental circumstances. Each form of structure is employed for a specific purpose, aligned with crop needs, seasonal conditions, and investment capacity. The primary categories of protected cultivation structures are defined below:

11.2.1. Greenhouse or Polyhouse

A polyhouse or greenhouse is a structured facility with translucent or semi-transparent coverings, such as polyethylene film or polycarbonate sheets, that allow sunlight to penetrate while controlling internal temperature and humidity levels. These structures are highly suitable for cultivating high-value crops such as capsicum, tomato, cucumber, exotic vegetables, and flowers. Equipped with irrigation systems, ventilation, heating, and occasionally automation, polyhouses enable comprehensive environmental management, facilitating year-round agricultural cultivation regardless of exterior weather conditions. These buildings are optimally designed for commercial and export-focused agriculture.

11.2.2. Shade Net House

A shade net house is a structure covered with synthetic netting material, typically offering varied degrees of shading (30 per cent, 50 per cent, 75 per cent) according to crop needs. This structure is mostly employed to acclimatize seedlings cultivated in nurseries, propagate ornamental plants, and cultivate semi-shade-tolerant or shade-preferring vegetables, including leafy greens and herbs. A shade net house mitigates excessive light intensity, provides wind protection, reduces heat stress, and minimizes evapotranspiration, so ensuring enhanced plant development under extreme summer conditions. It is highly economical and suitable for small- and medium-sized agricultural producers.

11.2.3. Low Tunnels or Plastic Tunnels

Low tunnels are diminutive, arc-shaped structures formed by bending flexible rods or tubes and enveloping them with translucent plastic sheeting. These are cost-effective methods for the early-season cultivation of cucurbits, including muskmelon, bottle gourd, and cucumber. Plastic sheeting retains solar heat by establishing a greenhouse effect, thus protecting young plants from cold, frost, or excessive precipitation in late winter or early spring. Low tunnels are frequently utilized by smallholder farmers to accelerate crop maturation, obtain early market advantage, and enhance income generation.

11.2.4. Walk-In Tunnels

Walk-in tunnels are of moderate dimensions, generally ranging from 6 to 8 feet in height, providing adequate space for agricultural personnel to navigate and engage in tasks such as pruning, harvesting, and trellising within the structure. They establish a semi-controlled cultivation environment, proving especially advantageous for crops such as tomatoes, chilies, French beans, and gourds. These structures enable early planting, mitigate crop vulnerability to abrupt weather fluctuations, and improve overall crop output and quality. Walk-in tunnels are more controllable and easier to maintain than fully climate-controlled greenhouses, offering a cost-effective solution for marginal and small-scale growers.

11.2.5. Insect-Proof Net Houses

Insect-proof net houses are protective structures covered with fine-mesh nets (40–60 mesh) that act as a barrier against insect pests like whiteflies, aphids, and fruit borers. They are particularly useful for cultivating crops such as brinjal, okra, chili, and leafy vegetables, which are highly vulnerable to insect-transmitted viral diseases. By physically excluding pests, these structures significantly reduce the need for chemical pesticides, supporting safe, residue-free, and eco-friendly vegetable production. Insect-proof net houses enhance crop health, productivity, and sustainability, making them highly suitable for organic farming and Integrated Pest Management (IPM) practices.

11.3 Off-season Production of Flowers

Off-season production of flowers has gained importance as it ensures continuous supply, higher returns, and better-quality blooms when market demand is high. Protected cultivation of ornamentals involves the use of specialized structures like greenhouses, polyhouses, shade nets, and walk-in tunnels to create favourable growing conditions. These structures help regulate temperature, humidity, light, and protect crops from adverse weather. They also minimize pest and disease incidence, ensuring healthier plants and superior quality flowers. Such systems make it possible to cultivate sensitive ornamentals year-round. Overall, protected structures play a vital role in sustainable and profitable off-season floriculture.

11.3.1. Greenhouses/Polyhouses

Greenhouses and polyhouses are advanced protected structures designed to regulate temperature, humidity, and light for optimum plant growth. They create a stable microclimate that supports the year-round cultivation of high-value flowers. By shielding crops from adverse weather and pests, they ensure better quality and higher yields. Such systems are widely adopted for premium ornamentals like rose, carnation, gerbera, lily, and orchids.

- ☆ **Rose (*Rosa* spp.):** Roses are the most commercially important cut flowers, requiring a controlled climate for uniform growth and continuous blooming. Greenhouses ensure the right balance of temperature, humidity, and light to enhance colour and fragrance. Protected cultivation also minimizes pest and disease attacks. This makes year-round high-quality rose production possible, catering to domestic and export markets.
- ☆ **Carnation (*Dianthus caryophyllus*):** Carnations are highly sensitive to extreme temperatures and grow best under cool, controlled greenhouse environments. Proper regulation of climate helps maintain longer stems, vibrant colours, and extended vase life. They are in high demand for floral arrangements and exports. Off-season production in polyhouses significantly increases profitability for growers.
- ☆ **Gerbera (*Gerbera jamesonii*):** Gerbera thrives in polyhouses under diffused light conditions that prevent scorching and maintain bright petal colours. Controlled humidity ensures a longer flowering season and higher productivity. Protected structures help achieve uniform flower size and stem strength. These qualities make gerbera one of the most preferred flowers for bouquets and decorative purposes.
- ☆ **Orchid (*Dendrobium, Phalaenopsis*):** Orchids demand high humidity, filtered light, and warm conditions, which are best provided in greenhouses or mist chambers. Protected structures mimic their natural tropical environment, ensuring continuous flowering and exotic blooms.

With proper care, orchids produce vibrant, long-lasting flowers. They are highly valued in the international cut flower trade.

- ☆ **Anthurium (*Anthurium andreanum*):** Anthurium prefers warm, humid greenhouse conditions where temperature and shade can be carefully regulated. Protected cultivation enhances the glossiness, size, and longevity of its attractive spathes. Humidity control also ensures better foliage and disease resistance. This makes anthurium a premium flower in both domestic and export markets.
- ☆ **Lilium (*Lilium* spp.):** Lilium is a high-value bulbous crop that performs best in cool polyhouse environments. Controlled conditions support straight stems, larger buds, and vibrant-coloured flowers. Protection from excessive heat and rainfall ensures better bulb development and repeat cycles. Its popularity as a luxury cut flower makes lilium cultivation highly profitable under protected systems.

11.3.2. Shade Net Houses

Shade net houses are protective structures designed to filter sunlight, thereby reducing heat stress on plants and maintaining an optimal microclimate. They provide partial shade that supports healthy growth, vibrant colours, and extended flowering periods in ornamentals. By lowering the risk of leaf scorch and sun damage, they improve both quality and yield. Shade nets are particularly effective for crops like chrysanthemum, gladiolus, tuberose, and various foliage plants.

- ☆ **Chrysanthemum (*Chrysanthemum morifolium*):** Chrysanthemum is highly responsive to photoperiod, and shade net houses allow precise regulation of day length for off-season flowering. Partial shading reduces stress during hot periods and promotes uniform bud initiation. Protected cultivation ensures larger blooms with brighter colors. This makes chrysanthemum a reliable flower for both domestic use and export markets.
- ☆ **Gladiolus (*Gladiolus grandiflorus*):** Gladiolus produces superior spikes when grown under shade nets that reduce excessive heat and strong light. Moderate shading helps maintain soil moisture, improves stem length, and enhances flower colour intensity. Protection from harsh conditions results in straight, market-quality spikes. As a cut flower, gladiolus fetches premium prices when grown off-season.
- ☆ **Tuberose (*Polianthes tuberosa*):** Tuberose is a fragrant cut flower that flourishes in shaded environments where heat stress is minimized. Shade nets promote prolonged flowering and increase the number of spikes per plant. Flowers harvested under protected conditions have improved fragrance retention and vase life. This makes tuberose cultivation highly profitable for local and export markets.

- **Foliage Plants (*Dracaena, Areca palm, Dieffenbachia*):** Foliage ornamentals thrive under shade net houses as they require filtered light instead of direct sun. Controlled shading enhances leaf size, colour, and glossiness, improving their decorative value. These conditions also reduce leaf scorching and pest infestations. Such plants are in great demand for landscaping, indoor decoration, and the nursery trade.

11.3.3. Walk-in Tunnels/Low-cost Poly Tunnels

Walk-in tunnels or low-cost poly tunnels are simple and economical protected structures that help small and medium farmers extend the growing season of flowers. They shield crops from heavy rain, wind, and temperature extremes, ensuring better plant survival and quality blooms. By creating a favorable microclimate, these tunnels make it possible to advance or delay flowering for market advantage. They are especially useful for short-duration ornamentals and seasonal flowers, making off-season production more profitable.

- **Marigold (*Tagetes* spp.):** Marigolds adapt well to walk-in tunnels, allowing early or late-season cultivation beyond natural outdoor conditions. The structure protects plants from heavy rain, strong winds, and temperature fluctuations. This results in uniform flowering, larger blooms, and longer spike retention. Off-season production increases market availability and profitability for farmers.
- **Gaillardia and Zinnia:** Gaillardia and zinnia benefit from tunnel protection that shields them from adverse weather, reducing flower damage and stem breakage. Controlled conditions help maintain vibrant petal colours and uniform flower size. These annuals also show improved survival rates and growth under tunnels. As a result, growers can achieve higher-quality blooms for cut flowers or landscaping purposes.
- **Aster (*Callistephus chinensis*):** Aster thrives under low-cost tunnels, which provide shelter during off-season periods and protect young plants from cold or rain. The controlled microclimate promotes healthy vegetative growth and robust flowering. Tunnel-grown asters produce larger, more uniform blooms suitable for cut flower markets. This method allows growers to extend production windows profitably.
- **Seasonal Annuals (*Petunia, Calendula, Salvia*):** Seasonal annuals like petunia, calendula, and salvia respond well to tunnel cultivation, enabling early or staggered production. Protection from harsh sunlight, rain, and wind enhances flowering, colour intensity, and plant health. These structures also allow for better management of irrigation and nutrients. Consequently, growers can supply high-quality flowers to markets during off-peak seasons.

11.3.4. Mist Chambers/Humidity-Controlled Structures

Mist chambers and humidity-controlled structures are specialized systems designed to provide consistently high humidity and moderate temperatures. They are ideal for delicate and tropical ornamentals that cannot tolerate dry or fluctuating conditions. By maintaining optimal moisture levels, these chambers promote healthy growth, vibrant foliage, and prolonged flowering. Such systems are especially useful for high-value ornamental crops grown for landscaping and export markets.

- ☆ **Orchids (*Vanda, Cattleya, Dendrobium*):** Orchids flourish in mist chambers where humidity, temperature, and light are carefully controlled to mimic their natural tropical environment. High humidity promotes continuous flowering, vibrant colours, and robust plant growth. Protection from direct sunlight and pests ensures healthier blooms. These conditions make orchids highly desirable in both domestic and international cut flower and potted plant markets.
- ☆ **Fern Varieties (*Nephrolepis, Adiantum*):** Ferns require constant moisture and high humidity for lush fronds and vigorous growth. Mist chambers provide fine water sprays and moderate temperatures that prevent leaf desiccation and stress. This controlled environment results in fuller, greener, and more decorative foliage. Ferns grown in such systems are highly sought after for interior decoration and landscaping.
- ☆ **Bromeliads and Exotic Ornamentals:** Bromeliads and other exotic ornamentals benefit from the stable, humid environment of mist chambers, which promotes vibrant foliage and prolonged flowering. These conditions reduce stress, pest attacks, and leaf damage. The controlled environment supports propagation and ensures consistent quality for landscaping, gifting, and export. High-value exotic plants thrive under such carefully managed conditions, making them commercially profitable. Ornamental flowers for off-season protected cultivation is summarised in Table 11.1.

Table 11.1: Ornamental Flowers for Off-season Protected Cultivation

Flower/Crop	*Best Protected Structure*	*Advantage*
Rose (Rosa spp.)	Greenhouse/Polyhouse	Year-round high-quality blooms, premium market value
Carnation (Dianthus spp.)	Greenhouse/Polyhouse	Sensitive to climate, thrives in controlled cool environment
Gerbera (Gerbera jamesonii)	Polyhouse	Uniform size, better color intensity, higher yield
Orchid (Dendrobium, Phalaenopsis)	Greenhouse/Mist Chamber	Requires humidity and shade, high export potential

Flower/Crop	*Best Protected Structure*	*Advantage*
Anthurium (Anthurium andreanum)	Greenhouse/Polyhouse	Prefers warm humid climate, glossy long-lasting blooms
Lilium (Lilium spp.)	Polyhouse	Bulbous crop, protected cool climate ensures bigger flowers
Chrysanthemum	Shade Net House	Photoperiod control, off-season production possible
Gladiolus	Shade Net House	Protection from heat stress, straight spikes
Tuberose (Polianthes tuberosa)	Shade Net House	Longer spikes and extended flowering season
Foliage Plants (Dracaena, Areca palm)	Shade Net House	Thrive under filtered light, better leaf quality
Marigold (Tagetes spp.)	Walk-in Tunnel/Low-cost Polyhouse	Early/late season production, weather protection
Aster (Callistephus chinensis)	Walk-in Tunnel	Healthy plants, better flower size under protection
Zinnia, Gaillardia, Petunia	Walk-in Tunnel	Off-season cultivation of annuals, higher survival
Ferns (Nephrolepis, Adiantum)	Mist Chamber	Require humidity, lush ornamental foliage
Bromeliads and Exotics	Mist Chamber	Niche crops, landscaping and export markets

11.4 Common Practices in Protected Off-Season Flowers and Vegetable Production

Successful off-season vegetable production under protected conditions depends on the careful integration of environmental regulation, scientific crop management, and effective agronomic practices. It involves maintaining optimal temperature, humidity, and ventilation, choosing high-yielding and stress-tolerant varieties, and ensuring balanced nutrition and irrigation. Timely pest and disease management using physical, biological, and minimal chemical interventions further protects crop health. Implementing these strategies not only enhances yield and quality but also supports sustainable, resource-efficient, and profitable cultivation systems. The key practices that promote higher yields, premium-quality produce, and sustainable off-season production under controlled conditions include:

11.4.1. Selection of Suitable Varieties

The primary and most essential phase in protected agriculture is the selection of crop varieties or hybrids appropriate for regulated conditions. These types must possess traits like as tolerance to elevated humidity, adaptability to low or dispersed light conditions, and resistance to common diseases and pests. Indeterminate tomato and colored capsicum hybrids exhibiting robust fruit set in greenhouse conditions are advantageous. The selection of cultivars also considers the season, market demand, and the intended duration of planting.

11.4.2. Soil Sterilization

Soil sterilization is essential before planting to prevent the emergence of soil-borne illnesses such as damping-off, Fusarium wilt, and nematode infestations. This can be achieved through solarization (using clear polythene as plastic mulch over moist soil for 4–6 weeks during summer), the application of formalin or biological agents (such as Trichoderma harzianum), or steam sterilization in advanced operations. In hydroponic or soilless systems, the growing medium (*e.g.*, cocopeat, perlite) requires disinfection before reuse.

11.4.3. Management of Water and Nutrients via Drip Irrigation and Fertigation

Drip irrigation is essential for sheltered horticulture since it delivers water directly to the root zone, minimizes evaporation loss, and maintains ideal soil moisture levels. When combined with fertigation systems, water-soluble nutrients can be delivered in precise quantities according to the crop's growth stage and nutritional requirements. It enhances nutrient utilization efficiency, reduces leaching losses, and ensures uniform crop development. It is advisable to conduct regular monitoring of electrical conductivity (EC) and pH to maintain balanced nutritional solutions.

11.4.4. Pest and Disease Management through Integrated Pest Management (IPM)

In protected cultivation, closed environments can quickly become hotspots for pest outbreaks if not properly managed, making IPM essential. Effective strategies include the use of yellow and blue sticky traps to monitor and control flying pests like whiteflies and thrips, application of biological control agents such as *Trichogramma*, *Beauveria bassiana*, and neem-based extracts, and maintaining strict sanitation by removing plant waste and preventing waterlogging. Additionally, insect-proof netting and physical barriers help keep pests out. These eco-friendly measures reduce pesticide residues, promote safer vegetable production, and support sustainable cultivation practices.

11.4.5. Soil Health Management and Crop Rotation

Continuous monocropping in protected environments can lead to the accumulation of soil-borne diseases and depletion of essential nutrients. To prevent this, crop rotation with non-host or leguminous crops should be practiced, breaking pest and disease cycles and maintaining soil fertility. Additionally, the application of organic amendments such as vermicompost, neem cake, and biofertilizers like Azospirillum and Phosphobacteria supports beneficial microbial activity, improves soil structure, and sustains long-term productivity. These practices collectively enhance soil health and ensure higher, more sustainable yields.

11.4.6. Training and Pruning

Training and pruning are essential cultural practices for indeterminate crops such as tomato, capsicum, and cucumber. Plants are typically guided along trellises or vertical strings to encourage upright growth, improve air circulation, and maximize light interception for optimal photosynthesis. Regular pruning of side shoots and excess foliage directs the plant's energy toward fruit development, reduces disease incidence, and maintains the desired plant structure. Additionally, removing lower leaves improves field hygiene, facilitates timely harvesting, and ensures better overall crop management and yield quality.

11.5. Advantages of Protected Off-Season Production Flowers and Vegetables

Protected off-season cultivation offers numerous benefits by creating controlled environments that optimize plant growth and productivity. Key advantages include:

- ☆ **Year-round Production:** Enables continuous cultivation irrespective of seasonal limitations, ensuring steady supply to markets.
- ☆ **Higher Yields and Quality:** Controlled temperature, humidity, and light enhance flower size, color, shelf life, and vegetable yield.
- ☆ **Early or Extended Harvest:** Crops can be harvested earlier or later than open-field cultivation, capturing premium market prices.
- ☆ **Reduced Pest and Disease Pressure:** Enclosed structures limit pest entry and, combined with IPM, reduce chemical usage.
- ☆ **Resource Efficiency:** Optimized water, nutrient, and space management improves overall input-use efficiency.
- ☆ **Market Competitiveness:** Off-season produce often commands higher prices due to limited availability.

These advantages make protected cultivation a profitable and sustainable approach for both flowers and vegetables.

11.6. Challenges in Off-Season Flowers and Vegetables Production under Protected Cultivation

- ☆ **High Initial Investment:** Structures like greenhouses, polyhouses, and mist chambers require substantial capital.
- ☆ **Skilled Labor Requirement:** Trained personnel are needed for climate control, crop management, and pest/disease handling.
- ☆ **Continuous Monitoring:** Maintaining optimal temperature, humidity, and light conditions is labour- and knowledge-intensive.

- ☆ **Pest and Disease Management:** Enclosed environments can favour rapid pest and disease outbreaks if not properly managed.
- ☆ **Market Fluctuations:** Off-season produce prices can be unpredictable, affecting profitability.
- ☆ **Post-Harvest and Value Addition Constraints:** Lack of proper storage, packing, and processing facilities limits returns.
- ☆ **Certification Requirements:** Compliance with organic, GAP, or fair-trade standards adds operational complexity.

Insect Pest and Disease Management in Protected Cultivation

12.1 Introduction: An Overview

Protected cultivation, such as greenhouses, polyhouses, and shade nets, has emerged as a key approach for year-round production of high-value crops. These controlled environments allow optimal regulation of temperature, humidity, light, and irrigation, enhancing crop growth and yield. However, the enclosed nature of protected structures also creates favourable conditions for the rapid proliferation of insect pests and diseases. Pests such as aphids, whiteflies, thrips, mites, and caterpillars can cause severe damage, while pathogens including fungi, bacteria, and viruses threaten plant health. Unlike open-field cultivation, natural predators and environmental checks are limited in protected systems, making pest and disease outbreaks more intense. Effective management strategies are therefore critical to maintain crop quality and economic viability. Integrated Pest Management (IPM) is the preferred approach, combining cultural, biological, and chemical methods to minimize losses. Regular monitoring, early detection, and timely intervention form the cornerstone of effective pest and disease control. Sanitation practices, resistant varieties, and optimal plant spacing further reduce infection risks. Biological agents, such as predatory insects and microbial formulations, provide eco-friendly alternatives to chemical pesticides. Cultural practices, including crop rotation, pruning, and proper irrigation, help maintain a healthy growing environment. Chemical interventions are used judiciously, targeting pests at critical stages while avoiding resistance development. Advances in technology, such as pheromone traps, UV traps, and sensor-based monitoring, enhance precision in pest management. Successful pest and disease management in protected cultivation not only ensures sustainable production but also reduces post-harvest losses and improves market value. Overall, a holistic and integrated approach is essential to achieve high productivity while safeguarding environmental and human health.

12.2 Preventive Measures in Pest and Disease Management

12.2.1. Environmental Control

A crucial preventive measure in pest and disease management is the maintenance of a meticulously regulated and sanitary polyhouse environment, which substantially diminishes the likelihood of pest infestations and disease occurrences. This entails regular cleaning, removal of plant detritus, and eradication of weeds or algae that may hold pathogens or insect pests. Adequate ventilation and air movement are crucial to mitigate excessive humidity, which frequently fosters fungal and bacterial infections. Regulating temperature and humidity with controlled watering methods enhances plant health and resilience. Contemporary protected cultivation systems provide sophisticated environmental control mechanisms, encompassing automatic climate regulation, ventilation apparatus, misting or fogging devices, and tailored shade materials, all intended to establish an ideal microclimate for crop development. These techniques not only inhibit the proliferation of pests and pathogens but also augment overall crop vitality and production. Through the integration of hygiene, environmental regulation, and technology assistance, farmers may create and sustain a polyhouse environment that is both productive and resilient to biotic stresses. The consistent use of these tactics constitutes the foundation of integrated pest and disease management in protected agriculture systems.

12.2.2. Resistant Crop Varieties

Choosing crop types with intrinsic resistance to particular pests and diseases is an exceptionally successful approach for reducing the likelihood of infestations and outbreaks. Resistant cultivars inherently endure or tolerate assaults from prevalent pests and diseases, diminishing the incidence and intensity of damage. In protected farming, partnership with crop breeding programs and certified seed suppliers guarantees the availability of resistant types suited for greenhouse and polyhouse environments. The utilization of these cultivars enables farmers to diminish reliance on chemical pesticides, so fostering safer and more sustainable agricultural methods. Moreover, resistant crops frequently demonstrate enhanced growth, yield stability, and quality under stress circumstances, rendering them particularly suitable for intense protected situations. Incorporating resistant cultivars into cropping systems enhances other preventive and integrated pest control strategies, fostering a comprehensive approach to crop protection. By emphasizing resistant cultivars, farmers can improve productivity, reduce production costs, and promote ecologically sustainable agriculture. This proactive method constitutes an essential element of contemporary protected farming practices focused on long-term sustainability.

12.2.3. Exclusion Techniques

Employing exclusion techniques, such as insect netting and screens, constitutes an efficient preventive strategy to diminish pest ingress inside polyhouses, so substantially mitigating the danger of infestations. These physical barriers serve as the initial line of defence, obstructing insects such as whiteflies, aphids, and thrips from accessing the crops. Contemporary protected cultivation systems provide several high-quality netting and screening options that optimize pest exclusion while ensuring sufficient ventilation, light infiltration, and microclimate control. Correctly fitted screens facilitate airflow while preserving conditions favourable for optimal plant growth. Moreover, integrating netting with structural elements like double-door entrances or airlocks significantly improves defence against external pests. This method lowers reliance on chemical treatments and enhances other integrated pest management techniques. Consistent inspection and maintenance of screens are crucial to preserve their efficacy over time. Through the efficient integration of exclusion measures, farmers may protect crops, increase productivity, and foster environmentally sustainable agricultural practices. Consequently, exclusion techniques constitute a fundamental element of preventive pest management in protected agriculture systems.

12.3 Monitoring and Early Detection

12.3.1. Routine Inspections

Consistent and methodical inspections, along with diligent surveillance of the polyhouse environment and crops, are crucial for the prompt identification of pests and diseases. Agriculturists must regularly inspect vegetation for indicators of infestation, such leaf discoloration, wilting, stunted growth, or observable insect activity, and implement prompt corrective measures upon detection of issues. Timely detection averts minor problems from developing into significant outbreaks that may jeopardize yield and quality. In protected cultivation systems, monitoring include not only eye examination but also the utilization of traps, sensors, and environmental data logging to assess insect populations and microclimatic conditions. Agricultural experts in Protected Cultivation provide instruction and training on optimal procedures for efficient scouting, documentation, and diagnostic methods. Equipping farmers with these abilities enables them to immediately address emerging challenges and execute tailored management strategies. Ongoing monitoring facilitates the assessment of the efficacy of integrated pest management strategies, encompassing biological controls and cultural activities. Proactive inspections and careful monitoring are essential elements of an effective and sustainable pest and disease management program in protected horticulture.

12.3.2. Use of Traps and Lures

Establishing traps and lures is an efficient method for surveillant insect populations and identifying pest infestations at an early stage. These instruments enable farmers to monitor the occurrence, density, and behavioural patterns of pests, so permitting prompt actions prior to substantial harm. Protected Cultivation provides an extensive array of pest-specific monitoring options, such as yellow sticky traps for aerial insects, pheromone lures designed to attract particular pest species, and insect light traps aimed at nocturnal pests. The strategic positioning and routine examination of these traps yield critical information regarding pest prevalence, enabling farmers to make informed management choices. These monitoring instruments can be amalgamated with additional preventive and control strategies, such as biological agents and exclusion techniques, to augment the overall efficacy of pest management. The utilization of traps diminishes the necessity for widespread chemical applications, hence promoting environmentally sustainable agricultural practices. Consistent rotation and upkeep of traps guarantee sustained efficacy in capturing intended pests. Utilizing these sophisticated monitoring systems, growers may ensure healthier crops, enhance production quality, and preserve a healthy ecology within the polyhouse. Consequently, traps and lures function as both early-warning mechanisms and management instruments within integrated pest control programs.

12.4 Control Methods

12.4.1 Biological Controls

Biological control is one of the most sustainable and eco-friendly approaches in Integrated Pest Management (IPM). It involves the deliberate use of living organisms to suppress pest populations below economic thresholds. Beneficial insects and mites play a vital role by preying on harmful pests or parasitizing them at specific life stages. Predatory mites, for example, target spider mites and thrips, reducing their damage without chemical intervention. Parasitic wasps lay eggs inside or on pest larvae, eventually killing them and preventing further reproduction. Nematodes can infect soil-dwelling pests, including root grubs and larvae, causing mortality and reducing pest pressure. Entomopathogenic fungi, such as Beauveria bassiana and Metarhizium anisopliae, infect and kill insects while being safe for plants, humans, and beneficial insects. Biological controls are compatible with other IPM strategies and reduce dependence on chemical pesticides. Proper timing, release rates, and environmental conditions are critical for their effectiveness. These agents can be introduced preventively or in response to pest monitoring data. Conservation of naturally occurring beneficials within the greenhouse environment enhances long-term pest suppression. In addition, combining multiple biological agents often provides a synergistic effect. Regular scouting and monitoring help determine the right moment to introduce beneficial organisms. Maintaining suitable temperature, humidity, and vegetation diversity improves their survival and efficacy.

Overall, biological controls contribute to sustainable pest management, reducing environmental contamination and pesticide residues.

Botanical Pesticides

Biological management is among the most sustainable and environmentally benign methods in Integrated Pest Management (IPM). It entails the intentional utilization of live creatures to reduce insect populations beneath economic thresholds. Beneficial insects and mites are essential as they feast on dangerous pests or parasitize them throughout specific life stages. Predatory mites, for instance, prey on spider mites and thrips, mitigating their damage without chemical intervention. Parasitic wasps deposit eggs within or upon pest larvae, ultimately resulting in their demise and inhibiting further reproduction. Nematodes can infest soil-dwelling pests, such as root grubs and larvae, leading to mortality and diminished pest pressure. Entomopathogenic fungi, like Beauveria bassiana and Metarhizium anisopliae, infect and exterminate insects while remaining harmless to plants, humans, and beneficial insects. Biological controls are congruent with other Integrated Pest Management techniques and diminish reliance on chemical pesticides. Optimal timing, release rates, and environmental conditions are essential for their efficacy. These agents may be applied either prophylactically or in reaction to pest surveillance data. The preservation of naturally occurring beneficial organisms in the greenhouse environment improves sustained pest control. Moreover, the amalgamation of many biological agents frequently yields a synergistic impact. Consistent scouting and monitoring facilitate the identification of the optimal time to introduce beneficial organisms. Regulating optimal temperature, humidity, and plant diversity enhances their survival and effectiveness. Biological controls enhance sustainable pest management by minimizing environmental pollution and chemical residues.

Horticultural Oils and Soaps

Horticultural oils and insecticidal soaps are effective contact agents for controlling soft-bodied pests, including aphids, whiteflies, mealybugs, and mites. These compounds operate by covering the insect cuticle, impeding respiration, or inducing dehydration, resulting in swift mortality. Horticultural oils are often either petroleum-based or plant-derived, whereas insecticidal soaps consist of fatty acid salts that disrupt the insect's protective wax coating. Both are regarded as low-toxicity substitutes for traditional pesticides, ensuring safety for workers and non-target creatures. Comprehensive application to all plant surfaces, especially the undersides of leaves, is crucial for efficient management. Oils and soaps are most effective when pests are actively eating or in their juvenile stages, as they have reduced efficacy against eggs or pupae. These therapies exhibit negligible residual activity, hence diminishing the likelihood of resistance emergence. They are compatible with biological control agents and can be safely incorporated into Integrated Pest Management systems. The time of application is crucial; refrain from spraying during severe temperatures to avert phytotoxicity. Multiple applications may be required in instances of severe infestation. Horticultural oils has fungicidal activity against certain plant diseases. Appropriate blending, concentration, and application strategies improve efficacy and minimize plant stress. The products' environmental safety, minimal residue, and user-friendliness render them optimal for greenhouse cultivation. They are extensively utilized in organic and protected agricultural systems for sustainable pest control.

12.4.2 Chemical Controls

Chemical pesticides are a crucial instrument for pest management in protected farming; however, they should be employed sparingly and only as a final option. The selection of an appropriate pesticide necessitates precise identification of the pest species and comprehension of its life cycle. Focusing on particular developmental stages enhances efficacy and minimizes superfluous chemical exposure. Employing a rotation of pesticides with varying modes of action aids in averting the emergence of resistance within pest populations. Consistently adhere to the manufacturer's label directives for dosage, application technique, and safety measures. Enclosed areas such as greenhouses necessitate careful consideration of ventilation and accumulation of residues. Excessive use or improper application can adversely affect beneficial insects, pollinators, and human health. Incorporating IPM strategies, such as biological agents, cultural techniques, and monitoring, guarantees less dependence on chemicals. Localized applications and spot treatments diminish the total pesticide burden. Pre-harvest intervals must be rigorously adhered to in order to guarantee food safety. The application of systemic pesticides in soil or on foliage effectively targets elusive bugs. The appropriate disposal of residual chemicals and vacant containers mitigates environmental contamination. Maintaining records of applications facilitates the assessment of effectiveness and the planning of further initiatives. Regular resistance monitoring facilitates the prompt rotation of pesticide categories. Personnel training in safe handling, personal protection equipment, and emergency protocols is essential. Prudent chemical control, when integrated with other IPM tactics, guarantees effective, safe, and sustainable pest management in protected horticulture.

12.4.3. Integrated Pest Management (IPM)

Integrated Pest Management (IPM) is a holistic and sustainable methodology for pest and disease control, integrating preventative, cultural, biological, and chemical tactics to reduce crop losses while maintaining pest populations at economically viable levels. In protected agriculture systems, Integrated Pest Management (IPM) prioritizes early identification, monitoring, and the strategic application of control techniques customized for specific crops and environmental conditions. Protected Cultivation assists farmers in implementing Integrated Pest Management (IPM) by providing an extensive array of tools, technology, and expert advice to develop and execute programs tailored to their polyhouse operations. An efficient Integrated Pest Management program depends on a comprehensive strategy that incorporates cultural practices, resistant cultivars, biological controls, and, where required, specific chemical interventions. Consistent monitoring, documentation, and environmental stewardship are essential elements that improve decision-making and minimize superfluous pesticide usage. IPM fosters healthier crops, increased yields, and environmentally sustainable production through a focus on prevention, ecological equilibrium, and prudent intervention. Ultimately, IPM offers a paradigm for sustainable management of pests and diseases that safeguards both productivity and the surrounding ecology.

- **Prevention:** Commence with a meticulously cleansed greenhouse and verify that any plants, seeds, or cuttings are devoid of pests and diseases. Effective cleanliness diminishes the entry of germs and lowers the probability of infestations. Creating a sanitary atmosphere constitutes the basis for all subsequent pest control strategies.
- **Monitoring:** Begin with a thoroughly sanitized greenhouse and ensure that all plants, seeds, or cuttings are free from pests and illnesses. Thorough cleanliness reduces the ingress of pathogens and decreases the likelihood of infestations. Establishing a hygienic environment is the foundation for all subsequent pest management measures.
- **Identification:** Commence with a meticulously cleansed greenhouse and verify that any plants, seeds, or cuttings are devoid of pests and diseases. Meticulous hygiene decreases the entry of diseases and lowers the probability of infestations. Creating a sanitary atmosphere is essential for all subsequent pest control strategies.
- **Thresholds for Action:** Initiate action solely when pest numbers above thresholds capable of inflicting economically substantial harm. This method curtails superfluous pesticide usage and reduces ecological consequences. Setting thresholds reconciles crop protection with sustainable management approaches.
- **Control Tactics:** Employ a synthesis of techniques, encompassing cultural norms, physical obstacles, biological restrictions, and targeted pharmacological interventions. A comprehensive strategy improves efficacy while diminishing dependence on any one technique. The integration of several strategies aids in preserving ecological equilibrium and ensuring sustained pest control.
- **Record-keeping:** Keep comprehensive records of monitoring outcomes, pest numbers, and any implemented measures. Documentation facilitates the monitoring of trends, assessment of management effectiveness, and enhancement of future initiatives. Efficient record-keeping facilitates informed decision-making and ongoing enhancement in pest management initiatives.

Preventive Measures

- **Sanitation:** Sanitation is a highly efficient method for preventing pest and disease outbreaks in controlled environments. Eliminating weeds, algae, and plant detritus from both the interior and exterior of the greenhouse eradicates possible breeding grounds for insects and reservoirs for infections. Decomposing or deceased plant matter frequently contains fungal spores, bacteria, or viruses that can readily disseminate to healthy crops. Sanitizing benches, trays, pots, and containers between cropping cycles effectively eliminates any residual microorganisms. Instruments such as pruners, knives, and scissors must be routinely disinfected to avert cross-contamination. Floors, walls, and irrigation systems require regular cleaning to diminish microbial burdens. Appropriate disposal of infected plant waste inhibits disease spread. A regular sanitation plan ensures a hygienic environment and reduces bug infestations. Sanitation is a fundamental element of integrated pest management, safeguarding crop health and productivity.

- ☆ **Quarantine:** Quarantine is a preventive measure that secludes new or incoming plants to avert the entry of pests and illnesses into the primary greenhouse. Recently obtained plants should be positioned in a designated area for several weeks for careful observation. This timeframe facilitates the identification of observable symptoms, including leaf spots, wilting, reduced growth, or pest infestation. Infected plants can thereafter be eliminated or treated before to posing a risk to good harvests. Quarantine zones must possess distinct tools, containers, and watering systems to prevent inadvertent contamination. Consistent monitoring during this interval, along with the deployment of adhesive traps for insects, improves early identification. Quarantining mitigates the possibility of importing exotic pests devoid of natural predators in the greenhouse. Personnel managing confined plants must adhere to stringent hygiene measures. The documentation of observations facilitates informed decision-making and prompt interventions. Quarantine aids in preserving overall crop health and diminishing long-term pest management expenses by isolating potentially hazardous plants.
- ☆ **Environmental Control:** Environmental control is essential in averting the emergence of pests and illnesses by establishing conditions that are detrimental to pathogen proliferation. Regulating temperature, humidity, and air circulation effectively diminishes illness prevalence. Excess humidity and condensation on foliage facilitate fungal infections, including Botrytis and powdery mildew. Horizontal airflow fans facilitate uniform air circulation, mitigating moisture buildup on leaf surfaces. Effective ventilation, adequate spacing, and optimal light management enhance plant health and reduce insect attraction. Drip irrigation or subirrigation methods diminish leaf moisture, hence reducing the danger of fungal infections. Ongoing surveillance utilizing thermometers, hygrometers, and data recorders guarantees consistent environmental conditions.

Environmental regulation enhances plant immunity and diminishes dependence on chemical treatments. Integrating environmental management with additional Integrated Pest Management strategies improves comprehensive pest and disease control. A meticulously controlled greenhouse environment is essential for sustainable, high-quality crop production.

Table 12.1: Common Greenhouse Pests and Control Methods

Pest	*Identification*	*Control Methods*
Aphids	Small, soft-bodied insects that cluster on new growth. They excrete sticky "honeydew," which can lead to black sooty mold.	Release natural predators like ladybugs and lacewings, or spray with insecticidal soap or neem oil for mild infestations.
Spider Mites	Tiny, almost microscopic pests that thrive in hot, dry conditions. They cause stippling on leaves and may produce fine webbing in severe infestations.	Increase humidity, regularly misting plants. Introduce predatory mites, like Phytoseiulus persimilis.
Whiteflies	Small, white, moth-like insects that feed on the undersides of leaves, causing yellowing. They also produce sticky honeydew.	Use yellow sticky traps to catch adults. Release parasitic wasps (Encarsia formosa) or spray with insecticidal soaps or neem oil.
Fungus Gnats	Small, black, fly-like insects attracted to damp soil. Larvae feed on plant roots, damaging seedlings.	Let the growing medium dry out between waterings. Apply beneficial nematodes (*Steinernema feltiae*) to the soil.
Thrips	Tiny, slender insects that feed on plant tissue. Leaves may show silvering or mottling, and flowers can become distorted.	Use blue sticky traps, which are more attractive to thrips than yellow ones. Release predatory mites or introduce the pirate bug (Orius laevigatus).
Mealybugs	White, cottony pests found on stems and leaf joints. They suck plant sap and secrete honeydew.	Treat small infestations by dabbing with a cotton swab soaked in rubbing alcohol. For larger issues, use insecticidal soap or neem oil

☆ **Entry Barriers:** Entry barriers are tangible and procedural strategies implemented to inhibit the ingress of pests and diseases into the greenhouse. Insect-resistant screens on vents, windows, and doors restrict the entry of airborne pests like whiteflies, aphids, and thrips. Double-door entrances, or "airlocks," diminish the likelihood of inadvertent pest ingress during door openings. Disinfectant footbaths at entry points inhibit the transmission of infections via footwear. Personnel and guests must don clean attire, gloves, and additional safety equipment when within the greenhouse. External tools, containers, and equipment must be meticulously sterilized. Consistent examination and upkeep of screens and access barriers guarantee their continued efficacy. Mitigating external contamination diminishes the necessity for reactive pest management strategies. Entry barriers constitute the initial line of defense in an integrated pest management system. When consistently implemented, they markedly diminish the likelihood of pest and disease occurrences and safeguard overall crop vitality.

Table 12.2: Common Greenhouse Diseases and Control Methods

Disease	*Identification*	*Control Methods*
Botrytis Blight (Grey Mold)	A fungal disease favored by high humidity and poor air circulation. It appears as tan-to-brown lesions that develop a fuzzy grey mold.	Improve air circulation and reduce humidity by heating and venting. Remove infected plant material immediately.
Damping-Off	Caused by soil-borne fungi like Pythium and Rhizoctonia, it causes seeds or seedlings to rot before or after they emerge from the soil.	Use pasteurized growing media and clean pots. Avoid overwatering and provide adequate heat and drainage.
Powdery Mildew	A fungal disease that produces a white, powdery coating on leaves and stems. It thrives in humid conditions.	Ensure good ventilation and air circulation. Remove infected foliage and apply appropriate fungicides or botanical sprays.
Viruses	Viruses cause various symptoms, including mosaic patterns, stunting, or leaf distortion. They often spread via insect vectors like thrips.	There is no cure for viruses. Focus on prevention by using virus-indexed stock and controlling insect vectors. Destroy all infected plants and weeds.

Colour Plates

Figure 1.1: Protected Cultivation. (Pg. 2)

Figure 2.1: Low-Cost Greenhouse. (Pg. 25)

Figure 2.2: Hi-Tech Greenhouse (Pg. 26)

Figure 2.3: Medium Cost Greenhouse. (Pg. 26)

Figure 4.1: Smart Greenhouse System. (Pg. 66)

Figure 4.2: Benefits of Protected Cultivation: Maximizing Yield with Polyhouse. (Pg. 68)

Figure 4.3: Protected Cultivation of Vegetables, Flowers, and Fruits. (Pg. 69)

Figure 4.4: Naturally Ventilated Greenhouse. (Pg. 70)

Figure 4.5: Climate-Controlled or Hi-Tech Greenhouse. (Pg. 71)

Figure 4.6: Naturally Ventilated Polyhouse. (Pg. 84)

Figure 4.7: Shade Nets to Protect Plants from Sun Damage. (Pg. 85)

Figure 4.8: Heat Greenhouse in Winter (14 Proven Methods) (Pg. 85)

Figure 4.9: Thermal Screens. (Pg. 86)

Figure 4.10: Polycarbonate Solid Sheet (Pg. 86)

Figure 4.11: Drip Irrigation in Protected Cultivation. (Pg. 92)

Figure 4.12: Sprinkler Irrigation (Pg. 93)

Figure 4.13: Subsurface Irrigation (Pg. 93)

Figure 4.14: Rainwater Harvesting. (Pg. 94)

Figure 4.15: Soil Sterilization. (Pg. 96)

Figure 4.16: Greenhouse Automation Systems. (Pg. 109)

Figure 4.17: Robotics in 21st Century Agriculture. (Pg. 111)

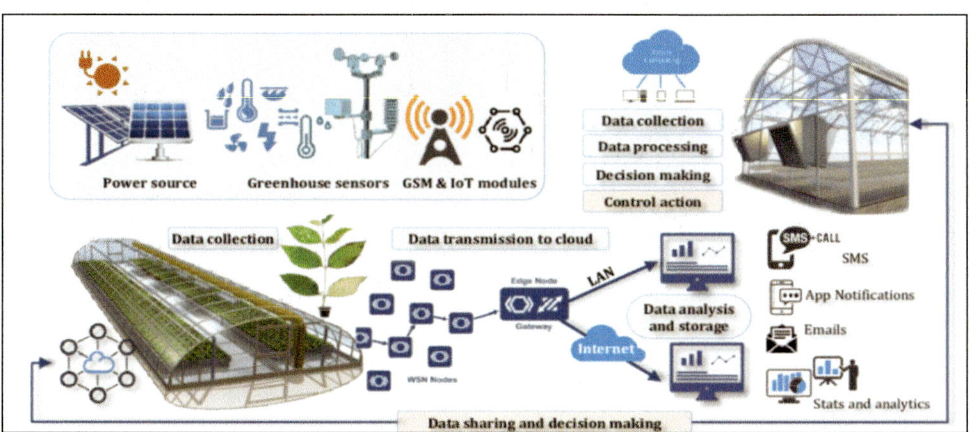

Figure 4.18: IoT Based Automated Greenhouse Monitoring System. (Pg. 112)

Figure 5.1: Soil and Substrate Management (Pg. 117)

Figure 42: Soil Levelling. (Pg. 124)

Figure 5.3: Bed Formation (Pg. 125)

Figure 5.4: Soil Solarization in Greenhouse. (Pg. 126)

Figure 5.5: Soil Sterilization. (Pg. 127)

Figure 5.6: Enriching Soil with Organic Matter in Greenhouse. (Pg. 128)

Figure 5.7: Soil Conditioners. (Pg. 129)

Figure 5.8: Soil Mulching. (Pg. 131)

Figure 5.9: Nursery Seed Bed Preparation. (Pg. 133)

Figure 5.10: Methods of Soil Conservation (Pg. 134)

Reduce Soil Erosion

Increase water Infiltration and Storage

Improve Soil Fertility

Providing Wildlife Habitat

Increase Crop Yield

Figure 5.11: Benefits of Soil Conservation. (Pg. 135)

Figure 5.12: Soil-Based Substrates. (Pg. 136)

Figure 5.13: Coir Pith or Coir Dust (Pg. 139)

Figure 5.14: Vermiculite. (Pg. 140)

Figure 5.15: Perlite (Pg. 142)

Figure 5.16: Rockwool (Mineral Wool) as a Hydroponic Substrate. (Pg. 143)

Figure 5.17: Sand Used Inert Growing Media in Horticulture and Protected Cultivation Systems. (Pg. 145)

Figure 5.18: Gravel and Expanded Clay Pebbles are Widely Used... (Pg. 146)

Figure 5.19: Sawdust and Wood Chips.. (Pg. 148)

Figure 5.20: Peat Moss is Widely Used Soilless Substrates in Horticulture and Hydroponics. (Pg. 149)

Figure 5.20: Organic Composts. (Pg. 151)

Figure 6.1: Walk-in Tunnels. (Pg. 161)

Figure 6.2: Plastic Low Tunnels/Row Covers. (Pg. 163)

Figure 6.3: Net Houses (Pg. 164)

Figure 6.4: Plastic Mulch. (Pg. 165)

Figure 6.5: Fresh Produce in a Plastic Bag Tomatoes and Greens. (Pg. 166)

Figure 6.6: Soil Trenches (Pg. 167)

Figure 6.7: Hot Beds (Pg. 168)

Figure 6.8: Benches of Greenhouse. (Pg. 173)

Figure 6.9: Dendrobium Orchid Cultivation on Metal Bench. (Pg. 174)

Figure 6.10: Vertical Cultivation of Leafy Vegetables on Shelves (Pg. 175)

Figure 6.11: Floor as Bench. (Pg. 178)

Figure 6.12: Raised Benches. (Pg. 179)

Figure 6.13: Growing Leafy Vegetables in Plastic Containers. (Pg. 182)

Figure 6.14: Ceramic Containers (Pg. 183)

Figure 6.15: Wood Containers. (Pg. 184)

Figure 6.16: Fiberglass Containers. (Pg. 185)

Figure 6.17: Metal Containers. (Pg. 186)

Figure 6.18: Metal Containers. (Pg. 187)

Figure 7.1: Hand Watering. (Pg. 193)

Figure 7.3: Overhead Sprinklers in Greenhouse Irrigation. (Pg. 197)

Figure 7.4: Boom Watering Irrigation. (Pg. 199)

Figure 7.5: Boom Watering Irrigation System. (Pg. 200)

Figure 7.6: Drip Irrigation System. (Pg. 202)

Figure 7.7: Greenhouse Irrigation Misting System. (Pg. 205)

Figure 7.8: Fogging System. (Pg. 205)

Figure 7.9: High Pressure Fog Systems in Horticulture (Pg. 206)

Figure 8.1: Propagation by Seed (Pg. 225)

Figure 8.3: Propagating Plants from Cutting (Pg. 227)

Figure 8.4: Layering (Pg. 227)

Figure 8.5: Grafting and Budding. (Pg. 228)

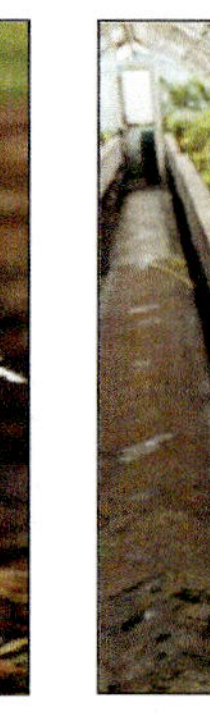

Figure 8.6: Tissue Culture (Micropropagation) (Pg. 228)

Figure 9.1: Flower Crops their Varieties Cultivated under Greenhouses (Pg. 250)

Figure 9.2: Rose Cultivation in Greenhouse. (Pg. 253)

Figure 9.3: Carnation (*Dianthus caryophyllus*) in Greenhouse (Pg. 258)

Figure 9.4: Chrysanthemum (*Dendranthema grandiflora*) in Greenhouse. (Pg. 264)

Figure 9.5: Gerbera in Greenhouse. (Pg. 270)

Figure 9.8*:* Dendrobium in Greenhouse. (Pg. 274)

Figure 9.9: Spike Ready for Harvest. (Pg. 278)

Figure 9.10:Anthurium Under Shade Net. (Pg. 279)

Figure 9.11: Lilium in Greenhouse. (Pg. 283)

Figure 9.12: Tulips in Greenhouse (Pg. 287)

Figure 9.13: Protected Cultivation of Tomato in Open Ventilated Oolyhouse (Pg. 291)

Figure 9.14: Cherry Tomato grown under Protected Cultivation (Pg. 291)

Figure 9.15: Capsicum Farming in Polyhouse (Pg. 295)

Figure 9.16: Green Capsicum. (Pg. 297)

Figure 9.17: Cucumber in Greenhouse. (Pg. 300)

Figure 9.18: Strawberry in Greenhouse (Pg. 305)

Figure 9.19: Pot Plants (Pg. 310)

Glossary

Aeroponics: Aeroponics grows plants with roots suspended in air and periodically misted with nutrient solutions. This system provides high oxygen availability to roots, promoting faster growth and efficient nutrient uptake. Aeroponics is used for rapid propagation and space-saving cultivation.

Air Circulation Fans: Air circulation fans promote uniform temperature and humidity inside greenhouses by moving air. Good air circulation prevents hot spots and reduces disease risks. Fans improve plant transpiration and overall microclimate stability.

Air Exchange Rate: Air exchange rate is the frequency of complete air replacement inside a greenhouse per hour. Adequate exchange controls temperature, humidity, and CO2 levels. It prevents buildup of pests, diseases, and harmful gases.

Air Humidification: Air humidification increases moisture levels in greenhouse air, beneficial for crops sensitive to dry conditions. Techniques include misting and fogging systems. Maintaining optimal humidity improves plant physiology and reduces stress.

Anthurium Cultivation: Requires shaded conditions, warm temperatures, and controlled moisture levels for optimum flower and leaf development.

Aquaponics: Aquaponics combines aquaculture (fish farming) with hydroponics, creating a symbiotic environment where fish waste provides nutrients for plants. This sustainable system recycles water and nutrients efficiently. It is gaining interest in protected cultivation.

Aromatic Plants Cultivation: Essential oils quality and yield improve when cultivated under controlled environment with adequate light and moisture.

Artificial Lights: Supplemental lighting using LEDs or high-pressure sodium lamps to extend daylight, improve photosynthesis, and enable off-season production.

Automated Climate Control: Automated climate control systems use computers and sensors to regulate temperature, humidity, ventilation, and lighting. Automation ensures optimal growing conditions with minimal human intervention. It enhances crop uniformity and resource efficiency.

Automated Fertigation: Automated fertigation combines irrigation and fertilizer application controlled by sensors and timers. It provides precise nutrient delivery tailored to crop needs. Automation reduces labour and enhances nutrient use efficiency.

Automation in Protected Cultivation: Integration of sensors, controllers, and actuators to automate irrigation, fertigation, ventilation, and lighting for precise environment control.

Automation: Technology that automatically controls greenhouse conditions like temperature, humidity, and irrigation, improving efficiency and crop consistency. It reduces manual work and responds quickly to environmental changes.

Base/Foundation: The bottom support structure of the greenhouse ensuring stability and anchoring it to the ground.

Bell Pepper Cultivation: Requires stable temperature and humidity levels, along with fertigation for high yield and fruit quality.

Benching System: Raised platforms inside greenhouses for holding pots or trays, improving space use and airflow.

Biocontrol Agents: Biocontrol agents are beneficial organisms like predatory insects, nematodes, or fungi used to naturally control pests and diseases. They reduce chemical pesticide use and promote sustainable crop protection. Incorporating biocontrol enhances ecological balance in greenhouses.

Biodegradable Mulch: Biodegradable mulch is an eco-friendly alternative to plastic mulch that decomposes naturally. It conserves soil moisture and suppresses weeds without environmental pollution. Its use supports sustainable protected cultivation.

Biodegradable Pots: Biodegradable pots are containers made from organic materials that decompose naturally in the soil. They reduce plastic waste and improve root aeration when planted directly. These pots are popular for nursery and propagation.

Biological Control: Biological control uses natural enemies like predators, parasites, or pathogens to suppress pests. It reduces chemical pesticide use and environmental harm. Biological agents are key components of integrated pest management.

Biological Fertilizers: Biological fertilizers contain living microorganisms that enhance nutrient availability by fixing nitrogen or solubilizing phosphates. They improve soil fertility and plant health. Using biofertilizers reduces chemical fertilizer dependence.

Biological Filters: Biological filters use beneficial microbes to treat wastewater or recirculate nutrient solutions in hydroponics. They reduce pollutants and maintain solution quality. These filters support sustainable water management.

Biopesticides: Biopesticides are natural agents derived from microorganisms, plants, or minerals used to control pests. They offer environmentally friendly alternatives to chemical pesticides. Biopesticides are integral to sustainable pest management.

Biostimulants: Biostimulants are natural or synthetic substances that enhance plant growth and stress tolerance. They improve nutrient uptake, root development, and crop quality. Using biostimulants supports sustainable protected cultivation.

Capillary Mat Irrigation: Capillary mat irrigation involves placing plants on a water-retentive mat that delivers moisture via capillary action. It provides uniform water supply to container-grown plants. This method conserves water and reduces irrigation labour.

Carbon Dioxide Enrichment: Increasing CO2 concentration inside greenhouses to enhance photosynthesis and growth rates.

Carnation Cultivation: Protected cultivation of carnations requires well-drained substrates, balanced nutrition, and regulated environment for flower quality.

Chrysanthemum Cultivation: Requires precise photoperiod management and temperature control in greenhouses for continuous flower production.

Cladding Material: The outer covering used in greenhouses/polyhouses to protect crops and maintain internal climate; examples include glass, polyethylene, and polycarbonate.

Climate Adaptive Structures: Climate adaptive structures are greenhouses designed to adjust to varying weather conditions automatically. Features include adjustable vents, shading, and insulation. These structures optimize crop environment and energy use.

Climate Control System: An integrated system managing temperature, humidity, ventilation, and lighting to maintain ideal growing conditions inside a greenhouse. It helps optimize plant growth by keeping the environment stable.

Climate Sensors: Climate sensors monitor temperature, humidity, light intensity, and CO2 levels inside the greenhouse. Data from sensors help automate environment control systems for optimal crop growth. Real-time monitoring enhances precision and efficiency in protected cultivation.

Climate Zones: Climate zones classify regions based on temperature, humidity, and rainfall patterns. Greenhouse designs and crop choices are adapted to local climate zones. Understanding these zones helps optimize protected cultivation strategies.

Climatic Data Logging: Climatic data logging involves recording environmental parameters over time for analysis. This information aids in adjusting greenhouse management practices. Data logging improves crop

Climatic Stress Indicators: Indicators such as leaf curling, chlorosis, or slowed growth signal environmental stresses like temperature extremes or humidity imbalance. Early detection allows corrective measures. Managing stress ensures consistent crop performance.

Closed-loop Irrigation: Closed-loop irrigation recycles runoff water, treating and reusing it within the system to conserve resources. It reduces water demand and environmental impact. This system is common in hydroponic and soilless cultivation.

CO_2 Enrichment Systems: CO_2 enrichment systems inject carbon dioxide into greenhouses to increase levels above ambient concentrations. Elevated CO_2 enhances photosynthesis, growth, and yield. Proper management prevents wastage and ensures safety.

CO_2 Monitoring: CO_2 monitoring tracks carbon dioxide levels to optimize photosynthesis and growth. Sensors provide real-time data for adjusting enrichment systems. Proper CO_2 levels can increase yield and crop quality.

Controlled Environment Agriculture (CEA): CEA refers to fully or partially controlled growing environments that optimize crop production year-round. It includes greenhouses, vertical farms, and growth chambers. CEA enables high-efficiency, resource-conserving agriculture.

Controlled Release Fertilizers: Controlled release fertilizers slowly release nutrients over time, matching crop uptake patterns. They reduce nutrient losses and improve fertilizer efficiency. These fertilizers are beneficial in protected cultivation to maintain steady nutrient supply.

Cooling System: Cooling systems like evaporative pads, misting, and fans reduce excessive heat during hot weather. These systems lower the greenhouse temperature to prevent heat stress and promote healthy crop development. Effective cooling is essential, especially in tropical and subtropical climates.

Crop Canopy Management: Crop canopy management involves pruning, training, and spacing plants to optimize light interception and air circulation. Proper canopy structure improves photosynthesis and reduces disease incidence. It enhances fruit quality and yield in protected cultivation.

Crop Evapotranspiration (ETc): *etc.* is the total water loss from a crop through evaporation and transpiration. Understanding *etc.* helps optimize irrigation scheduling. Managing *etc.* reduces water waste and stress on plants.

Crop Load Management: Crop load management controls the number of fruits or flowers per plant to balance growth and yield quality. Proper load improves fruit size, flavour, and shelf life. Techniques include pruning and thinning.

Crop Load Thinning: Crop load thinning is the removal of excess flowers or fruits to balance vegetative growth and fruit quality. It prevents overburdening the plant and improves size and taste of remaining fruits. Thinning is practiced in fruits like tomato and bell pepper.

Crop Modelling: Crop modelling uses computer simulations to predict plant growth under various environmental conditions. It helps optimize greenhouse management by forecasting yields and resource needs. Modelling supports decision-making and precision agriculture.

Crop Monitoring Systems: Crop monitoring systems use sensors and imaging technologies to track plant health, growth, and stress factors. Real-time data enables timely interventions, improving crop management. These systems support precision agriculture in protected cultivation.

Crop Phenology: Crop phenology studies developmental stages such as germination, flowering, and fruiting. Monitoring phenology guides management practices and harvest timing. It is important for synchronized and optimized production.

Crop Residue Management: Crop residue management involves handling leftover plant materials after harvest, either by removal or incorporation into the substrate. Proper management prevents pest and disease buildup. It also recycles nutrients and improves substrate health.

Crop Rotation in Protected Cultivation: Crop rotation involves growing different crops sequentially in the same space to improve soil health. It reduces buildup of pests and diseases while replenishing nutrients. This practice supports sustainable greenhouse production.

Crop Scheduling: Crop scheduling plans planting and harvesting to optimize greenhouse utilization and market supply. It allows continuous production and reduces downtime. Effective scheduling aligns production with demand and climatic conditions.

Crop Sensors: Crop sensors measure plant parameters like leaf temperature, chlorophyll content, or moisture stress. They provide data to improve irrigation, nutrition, and pest management. These sensors enable precision agriculture inside greenhouses.

Crop Spacing: Crop spacing is the optimal distance between plants to maximize light penetration and air circulation. Proper spacing reduces competition for nutrients and water. It also helps prevent disease by improving airflow around plants.

Crop Stress Indicators: Indicators like leaf wilting, colour changes, or stomatal closure signal plant stress from water, nutrients, or temperature. Monitoring these signs helps adjust management practices promptly. Early intervention prevents yield loss.

Crop Uniformity: Crop uniformity refers to the consistency in size, growth rate, and development of plants within a greenhouse. Achieving uniformity improves harvesting efficiency and market quality. It depends on genetics, environment, and management practices.

Crop Yield: Crop yield refers to the amount of produce harvested per unit area. Protected cultivation techniques aim to maximize yield by optimizing environmental conditions. Higher yields improve profitability and resource use.

Cucumber Cultivation: Needs good ventilation, humidity control, and disease management in polyhouses for uniform fruit development.

Cuttings Propagation: Cuttings propagation involves rooting plant parts like stems or leaves to produce new plants. It allows cloning of desired varieties and quicker crop establishment. Controlled environments improve rooting success and quality.

Data Logger: Device that records environmental parameters over time for monitoring and analysis to improve management.

Data Logger: Electronic device recording environmental data over time to monitor and analyse greenhouse conditions. It aids precision management and informed decision-making.

Disease Management: Disease management in protected cultivation involves practices such as sanitation, using resistant varieties, and applying biological controls. These methods help prevent outbreaks of fungal, bacterial, and viral diseases inside greenhouses. Maintaining a clean environment is key to protecting crops and ensuring healthy growth.

Disease-resistant Varieties: Disease-resistant varieties possess genetic traits that reduce susceptibility to specific pathogens. Using these varieties in protected cultivation lowers chemical use and crop losses. They are essential for sustainable and profitable production.

Drainage in Greenhouse Beds: Good drainage removes excess water from beds or containers to prevent waterlogging. Proper drainage helps avoid root diseases caused by anaerobic conditions. It maintains oxygen availability in the root zone for healthy plant growth.

Drip Fertigation: Drip fertigation combines drip irrigation with fertilization, providing nutrients and water simultaneously. It supports targeted nutrient delivery and minimizes leaching. Drip fertigation is widely adopted in greenhouse vegetable production.

Drip Irrigation: Drip irrigation delivers water directly to the plant root zone through emitters, reducing wastage and evaporation. It provides precise moisture control, improving water use efficiency in protected cultivation. Drip systems also facilitate fertigation by combining irrigation with nutrient supply.

Drip Tape: Drip tape is a flat, thin irrigation tubing with built-in emitters used for drip irrigation systems. It efficiently delivers water and nutrients directly to plant roots. Drip tape is widely used in greenhouse

Electrical Conductivity (EC): EC measures the salinity or nutrient concentration in soil or nutrient solutions. Maintaining appropriate EC ensures balanced nutrient availability. High EC can cause toxicity, while low EC leads to deficiencies.

Energy Curtains: Energy curtains are retractable screens that conserve heat and reduce energy consumption in greenhouses. They help maintain stable temperatures during nights or cold periods. Energy curtains improve crop quality and reduce costs.

Environment Control in Greenhouses: Use of ventilation, heating, cooling, shading, and humidity control systems to maintain optimal growth conditions inside the structure.

Evaporative Cooling: Evaporative cooling lowers greenhouse temperature by passing air through wet pads, causing water evaporation and heat loss. It is an energy-efficient cooling method for hot climates. Cooling prevents heat stress and improves crop performance.

Fertigation Management: Combined application of fertilizers through irrigation systems to deliver nutrients efficiently and uniformly to crops.

Fertigation System: Fertigation delivers fertilizers dissolved in irrigation water directly to plant roots. It allows precise nutrient management tailored to crop growth stages. This system improves nutrient use efficiency and reduces fertilizer wastage.

Fertilizer Schedule: A fertilizer schedule plans the timing and amount of nutrient application to meet crop requirements. It optimizes nutrient availability throughout growth stages. Proper scheduling prevents nutrient deficiencies and excesses, enhancing crop yield and quality.

Fogging System: Fogging systems produce fine water droplets to cool and humidify greenhouse air. They help maintain optimal microclimate during hot and dry conditions. Fogging reduces plant heat stress and improves growth.

Foliar Feeding: Foliar feeding supplies nutrients directly to plant leaves through sprays. It provides rapid correction of deficiencies and supplements soil fertilization. This method is effective in protected cultivation for quick nutrient uptake.

Forced Ventilation: Using fans or blowers to mechanically move air for temperature and humidity control.

Gerbera Cultivation: Demanding in terms of substrate quality, light intensity, and temperature regulation for producing vibrant flowers.

Glass Cladding: Transparent, durable material offering high light transmission but expensive and fragile; common in traditional greenhouses.

Grafting: Grafting joins two plant parts, typically a rootstock and scion, to combine desired traits like disease resistance and high yield. It is used in protected cultivation to improve crop quality and resilience. Grafting accelerates propagation of superior plants.

Greenhouse Automation: Greenhouse automation uses sensors, controllers, and actuators to regulate climate, irrigation, and fertigation. This reduces manual labour and ensures precise environmental control. Automation improves consistency and crop productivity.

Greenhouse Design: Involves planning size, shape, orientation, and ventilation to optimize light, temperature, humidity, and CO2 levels for crop growth.

Greenhouse Orientation: Greenhouse orientation refers to aligning the structure to maximize sunlight exposure and ventilation. Proper orientation improves light interception and climate control. It is crucial for energy efficiency and crop performance.

Grow Bags: Grow bags are plastic or fabric containers filled with substrate used for growing plants in protected cultivation. They offer flexibility, good drainage, and easy transport. Grow bags are used for flowers, vegetables, and medicinal plants.

Growing Media: Growing media are materials like cocopeat, perlite, and vermiculite that support plant roots in container and soilless cultures. They influence water retention, aeration, and nutrient availability. Choosing the right media is vital for crop success.

Growth Regulators: Growth regulators are chemical substances that influence plant development stages like flowering, fruiting, and rooting. They are used in protected cultivation to manipulate crop growth and improve yields. Proper application enhances crop uniformity and quality.

Gutter-connected Greenhouse: Multiple greenhouse bays connected by gutters for efficient water management and increased growing area.

Heat Retention: Techniques and materials used to keep warmth inside the structure, especially during cold nights.

Heat Stress Management: Heat stress management involves strategies to protect plants from excessive temperatures such as shading, ventilation, and cooling systems. Managing heat stress prevents yield loss and physiological disorders. It is critical during hot seasons.

Heating System: Heating devices such as boilers, heaters, or heat mats maintain optimum temperatures during cold seasons. They protect sensitive plants from frost damage and ensure steady growth throughout the year. Proper heating improves flowering, fruit set, and overall yield in protected cultivation.

High-Pressure Sodium (HPS) Lamps: Common artificial lights emitting orange-yellow light, efficient for promoting flowering and fruiting.

Humidifier: A humidifier adds moisture to greenhouse air to maintain ideal relative humidity levels. It prevents excessive drying of plants, especially seedlings and tropical crops. Proper humidity promotes healthy growth and reduces stress.

Humidity Control: Humidity control techniques manage moisture in the greenhouse air to prevent fungal diseases and physiological disorders. High humidity encourages diseases like powdery mildew, while low humidity can stress plants. Maintaining balanced humidity promotes healthy crop growth and productivity.

Humidity Control: Techniques to regulate moisture in the air to prevent fungal diseases and ensure crop health.

Hydrogel: Hydrogel is a water-absorbing polymer added to substrates to improve water retention and availability to plants. It reduces irrigation frequency and drought stress. Hydrogels are especially useful in substrate management for container crops.

Hydroponic Nutrient Solution: A nutrient solution in hydroponics contains dissolved minerals tailored to crop needs. Its composition affects growth, yield, and quality. Monitoring and adjusting nutrient solutions is essential for successful hydroponic cultivation.

Hydroponics: Hydroponics is a soilless cultivation method where plants grow in nutrient-rich water solutions. It allows precise control over nutrient supply and reduces soil-borne diseases. Hydroponics is widely used in protected cultivation for high-value vegetables and ornamentals.

Insect Pest Management: Use of integrated pest management, including biological control agents, traps, and safe pesticides, to protect crops in greenhouses.

Integrated Disease Management (IDM): IDM combines cultural, biological, and chemical practices to control plant diseases effectively and sustainably. It minimizes pesticide use and resistance development. IDM is essential for maintaining healthy crops in protected cultivation.

Irrigation Automation: Automatic watering systems using timers or sensors to supply precise amounts of water based on plant needs. This conserves water and ensures consistent soil moisture.

Light Diffusion: Light diffusion spreads incoming light evenly throughout the greenhouse, reducing shadows and hotspots. This enhances photosynthesis by providing uniform illumination to all plant parts. Diffuse light improves crop growth and quality.

Light Emitting Diodes (LEDs): Energy-efficient lights with customizable spectra, increasingly popular for targeted plant growth.

Light Intensity Sensors: These sensors measure the amount of light available to plants, informing adjustments in shading or supplemental lighting. Maintaining optimal light levels maximizes photosynthesis. Sensor data improves energy use efficiency.

Light Intensity: Light intensity refers to the amount of light reaching plant surfaces, influencing photosynthesis and development. Adequate light promotes healthy growth, flowering, and fruiting, while insufficient light reduces crop quality and yield. Managing light intensity is critical in greenhouse design and operation.

Light Quality: Light quality refers to the spectral composition of light, including red, blue, and UV wavelengths. Different spectra influence plant morphology, flowering, and photosynthesis. Supplemental lighting can be tuned for desired effects.

Light Saturation Point: The light saturation point is the light intensity beyond which photosynthesis no longer increases. Understanding this helps optimize lighting to avoid energy waste. Managing light levels at or below saturation improves efficiency.

Light-emitting Diodes (LEDs): LEDs provide energy-efficient, customizable light spectra for supplemental lighting in greenhouses. They promote photosynthesis and control plant morphology. LED lighting reduces electricity costs and enhances crop quality.

Lilium Cultivation: Temperature management and disease control are critical for high-quality lily production in protected environments.

Medicinal Plants Cultivation: Protected environments help in producing high-value medicinal and aromatic plants with controlled quality and yield.

Microbial Inoculants: Microbial inoculants introduce beneficial microbes like nitrogen-fixing bacteria or mycorrhizae to substrates. They enhance nutrient availability and disease resistance. Inoculants support sustainable and productive cultivation.

Microclimate: Microclimate refers to the small-scale climate within the greenhouse, including temperature, humidity, light, and airflow. Managing microclimate is critical to optimize plant growth and development. Precise control reduces stress and improves crop quality.

Microcontroller: A microcontroller is a small computer embedded in greenhouse control systems. It processes data from sensors and automates operations like irrigation, ventilation, and lighting. This ensures precise and efficient management of the protected environment.

Mist Irrigation: Mist irrigation sprays a fine mist of water to maintain high humidity and cool the plant canopy. It is useful for delicate seedlings and during hot weather to prevent heat stress. This method helps create favourable microclimates inside greenhouses for sensitive crops.

Mulch Films: Mulch films cover soil surfaces to conserve moisture, suppress weeds, and regulate temperature. They improve water use efficiency and reduce labour. Biodegradable and coloured mulch films are used based on crop and climate.

Mulching: Mulching involves covering the soil or substrate surface with plastic sheets or organic materials to retain moisture and suppress weeds. It also regulates soil temperature and prevents nutrient leaching. Mulching improves water use efficiency and reduces labour in protected cultivation.

Natural Ventilation: Using openings like side vents and roof vents to allow airflow driven by wind and thermal buoyancy without mechanical aid.

Nutrient Deficiency Symptoms: Symptoms like yellowing leaves, stunted growth, or necrosis indicate nutrient deficiencies. Early identification allows corrective measures through fertilization or foliar feeding. Managing nutrient levels prevents yield losses.

Nutrient Film Recirculation: In hydroponics, nutrient film recirculation continuously cycles nutrient solution through plant roots, conserving water and nutrients. Recirculation reduces waste and maintains solution quality. It supports sustainable water and nutrient use.

Nutrient Film Technique (NFT): NFT is a hydroponic system where a thin film of nutrient solution flows continuously over plant roots in a sloped channel. This ensures constant nutrient and oxygen supply while minimizing water use. NFT is popular for growing leafy greens and small fruiting crops in greenhouses.

Off-season Production: Protected cultivation allows growing flowers and vegetables outside their natural season to meet market demand and fetch premium prices.

Orchid Cultivation: Specialized environment control including high humidity and indirect light is essential for the successful greenhouse cultivation of orchids.

Pest Exclusion Screens: Pest exclusion screens are fine mesh materials installed on greenhouse openings to prevent insect entry. They reduce pest infestations and pesticide use. These screens are essential in integrated pest management.

Pest Life Cycle Understanding: Knowledge of pest life cycles aids in timing control measures for maximum effectiveness. Interventions can target vulnerable stages like eggs or larvae. Understanding life cycles improves integrated pest management outcomes.

Pest Monitoring: Pest monitoring involves regular scouting and using traps to detect pest populations early. Early detection enables timely control measures, preventing major outbreaks. Monitoring is a key component of integrated pest management.

Pest-resistant Varieties: Pest-resistant varieties possess traits that deter or withstand insect attacks. Growing these varieties reduces pesticide use and crop damage. They are crucial for integrated pest management strategies.

Pheromone Traps: Pheromone traps use synthetic insect sex attractants to monitor or control pest populations. They enable early pest detection and reduce pesticide reliance. Traps are integral to integrated pest management in protected cultivation.

Photomorphogenesis: Photomorphogenesis is the process by which plants develop form and structure in response to light quality and intensity. Manipulating light spectra in greenhouses can influence plant height, leaf size, and flowering. Understanding this helps optimize growth conditions.

Photoperiod: Duration of light exposure that affects flowering and other developmental processes in plants.

Photoperiod: Photoperiod is the duration of daily light exposure affecting flowering and other physiological processes in plants. Manipulating photoperiod through artificial lighting can induce or delay flowering to meet market demands. Understanding crop-specific photoperiod requirements helps optimize production cycles.

Photoperiodism: Photoperiodism is a plant's physiological response to the length of day or night, affecting flowering and growth. Manipulating photoperiod in greenhouses can control crop cycles. Understanding photoperiodism helps achieve off-season production.

Photoselective Films: Photoselective films modify the light spectrum entering the greenhouse to enhance specific plant responses. They improve photosynthetic efficiency and crop quality. These films are an advanced tool in protected cultivation.

Photosynthetically Active Radiation (PAR): The spectrum of light (400–700 nm) plants use for photosynthesis; important for selecting grow lights.

Plant Growth Chambers: Plant growth chambers are fully controlled environments used for research or propagation. They regulate temperature, light, humidity, and CO2 precisely. Chambers enable year-round cultivation and experimental studies.

Plant Growth Promoters: Plant growth promoters are substances that stimulate physiological processes like cell division, elongation, and nutrient uptake. They include hormones, vitamins, and organic extracts. Application improves crop vigour, yield, and stress tolerance.

Plant Quarantine: Plant quarantine involves isolating new or imported plants to prevent introduction of pests and diseases. It ensures biosecurity in protected cultivation operations. Quarantine protocols protect existing crops and ecosystems.

Plasticulture: Plasticulture uses plastic materials like mulch films, drip tapes, and covers in agriculture to improve water use efficiency and crop yield. It is widely applied in protected cultivation for moisture conservation and weed control. Plasticulture enhances productivity and resource management.

Pollination Efficiency: Pollination efficiency measures the effectiveness of pollen transfer leading to fruit set. Improving efficiency through managed pollinators or manual methods increases yield. Efficient pollination is vital in protected fruit and vegetable crops.

Pollination in Greenhouses: Pollination inside greenhouses can be done manually or using mechanical aids such as bumblebees or vibrating devices. Proper pollination ensures fruit set and seed production, critical for many vegetable and fruit crops. Controlled pollination improves yield and fruit quality.

Pollinator Habitat: Pollinator habitats within or near greenhouses support beneficial insects like bees. These habitats improve pollination and crop yield naturally. Promoting biodiversity enhances ecosystem health.

Polycarbonate Sheets: Polycarbonate sheets are durable, lightweight greenhouse covers with high light transmission. They provide good insulation and UV protection, increasing energy efficiency. These sheets are popular for modern greenhouse construction.

Polyethylene Film: Polyethylene film is a common greenhouse covering material that is lightweight and flexible. UV-stabilized films protect crops and extend film life. It is cost-effective for polyhouses but requires periodic replacement.

Polyhouse vs Greenhouse: Polyhouses use polyethylene films for covering and are low-cost structures mainly for moderate climates. Greenhouses are more durable, often use glass or polycarbonate, and offer better environment control. Both protect crops from harsh weather.

Polyhouse: A low-cost greenhouse covered with UV-stabilized polyethylene film, mainly used for vegetable and flower production under moderate climate conditions.

Postharvest Cooling: Postharvest cooling rapidly reduces harvested crop temperature to preserve freshness and extend shelf life. Techniques include forced-air cooling and hydro-cooling. Proper cooling reduces spoilage and maintains market quality.

Post-Harvest Handling: Post-harvest handling includes cleaning, grading, packaging, and storage to maintain crop quality. Proper handling extends shelf life and reduces losses. It is essential for delivering fresh, market-ready produce.

Pot Plants: Ornamental plants grown in pots under protected conditions for better growth and market value.

Precision Irrigation: Precision irrigation uses data and technology to apply water exactly where and when needed. It minimizes waste and maximizes crop water use efficiency. Techniques include drip irrigation controlled by soil moisture sensors.

Propagation Techniques: Methods like seed sowing, cutting, grafting, and tissue culture used to produce quality planting material under protected conditions.

Protected Cultivation: A method of growing crops in controlled environments like greenhouses or polyhouses to enhance yield and quality by managing temperature, humidity, and pests.

Quality Planting Material: Healthy, disease-free, and genetically true-to-type seedlings or cuttings essential for successful crop establishment.

Rainwater Harvesting: Collecting and storing rainwater from greenhouse roofs for irrigation use, promoting water conservation and reducing dependence on groundwater. Useful in water-scarce areas.

Recycled Water Use: Recycled water in protected cultivation involves reusing treated wastewater to reduce freshwater demand. Proper treatment ensures safety and nutrient content. Recycling supports sustainability and cost savings.

Relative Humidity: Relative humidity is the percentage of moisture in the air compared to its maximum capacity at a given temperature. It affects transpiration, disease development, and plant growth. Maintaining ideal humidity reduces stress and disease risks.

Ridge and Furrow Greenhouse: Greenhouse design with alternating ridges and furrows to maximize rainwater drainage and sunlight exposure.

Root Aeration: Root aeration ensures oxygen availability in the root zone critical for respiration and nutrient absorption. Aerated substrates or irrigation practices prevent root suffocation. Proper aeration supports healthy root systems and plant vigor.

Root Development: Root development encompasses the growth and expansion of the root system for water and nutrient uptake. Healthy roots are vital for plant stability and productivity. Substrate quality and irrigation affect root development in protected cultivation.

Root Pruning: Root pruning involves trimming roots to control plant size and encourage healthy root branching. It is practiced in container-grown plants to prevent root circling. Proper root pruning improves transplant success and vigour.

Root Substrate Amendments: Amendments like compost, perlite, or vermiculite improve substrate properties including aeration, drainage, and nutrient availability. They tailor the medium to specific crop needs. Proper amendments promote root health and growth.

Root Substrate Interaction: This refers to how plant roots interact with the substrate, including water and nutrient uptake, aeration, and microbial activity. Good interaction supports healthy root development and crop growth. Monitoring this helps optimize substrate management.

Root Zone Heating: Root zone heating warms the soil or substrate around roots to promote nutrient uptake and growth in cool conditions. It enhances early crop development and reduces chilling injury. Root heating is common for flowers and vegetables.

Root Zone Moisture Sensors: These sensors measure moisture levels around plant roots to optimize irrigation scheduling. Maintaining ideal root zone moisture prevents water stress and root diseases. Sensor-based irrigation improves water use efficiency.

Root Zone Monitoring: Root zone monitoring involves tracking moisture, temperature, and oxygen levels around plant roots. It provides crucial data to optimize irrigation and substrate conditions. Proper monitoring prevents stress and promotes healthy root systems.

Root Zone Oxygenation: Root zone oxygenation ensures sufficient oxygen supply to roots, vital for respiration and nutrient uptake. Aeration can be improved by substrate choice or irrigation management. Adequate oxygen prevents root diseases and promotes growth.

Root Zone pH Management: Maintaining optimal pH in the root zone ensures nutrient availability and microbial activity. pH adjustments are done through substrate amendments or nutrient solution control. Proper pH management is key for healthy root function.

Root Zone Temperature: Root zone temperature is the temperature surrounding plant roots, influencing nutrient uptake and metabolism. Maintaining ideal root temperatures promotes healthy growth and development. Extreme root temperatures can cause stress or inhibit root function.

Rose Cultivation in Greenhouses: Growing roses under controlled conditions to produce high-quality flowers year-round with optimum temperature and humidity control.

Sanitation Practices: Sanitation includes cleaning tools, benches, and surfaces to minimize disease spread. Regular disinfection prevents contamination of planting material and substrates. Good sanitation is vital for maintaining a healthy greenhouse environment.

Seed Germination: Seed germination is the process where a seed develops into a seedling under favourable conditions like moisture, temperature, and oxygen. Controlled environment in greenhouses ensures uniform and rapid germination. This is critical for successful crop establishment.

Seed Propagation: Seed propagation is the reproduction of plants by sowing seeds under controlled conditions. It is cost-effective and widely used for vegetables and flowers. Controlled environments improve germination rates and seedling quality.

Seed Treatment: Seed treatment involves applying fungicides, insecticides, or bioagents to seeds before sowing. It protects seedlings from soil-borne diseases and pests. Seed treatment enhances germination rates and seedling vigour.

Seedling Hardening: Seedling hardening gradually exposes young plants to less favourable conditions to increase their tolerance before transplanting. This improves survival rates and establishment. Hardening includes reducing humidity and increasing light intensity.

Seedling Nursery: A seedling nursery is a protected area for raising young plants under controlled conditions before transplanting. It ensures healthy, uniform planting material. Proper nursery management reduces crop losses and improves yields.

Seedling Vigor: Seedling vigour describes the health and robustness of young plants, influencing establishment and growth. High vigour seedlings adapt better to transplanting and stress. Vigor is affected by seed quality and nursery management.

Sensors: Devices that continuously measure environmental factors such as temperature, humidity, and light inside the greenhouse. They provide data needed for monitoring and automated adjustments.

Shade House: A shade house is a protected structure covered with shade netting to reduce sunlight and protect plants from heat stress. It is used for crops that require partial shade or nursery plants. Shade houses improve plant quality and reduce water demand.

Shade Netting: Shade netting is used to reduce light intensity and protect plants from excessive sunlight and heat stress. It helps maintain cooler temperatures and prevents leaf scorch during peak sun hours. Shade nets come in various densities depending on crop requirements and climatic conditions.

Single-span Greenhouse: A standalone greenhouse structure without adjoining bays, suitable for small-scale cultivation.

Soil Moisture Sensors: Soil moisture sensors measure water content in soil or substrates to guide irrigation scheduling. They prevent over- or under-watering, optimizing plant health and water use. Sensor-based irrigation increases efficiency and reduces waste.

Soil Preparation: Tilling, levelling, and amending soil inside the greenhouse to ensure good aeration, moisture retention, and nutrient availability.

Soil Solarization: Soil solarization uses transparent plastic covers to trap solar heat, sterilizing the soil by killing pathogens and weed seeds. It is an eco-friendly method for disease management before planting. Solarization improves soil health and reduces reliance on chemical treatments.

Soil-less Culture Systems: These systems grow plants without soil, using substrates or nutrient solutions for support and nutrition. They reduce soil-borne diseases and allow precise nutrient management. Examples include hydroponics, aeroponics, and aquaponics.

Soil-less Culture: Soilless culture grows plants without soil, using substrates or nutrient solutions to support roots. It reduces soil-borne diseases and allows precise nutrient management. This method is widely used in hydroponics and container cultivation.

Solar Greenhouse: A solar greenhouse harnesses solar energy for heating and lighting to reduce external energy needs. It uses materials and designs that maximize sunlight capture. Solar greenhouses promote energy-efficient protected cultivation.

Strawberry Cultivation: Protected cultivation extends the growing season and improves fruit quality by regulating temperature and moisture.

Structural Frame: The skeleton of the greenhouse made of materials like steel, aluminium, or wood supporting the cladding.

Substrate pH: Substrate pH measures the acidity or alkalinity of the growing medium. It affects nutrient availability and microbial activity. Adjusting pH to crop-specific ranges ensures optimal nutrient uptake and plant health.

Substrate Porosity: Substrate porosity is the volume of pore spaces within the growing medium, affecting air and water movement. High porosity improves root aeration and drainage. Managing porosity is key to avoiding root diseases.

Substrate Reuse: Substrate reuse involves recycling growing media after proper sterilization or treatment. It reduces costs and environmental impact. Careful management is necessary to avoid pathogen carryover.

Substrate Sterilization: Substrate sterilization eliminates pathogens and weed seeds from growing media using heat, chemicals, or solarization. This practice prevents disease outbreaks in protected cultivation. Sterilized substrates support healthy root growth and crop establishment.

Substrate Water Holding Capacity: Water holding capacity is the amount of water a substrate can retain for plant use. It affects irrigation frequency and root health. Substrates with balanced water retention and drainage support optimal growth.

Substrate: Soilless growing media like cocopeat or perlite that support plant roots while providing aeration and moisture retention. It helps reduce soil-borne diseases and improves root health.

Supplemental CO2 Delivery: Supplemental CO2 delivery increases carbon dioxide levels to enhance photosynthesis and plant growth. Systems include tanks or generators releasing CO_2 inside greenhouses. Controlled delivery maximizes yield and resource efficiency.

Supplemental Lighting: Supplemental lighting extends day length or increases light intensity to boost photosynthesis and growth. LED and high-pressure sodium lamps are common types. This practice enables off-season and higher-quality production.

Temperature Gradient: Temperature gradient refers to the variation of temperature within different greenhouse zones or heights. Managing gradients ensures uniform crop conditions. It prevents stress caused by localized hot or cold spots.

Temperature Management: Temperature management maintains optimal day and night temperatures inside protected structures. It influences physiological processes like photosynthesis and flowering. Proper management prevents heat or cold stress.

Thermal Insulation: Thermal insulation materials reduce heat loss in greenhouses, maintaining stable internal temperatures. They include double glazing, insulating screens, and foam panels. Insulation decreases energy costs and protects crops from temperature extremes.

Thermal Mass: Thermal mass refers to materials that absorb, store, and release heat, helping stabilize greenhouse temperatures. Examples include water barrels or concrete floors. Thermal mass reduces temperature fluctuations and energy costs.

Thermal Screens: Thermal screens are retractable covers inside greenhouses that reduce heat loss during cold nights. They improve energy efficiency and maintain stable temperatures. Thermal screens help protect crops from chilling injury.

Tissue Culture: Tissue culture is a sterile, in-vitro technique of growing plants from small tissue samples. It enables rapid multiplication of disease-free and genetically uniform plants. Widely used for orchids, bananas, and ornamentals in protected cultivation.

Tomato Cultivation in Greenhouses: Protected cultivation enables off-season tomato production with controlled temperature, humidity, and pest management.

Transplanting: Transplanting moves seedlings from nursery beds or containers to the main growing area. It is timed to optimize plant establishment and growth. Careful handling during transplanting prevents stress and transplant shock.

Tray Culture: Tray culture involves growing seedlings or plants in shallow trays filled with substrate. It facilitates uniform growth and easy handling during propagation. Tray culture is common for vegetable seedlings and ornamentals.

Tulip Cultivation: Cool temperature requirements and proper chilling treatments are necessary for tulip flowering in greenhouses.

UV Radiation: Ultraviolet (UV) radiation affects plant growth and can cause stress or damage at high levels. UV-blocking films or nets protect crops while allowing beneficial light. Managing UV exposure improves plant health and product quality.

UV-Stabilized Films: UV-stabilized films are plastic coverings treated to resist ultraviolet radiation degradation. This treatment extends the lifespan of greenhouse films. They maintain transparency and protect crops from harmful UV rays.

Ventilation System: Ventilation systems, either natural or mechanical, regulate the exchange of air inside the greenhouse. Proper ventilation controls temperature, humidity, and carbon dioxide levels, creating an optimal microclimate for plants. It also helps reduce disease pressure by improving air circulation.

Ventilation: The process of exchanging inside air with outside air to control temperature, humidity, and carbon dioxide levels.

Vertical Farming: Vertical farming stacks crops in layers inside controlled environments to maximize space efficiency. It allows year-round production with precise control over climate and nutrients. Vertical farming is an emerging protected cultivation technique.

Water Use Efficiency (WUE): WUE measures the amount of crop produced per unit of water used. Improving WUE is crucial in protected cultivation to conserve water resources. Efficient irrigation and substrate management enhance WUE.

Yield Enhancement: Yield enhancement involves practices and technologies that increase crop quantity and quality. These include optimized environment control, nutrient management, and pest control. Enhanced yields improve profitability and resource use efficiency.

References

Afolami, C.A.; Obayelu, A.E.; Vaughan, I.I. Welfare impact of adoption of improved cassava varieties by rural households in South Western Nigeria. *Agric. Econ.* 2015, *3*, 18.

Anton, A.; Torrellas, M.; Montero, J.I.; Ruijs, M.; Vermeulen, P.; Stanghellini, C. Environmental impact assessment of Dutch tomato crop production in a Venlo glasshouse. *Acta Hortic.* 2012, *927*, 781–791.

Aryal, J.P.; Sapkota, T.B.; Khurana, R.; Khatri-Chhetri, A.; Rahut, D.B.; Jat, M.L. Climate change and agriculture in South Asia: Adaptation options in smallholder production systems. *Environ. Dev. Sustain.* 2020, *22*, 5045–5075.

Atheequlla GA, Manjula N, Balakrishna B, Bheemappa A. Constraints and strategies of smallholder farmers for successful protected cultivation of capsicum. J Hortic Sci. 2022;18(2):71–77.

BIS Standard (IS 14462: 1997), Indian Standard for Recommendation for Layout, Design and Construction of Greenhouse Structures. BIS, New Delhi.

BIS Standard (IS 14485: 1998), Indian Standard for Recommendations for Heating, Ventilating and Cooling of Greenhouses.

Bishta, D.R.; Kumar, P.; Mathur, V.C. Progress and performance of kisan credit card scheme with a case study of Bihar. *Agric. Econ. Res. Rev.* 2012, *25*, 125–135.

Bose, TK, Maiti, RG, Dhua, R S and Das P 1999. *Floriculture and Landscaping*. Naya Prokash, Kolkatta.

Chadha, KL and Bhattacharjee, SK 1995. *Advances in Horticulture, vol.12 Part I and II, Ornamental Plants.* Malhotra Publishing House, New Delhi.

Chahal, A.S.; Sharma, R.; Dev, K. Economic analysis of carnation under protective cultivation: A study of Mid Hills Zone of Himachal Pradesh. *Indian J. Ecol.* 2020, *47*, 369–377.

Chhogyel, N.; Kumar, L. Climate change and potential impacts on agriculture in Bhutan: A discussion of pertinent issues. *Agric. Food Secur.* 2018, *7*, 79.

Choudhary, A.K. Scaling-up of protected cultivation in Himachal Pradesh Indian. *Curr. Sci.* 2016, *111*, 272–277.

D.T. Santosh, K.N. Tiwari and Vikas Kumar Singh (2017) Influence of different protected cultivation structures on water requirements of winter vegetable. International Journal of Agriculture, Environment and Biotechnology (IJAEB): 10(1): 93-103.

Dadlani, NK 1999. Development of new production technologies for rose in India. J. Ornam. Hort. New Series 2(1): 25-31.

Desh Raj 2005. *Floriculture in Hills,* Agrotech Publishing Academy, Udaipur, Rajasthan.

Desh Raj 2010. *Floriculture at Glance.* Kalyani Publishers, Ludhiana.

Dr. N.V. Shende and R.R. Meshram (2015). Cost Benefit Analysis and Marketing of Tomato. American International Journal of Research in Formal, Applied and Natural Sciences, 11(1), pp. 46-54.

Espi, A Salmeron, A Fontecha, Y Garcia and AI Real (2006), Plastic films for Agricultural Applications, J. of Plastic Film and Sheeting 2006; 22; 85-102.

Evans, A 2009. Reflection of short-wave infrared. FloraCulture International, 19 (2):18-19.

Feder, G.; Just, E.R.; Zilberman, D. Adoption of Agricultural Innovations in Developing Countries: A Survey. *Econ. Dev. Cult. Chang.* 1985, *33*, 255–298.

Fernandez JA, Orsini F, Baeza E, Oztekin GB, Muñoz P, Contreras J, *et al.,* Current trends in protected cultivation in Mediterranean climates. Eur J Hortic Sci. 2018;83(5):294–305.

Franco, D.; Singh, D.R.; Praveen, K.V. Economic feasibility of vegetable production under polyhouse: A case study from Palakkad district of Kerala. *J. Crop Weed.* 2018, *14*, 134–139.

Franco, D.; Singh, D.R.; Praveen, K.V. Evaluation of adoption of precision farming and its profitability in banana crop. *Indian J. Econ. Dev.* 2018, *14*, 225–234.

Gamanagatti, P.; Patil, B. Economic evaluation of protected cultivation technology for horticultural crops. In Proceedings of the 2018 Conference International Association of Agricultural Economists, Vancouver, BC, USA, 28 July–2 August 2018.

Ghanghas BS, Malik JS, Yadav VPS. (2018) Sustainable vegetables and flowers production technology (Poly House): Problems and Prospects in Haryana. Indian Research Journal of Extension Education; 18(2):12-16.

Gittinger, J.P. *Economic Analysis of Agricultural Projects*; Published for the Economic Development Institute of the World Bank; Johns Hopkins University Press: Baltimore, MD, USA, 1982.

Government of India. *Horticultural Statistics at a Glance 2017*; Horticulture Statistics Division, Department of Agriculture, Cooperation and Farmers Welfare, Ministry of Agriculture and Farmers Welfare, Government of India: New Delhi, India, 2017.

Government of India. *Mission for Integrated Development of Horticulture, Operational Guidelines*; Horticulture Division, Department of Agriculture, and Cooperation, Ministry of Agriculture and Farmers Welfare, Government of India: New Delhi, India, 2014.]

Government of India. *Mission for Integrated Development of Horticulture*; Department of Agriculture, and Cooperation, Ministry of Agriculture and Farmers Welfare, Government of India: New Delhi, India, 2019.

Green, D.A.G.; Ng'ong'ola, D.H. Factors Affecting Fertilizer Adoption in Less Developed Countries: An Application of Multivariate Logistic Analysis in Malawi. *J. Agric. Econ.* 1993, *44*, 99–109.

Greene, W.H. *Econometric Analysis*, 6th ed.; Prentice Hall: Hoboken, NJ, USA, 2007.

Griliches, Z. Hybrid Corn: An Exploration in the Economics of Technological Change. *Econometrica* 1957, *25*, 501–522.

Gruda, N.; Bisbis, M.; Tanny, J. Impacts of Protected Vegetable Cultivation on Climate Change and Adaptation Strategies for Cleaner Production–A Review. *J. Clean. Prod.* 2019, *225*, 324–339.

Harisha N., Tulsiram J., Joshi Amritha T. (2019) Techno-Economic Analysis of Vegetable Production under Protected Cultivation in Kolar district of Karnataka: Agricultural Science Digest - A Research Journal, Volume : 39(3):224 – 227.

Itigi Prabhakar, K Vijayaragavan, Premlata Singh, Balraj Singh, Janakiram, B L Manjunatha, SeemaJaggi And I Sekar (2017). Constraints in adoption and strategies to promote polyhouse technology among farmers: A multi-stakeholder and multi-dimensional study. Indian Journal of Agricultural Sciences 87 (4): 485–90.

J.A. Fernández, F. Orsini, E. Baeza, G.B. Oztekin, P. Muñoz, J. Contreras and J.I. Montero (2018) Current trends in protected cultivation in Mediterranean climates. European Journal of Horticultural Science: 83(5), 294–305.

Jain, Shubham, Deepak S. Kore, Kishorkumar G. K., Amrita Mohapatra, Hari Baksh, Vipin Kumar, Sukirti Mohanty, and Songthat William Haokip. 2023. "A Comprehensive Review on Protected Cultivation of Horticultural Crops:

Present Status and Future Prospects". *International Journal of Environment and Climate Change* 13 (11):3521-31.

Jensen, M.H. Controlled environment agriculture in deserts tropics and temperate regions–A world review. *Acta Hortic.* 2002, *578*, 19–25.

Jethi R, Srinivas K, Bisht JK. Economics of production of tomato under open and protected field conditions in hills of Uttarakhand. Indian J Ext Educ. 2012;48:13–16.

Jethi, R.; Srinivas, K.; Bisht, J.K. Economics of production of tomato under open and protected field conditions in hills of Uttarakhand. *Indian J. Ext. Educ.* 2012, *48*, 13–16.

Kanika Mehta, Rajesh Kumar Thakur and J.S. Guleria (2020) Socio-Economic Impact of Protected Cultivation on Tomato Growers of Himachal Pradesh. Journal of Economic Affairs, 65(1):01-07.

Katinka Weinberger and Thomas A. Lumpkin, (2007), Diversification into Horticulture and Poverty Reduction: A Research Agenda, World Development, 35, (8), 1464-1480.

Krishna H, Chaurasia SNS, Bahadur A. Protected cultivation of vegetable crops for sustainable food production. Indian Hortic. 2023;68(2):39–44. 6. Kumar P, Kar A, Singh DR, Perumal A, Shivamurthy SGC, Reddy KV, *et al.,* Protected cultivation of horticultural crops in Uttarakhand: an economic analysis. Agronomy. 2021;11(4):692.

Krishna, V.R.; Paramesh, V.; Das, V.A.B.; Elansary, H.O.; Parab, A.; Reddy, D.D.; Shashidhar, K.S.; El-Ansary, D.O.; Mahmoud, E.A.; El-Sheikh, M.A. Assessment of sustainability and priorities for development of Indian west coast region: An Application of Sustainable Livelihood Security Indicators. *Sustainability* 2020, *12*, 8716.

Kumar P, Chauhan RS, Tanwar N, Grover RK. (2018), Status and constraints in vegetable cultivation under poly house in Haryana. Advances in Bioresearch, 9(2):10-15.

Kumar P, Chauhan RS, Tanwar N, Grover RK. (2018), Status and constraints in vegetable cultivation under poly house in Haryana. Advances in Bioresearch, 9(2):10-15.

Kumar S, Meena RS, Jakhar SR, Jangir CK, Gupta A, Meena BL. Adaptation strategies for enhancing agricultural and environmental sustainability under current climate. In: Sustainable Agriculture. Jodhpur: Scientific Publisher; 2019. p. 226–274.

Kumar Suman, Singh Amar, and Banyal D. K. (2015) First record of occurrence and distribution of bacterial canker of capsicum under protected cultivation in Himachal Pradesh. Plant Disease Research:30(1):61-66.

Kumar, P.; Chauhan, R.S.; Grover, R.K. Economic analysis of tomato cultivation under poly house and open field conditions in Haryana. *Int. J. Appl. Sci.* 2016, *8*, 846–848.

Kumar, P.; Kar, A.; Singh, D.R.; Perumal, A.; Shivamurthy, S.G.C.; Reddy, K.V.; Badal, P.S.; LalaKamble, A.; Kamalvanshi, V.; Jha, G.K.; *et al.,* Protected Cultivation of Horticultural Crops in Uttarakhand: An Economic Analysis. *Agronomy* 2021, *11*, 692.

Malik K. (2017) Economic viability of cucumber cultivation in greenhouses. International Journal for Innovative Research in Multidisciplinary Field; 3(6):366-368.

Malik YS, Kumar N, Sircar S, Kaushik R, Bhat S, Dhama K, Gupta P, Goyal K, Singh MP, Ghoshal U, El Zowalaty ME. Coronavirus disease pandemic (COVID19): challenges and a global perspective. Pathogens. 2020 Jun 28;9(7):519.

Manoranjan Kumar, Narendra Kumar and K. Srinivas (2019). Greenhouse Farming in High Altitude Areas of North-West Himalayan Region of India: A Success Story. International Journal of Agriculture Sciences, 11(5):7944-7949.

Mariano, M.J.; Vilano, R.; Fleming, E. Factors influencing farmer's adoption of modern rice technologies and good management practices in the Philippines. *Agric. Syst.* 2012, *110*, 41–53.

Mehta, K.; Thakur, R.K.; Guleria, J.S. Socio-economic impact of protected cultivation on tomato growers of Himachal Pradesh. *Econ. Aff.* 2020, *65*, 1–7.

Mishra, G.P.; Singh, N.; Kumar, H.; Singh, S.B. Protected cultivation for food and nutritional security at Ladakh. *Def. Sci. J.* 2010, *61*, 219–225.

Murthy, D.S.; Prabhakar, B.S.; Hebbar, S.S.; Srinivas, V.; Prabhakar, M. Economic viability of vegetable production under polyhouse: A case study of capsicum and tomato. *J. Hortic. Sci.* 2009, *4*, 148–152.

Ndondo JT. Review of the Food and Agriculture Organisation (FAO) Strategic Priorities on Food Safety 2023. InFood Safety-New Insights 2023 Nov 29. IntechOpen.

Negi, V.S.; Maikhuri, R.K.; Rawat, S.; Parshwan, D. Protected cultivation as option of livelihood in mountain region of central Himalaya, India. *Int. J. Sustain. Dev. World Ecol.* 2013, *5*, 416–425.

Nimbrayan PK, Chauhan RS, Mehta VP, Bhatia JK. A review on economic aspect of protected cultivation in India. Res Trends Hortic Sci. 2018;7:43–59.

Nirmal, K.; Malik, D.P.; Raj, K. An economic analysisof Mango (*Mangiferaindica* L.) cultivation in Haryana. *Indian J. Econ. Dev.* 2019, *15*, 282–288.]

Ntshangase, N.L.; Muroyiwa, B.; Sibanda, M. Farmers perceptions and factors influencing the adoption of no-till conservation agriculture by small scale farmers in Zashuke Kwazula Natal Province. *Sustainability* 2018, *10*, 555.

Owen, William G. and Behe, Bridget (2020) "A National Survey to Characterize Industrial Hemp (Cannabis sativa L.) Production Challenges Under Protected Cultivation," Journal of Agricultural Hemp Research:1(2):230-236.

Pachiyappan D, Alam MS, Khan U, Khan AM, Mohammed S, Alagirisamy K, Manigandan P. Environmental sustainability with the role of green innovation and economic growth in India with bootstrap ARDL approach. Frontiers in Environmental Science. 2022 Oct 14;10:975177.

Panwar, N.L.; Kothari, S.; Kaushik, S.C. Cost-benefit and systems analysis of passively ventilated solar greenhouses for food production in arid and semi-arid regions. *Environ. Syst. Decis.* 2014, *34*, 160–167.]

Parmar MN (2020) Special horticultural practices for vegetable under protected cultivation. The Pharma Innovation Journal; 9 (3): 425 - 430.

Pepijn Schleiermacher's; Mei-huey Wu; Md. Nasiruddin; Shahabuddin Ahmad and Peter Hanson, (2016) Farmer training in off-season vegetables: Effects on income and pesticide use in Bangladesh, Food Policy, 61, (C), 132-140.

Prabhakar, I.; Vijayaragavan, K.; Singh, P.; Singh, B.; Janakiram; Manjunatha, B.L.; Jaggi, S.; Sekar, I. Constraints in adoption and strategies to promote poly house technology among farmers: A multi-stakeholder and multi-dimensional study. *Indian J. Agric. Sci.* 2017, *87*, 485–490.

Prakash, P.; Kumar, P. Performance of kisan credit card scheme in Tamil Nadu. *Indian J. Agric. Econ.* 2016, *71*, 191–211.

Prakash, P.; Kumar, P.; Kar, A.; Singh, A.K. Status and impact of protected cultivation of horticultural crops in Maharashtra. *Indian J. Hortic.* 2020, *77*, 518–526.

Prakash, P.; Kumar, P.; Kar, A.; Singh, A.K.; Ankukkani, P. Economic analysis of carnation (*Dianthus caryophyllus*) under protected cultivation in Maharashtra. *Indian J. Agric. Sci.* 2022, *92*, 460–463.

Prakash, P.; Kumar, P.; Kar, K.; Singh, A.K.; Anbukkani, P. Progress and performance of protected cultivation in Maharashtra. *Indian J. Econ. Dev.* 2019, *15*, 555–563.

Prakash, P.; Kumar, P.; Niranjan, S.; Jaganathan, D.; Kishore, P.; Immanuel, S. Economic feasibility of protected cultivation of rose under polyhouse and its supply chain in Maharashtra. *J. Community Mobilization Sustain. Dev.* 2021, *16*, 939–946.

Punera B, Pal S, Jha GK, Kumar P. (2017), Economics and Institutional Aspects of Protected Cultivation of Carnation in Himachal Pradesh. Agricultural Economics Research Review; 30(1):73-80.

Punera, B.; Pal, S.; Jha, G.K.; Kumar, P. Economics and institutional aspects of protected cultivation of carnation in Himachal Pradesh. *Agric. Econ. Res. Rev.* 2017, *30*, 73–80.]

R.K.Yadav, P.Kalia, H. Choudhary, Zakir Husain and Brihama Dev (2014). Low-Cost Polyhouse Technologies for Higher Income and Nutritional Security.

International Journal of Agriculture and Food Science Technology. 5(3): 191-196.

RanjitChatterjee, Ravi KiranThirumdasu and Dipika Mal (2015). Off-Season French Bean (Phaseolus vulgaris L.) Cultivation Inside Agro Shade Net: Influence of Planting Dates and Nutrient Sources, Journal of Agriculture and Technology: 2(1 and 2) 87-91.

Rogers, E.M. *Diffusion of Innovations*, 4th ed.; The Free Press, A Division of Simon and Schuster Inc.: New York, NY, USA, 1995.

Rosenzweig C, Iglesius A, Yang XB, Epstein PR, Chivian E. Climate change and extreme weather events–implications for food production, plant diseases, and pests. Glob Chang Hum Health. 2001;2:90–104.

Rosenzweig, C.; Iglesias, A.; Yang, X.; Epstein, P.R.; Chivian, E. Climate change and extreme weather events; implications for food production, plant diseases, and pests. *Glob. Chang. Hum. Health* 2001, *2*, 90–104.

S H Wittwer and N Castilla (1995), Protected Cultivation of Horticultural Crops Worldwide, HortTechnology, 5, 1, pp. 6-23. [2] MH Jensen and AJ Malter (1994), Protected Agriculture: Global Review, World Bank Technical Paper number 253, pp 160.

Sabir N, Singh B. (2013) Protected cultivation of vegetables in global arena: A review. Indian Journal of Agricultural Sciences; 83(2):123-135.

Sabir N, Singh B. Protected cultivation of vegetables in global arena: A review. The Indian Journal of Agricultural Sciences. 2013 Feb 1;83(2).

Sabir, N.; Singh, B. Protected cultivation of vegetables in the global arena: A review. *Indian J. Agric. Sci.* 2013, *83*, 123–135.

Sengar RS, Rani V. Opportunities and prospective of integrated development of horticulture: a review. Ann Hortic. 2020;13(1):1–8.

Sindhu V and Ranjit Chatterjee (2020) Off-Season Vegetable Cultivation under Protected Structures: A Promising Technology for Doubling Farmers Income. Int. Arch. App. Sci. Tech. (IAAST).11(3):208-214.

Singh B. Exploring potential of protected cultivation in India–a review. Curr Hortic. 2024;12(2):3–11.

Singh Balraj (2014) 3rd International Conference on Agriculture and Horticulture, Hyderabad International Convention Centre, India.2(4): 51.

Singh, B. Protected cultivation of horticultural crops in India: Challenges and opportunities. In Proceedings of the 2014 Conference Third International Conference on Agriculture and Horticulture, Hyderabad, India, 27–29 October 2014.

Slathia D, Nisa MU, Reshi M, Dolkar T, Hussain S. Protected cultivation of ornamentals. Glob J Bio-Sci Biotechnol. 2018;7(2):302–311.

Spehia R. S. (2015) Status and impact of protected cultivation in Himachal Pradesh, India. Current Science. 108(12):2254-2257.

Spehia, R.S. Status and impact of protected cultivation in Himachal Pradesh Indian. *Curr. Sci.* 2015, *108*, 2254–2257.

Sreedhara, D. S.; Kerutagi, M. G.; Basavaraja, H.; Kunnal, L. B.; Dodamani, M. T. (2013) Economics of capsicum production under protected conditions in Northern Karnataka: Journal of Agricultural Sciences.26(2):217-219.

SS Bhati and Dr. BL Baheti (2020) Occurrence and population status of root-knot nematode, Meloidogyne incognita on cucumber under protected cultivation in Rajasthan: Journal of Entomology and Zoology Studies (JEZS); 8(4):1441-1444.

Stanghellini, C.; Montero, J.I. Resource use efficiency in protected cultivation: Towards the greenhouse with zero emissions. *Acta Hortic.* 2012, *927*, 91–100.

Sudhagar, S. Production and marketing of cut flower (Rose and Gerbera) in hosurtaluk. *Int. J. Bus. Manag. Invent.* 2013, *2*, 15–25.

Theurl, M.C.; Hortenhuber, S.J.; Lindenthal, T.; Palme, W. Unheated soil-grown winter vegetables in Austria: Greenhouse gas emissions and socio-economic factors of diffusion potential. *J. Clean. Prod.* 2017, *151*, 134–144.

Tuzel Y, Leonardi C. Protected cultivation in Mediterranean region: trends and needs. J Agric Fac Ege Univ. 2009;46(3):215–223.

Vijayasarathy, K.; Ashok, K.R. Climate adaptation in agriculture through technological option: Determinants and impact on efficiency of production. *Agric. Econ. Res. Rev.* 2015, *28*, 103–116.

Vikram S. Negi, R. K. Maikhuri, L. S. Rawat and D. Parshwan (2013) Protected cultivation as an option of livelihood in mountain region of central Himalaya, India, International Journal of Sustainable Development and World Ecology, 20(5):416-425.

Waghmare, M.N.; Shendage, P.N. Economic Analysis of Production of Cut Roses under Polyhouses in Western Maharashtra. *Int. J. Curr. Microbiol. Appl. Sci.* 2019, *8*, 234–241.

Walawande, S.R.; Rahane, R.K.; Mali, R.R.; Waghmare, M.N. State of floriculture in Maharashtra: A case study of floriculturists in Pune District. *Agric. Econ. Res. Rev.* 2013, *26*, 221.

Index

A

Absorbers 90
Acclimatization 235, 245
Acrylic sheets 52
Advanced cladding materials 58
Aeration 156
Aeroponics 319
Air and gas regulation 99
Aloe vera 326
Anthracnose 307
Anthurium 278, 336
Aromatic plants 327
Artificial lighting 87, 105
Asexual 225
Ashwagandha 325
Aster 337
Automated climate control 104
Automated machinery 110
Automation 103, 108

B

Basal fertilizer application 130
Bed formation 123
Bed preparation 297
Bed sanitization 132
Bell pepper 294
Bemisia tabaci 282
Benches 159, 173
Biological aspects 121
Biological controls 101 346
Biosecurity measures 242
Biotic factors 82
Bitter gourd 332
Boom watering 198
Botanical pesticides 347
Bottle gourd 332
Brinjal 332
Bromeliads 338
Budding 227, 243

C

Capsicum 332
Carnation 258, 335
Ceramic 183
Certification 237, 247
Certification agencies 248
Chaetanaphothrips orchid 282
Chemical controls 348
Chemical properties 119
Chili 332
Chrysanthemum 264, 336
Cladding materials 43
Clay 187
Climate 35, 295
Climate compatibility 63
Climate conditions 78
Climate control 8
Climate-controlled or hi-tech greenhouses 71
Coastal area structures 34
Cocopeat 138
Coir pith 138
Cold and snow-prone climate 37
Cold frames 32
Cold storage 242
Concrete benches 176
Containers 159, 179
Controlled environment 1, 315
Cowpea 333
Crop management 323
Crop monitoring 110, 220
Crop rotation 340
Crop spacing 90
Crop varieties 344
Crop yield 3
Cucumber 300, 332
Cucurbits 331
Cutting 226
Cutting management 243

D

Data analytics 104, 111
Dehumidifiers 90
Dendranthema grandiflora 264
Dendrobium orchid 273
Dianthus caryophyllus 258
Disease management 257
Diseases 256, 262, 277
Disinfestation 126
Distillation techniques 325
Diversification and innovation 6
Drip irrigation 89, 91, 201, 322
Drones 104
Drying methods 324

E

Electrical conductivity 218
Employment generation 11
Energy 112
Energy efficiency 107
Environmental control 65, 79, 109, 310, 344
Essential oil crops 324
Exclusion techniques 345
Exotic crop production 5
Exotic ornamentals 338
Export-oriented agriculture 5
Export-oriented production 10

F

Fenugreek 333
Fern varieties 338

Fertigation 110, 217, 232, 243
Fertigation management 189
Fertilizer 312
Fertilizer doses 216
Fertilizer scheduling 97
Fiberglass reinforced plastic 54
Fibre glass 184
Floating plastic covers 165
Flower crops 249
Flower harvesting 257
Flowers 252, 331
Fogger-equipped greenhouses 318
Foggers 205
Fogging 89
Foliage plants 337
Food and nutritional security 12
Framed or inflated structure 170
French bean 333
Fruit thinning 302

G

Gaillardia 337
Geranium 327
Gerbera 270, 335
Gladiolus 336
Glass 45
Government support 16
Grading 299
Grafting 227, 243, 319
Gravel and expanded clay pebbles 146
Greenhouse 38
Greenhouse cultivation 249
Greenhouse design 65
Greenhouses 24, 69, 168, 318
Growing media 232, 311

H

Hand watering 192
Hardening 235, 245
Harvesting 299, 302
Heating systems 84
High-tech greenhouse 25
High tunnels 77
High-value crop production 7
Horticultural crops 249
Horticultural oils 347
Hot and arid climate 36
Hot beds 167
Humid climate 36
Humidity control 89
Humidity-controlled structures 338
Humidity management 91
Hydroponic nutrient solutions 322
Hydroponics 319
Hydroponic substrate 142
Hygiene 233, 242, 246

I

Insect pest and disease 343
Insect-proof net houses 30
Insect-proof nets 57
Insulation materials 86
Integrated pest management 102, 313, 324, 348
Irrigation 110, 189, 232, 243, 256
Irrigation management 91
Irrigation system 131

K

Kalmegh 326

L

Layering 226, 319
Leaf spot 308
Leafy vegetables 332
Leguminous vegetables 333
Lettuce 333
Light 80
Light control 87
Light diffusion 60
Light quality management 88
Light spectrum 106
Light transmission 60
Lilium 283, 336
Linear low-density polyethylene 48
Low-cost polyhouse 24
Low-cost poly tunnels 337
Low-density polyethylene 48
Low tunnels 30, 76

M

Manures and fertilizers 306
Marigold 337
Market opportunities 79
Mechanical strength 62
Medicinal and aromatic plants 315
Medium cost greenhouse 26
Metal 185
Metal benches 176
Micropropagation 227, 255
Mineral wool 142
Mint 328
Mist chambers 32, 338
Misters 205
Misting 89
Moisture 81, 156
Monitoring 345
Mulching 85, 89, 323
Muskmelon 332

N

Naturally ventilated greenhouses 70
Net houses 30, 75, 163
Nursery management 232, 241
Nursery production 10
Nutrient and biological support 246
Nutrient and water management 322
Nutrient management 156
Nutrients 82

O

Ocimum sanctum 326
Off-season 5, 7
Off-season production 331
Orchids 273, 335, 338
Organic amendments 98, 321
Organic and gap production 318
Organic composts 150
Organic liquid manures 322
Organic matter 127
Overhead sprinklers 196
Oxygen supply 99

O

Packaging 257, 262
Pallets 177
Patchouli 327
Peat moss 149
Perimeter watering system 194
Perlite 141
Pest and disease 4, 9, 101, 103, 233, 243, 282, 298, 324, 340
Pest management 257
Photoperiod 107

Photoperiod manipulation 88
Physical properties 118
Physiological disorder management 257
Physiological disorders 263
Physiological maturity 324
Pinching 260
Plant growth regulators 307
Plastic 176, 181
Plastic low tunnels 162
Plastic mulch 164
Plastic tunnels 27, 31
Pollination management 323
Polycarbonate sheets 50
Polyethylene films 47
Polyhouses 29, 39, 73, 241, 318, 335
Pot plants 309
Powdery mildew 308
Preventive measures 244, 344
Propagation 223, 259, 284
Propagation methods 311
Protected cultivation 2, 108, 315, 343
Protected structures 43, 160
Pruning 256, 298, 323
Pseudomonas 286

Q

Quality improvement 9
Quality planting material 223
Quality standards 234

R

Rainfall and monsoon 37
Rainfall zone 35
Rain shelters 33
Rainwater harvesting 94
Raised benches 179
Record keeping 247
Repotting 276
Research and development 11, 17
Resource-use efficiency 4
Ridge gourd 332
Robotics 105, 110
Rockwool 142
Rose 252, 327, 335
Routine inspections 345
Row covers 30, 162

S

Safed musli 325
Sand 144
Sanitation 102, 233, 246
Sawdust 147
Scheduling 216
Scouting 244
Seasonal annuals 337
Seed-based production 231
Seedling 10
Seed propagation 224, 280
Seed sowing 243
Semi-arid zone 35
Sensors 103
Sexual propagation 224
Shade houses 30, 74
Shade net house 40, 334
Shade nets 56
Shading 84
Skill development 16
Smart technologies 103
Soil and nutrient 98
Soil and nutrient control 96
Soil and substrate management 115

Soil-borne pathogen 244
Soil conditioning 128
Soil enrichment 127
Soil health management 340
Soilless culture 97
Soilless media 320
Soil management 133, 320
Soil moisture sensors 95
Soil mulching 130
Soil pH management 96
Soil preparation techniques 117
Soil solarization 125
Soil sterilization 340
Soil testing 118
Soil trenches 166
Solanaceous crops 332
Solar energy 66
Solar radiation 80
Specialty films 58
Spinach 333
Sprinkler irrigation 92
Sterilization 153
Sterilized soil 321
Stevia 325
Strawberry 304
Substrate 152
Substrates 136
Subsurface irrigation 93
Surface covering 130

T

Temperate climate 37
Temperature 80
Temperature control 84
Terra cota 187
Tetranychus cinnabarinus 282
Thermal insulation 61
Tissue culture 227, 232, 281, 319
Tomato 290, 332
Training 323, 341
Tuberose 336
Tulip 287
Tulsi 326

U

UV stability 61

V

Vanilla 328
Vegetables 290, 331
Vegetative propagation 225, 231
Ventilation 66, 84 , 89
Vermiculite 139
Verticillium wilt 307
Vetiver 328

W

Walk-in-tunnels 31, 161
Waste management 246
Water 81
Water and nutrient efficiency 8
Water management 95
Water storage 242
Weather 4
Weeding 261, 285
Windbreaks 100
Wood 184
Wood chips 147
Wooden benches 175

Y

Year-round production 7
Yield and quality 13